Sustainable Development Goals Series

The **Sustainable Development Goals Series** is Springer Nature's inaugural cross-imprint book series that addresses and supports the United Nations' seventeen Sustainable Development Goals. The series fosters comprehensive research focused on these global targets and endeavours to address some of society's greatest grand challenges. The SDGs are inherently multidisciplinary, and they bring people working across different fields together and working towards a common goal. In this spirit, the Sustainable Development Goals series is the first at Springer Nature to publish books under both the Springer and Palgrave Macmillan imprints, bringing the strengths of our imprints together.

The Sustainable Development Goals Series is organized into eighteen subseries: one subseries based around each of the seventeen respective Sustainable Development Goals, and an eighteenth subseries, "Connecting the Goals", which serves as a home for volumes addressing multiple goals or studying the SDGs as a whole. Each subseries is guided by an expert Subseries Advisor with years or decades of experience studying and addressing core components of their respective Goal.

The SDG Series has a remit as broad as the SDGs themselves, and contributions are welcome from scientists, academics, policymakers, and researchers working in fields related to any of the seventeen goals. If you are interested in contributing a monograph or curated volume to the series, please contact the Publishers: Zachary Romano [Springer; zachary.romano@springer.com] and Rachael Ballard [Palgrave Macmillan; rachael.ballard@palgrave.com].

Henny Warsilah · Agus Heri Purnomo
Editors

Approaching Natural Resources for Sustainable Development in Indonesia

Confronting the Socio-Ecological Crisis

BRIN
BADAN RISET
DAN INOVASI NASIONAL

Springer

Editors
Henny Warsilah
Research Center for Society and Culture
National Research and Innovation
Agency (BRIN)
Jakarta, Indonesia

Agus Heri Purnomo
Research Center for Society and Culture
National Research and Innovation
Agency (BRIN)
Jakarta, Indonesia

ISSN 2523-3084 ISSN 2523-3092 (electronic)
Sustainable Development Goals Series
ISBN 978-981-96-6912-7 ISBN 978-981-96-6913-4 (eBook)
https://doi.org/10.1007/978-981-96-6913-4

Jointly published with BRIN Publishing

This Springer imprint is published by the registered company Springer Nature Singapore Pte Ltd.
The registered company address is: 152 Beach Road, #21-01/04 Gateway East, Singapore 189721, Singapore

Preface

The Southeast Asian region, particularly Indonesia, has experienced high economic growth, resulting in the massive capitalization of natural resources. This growth has, in turn, created land-use conflicts in forest areas, peatlands, rural-urban regions, and karst rocks, as well as increased access to other natural resources. For instance, the coal, tin, and oil and gas mining industries have witnessed significant capitalization, leading to extensive mining excavations and environmental damage, thus contributing to the ongoing socio-ecological crisis.

This edited volume describes the socio-ecological crisis that has emerged, widening the gap between environmental preservation and economic development. The authors of this book emphasize that economic and environmental aspects cannot coexist harmoniously; a choice must be made, prioritizing one over the other.

The book adopts the paradigm of sustainable development and delves into its three main aspects – ecology, economics, and society – along with politics and policies. Researchers from various disciplines have contributed to this book, presenting a comprehensive analysis. As a result, a crucial conclusion is drawn: the socio-ecological crisis is not merely a 'result' of circumstance but a 'condition' inherent in the presence of the capitalist regime. In other words, the crisis itself is an integral part of every round of capitalist development.

This book draws attention to the dilemma of sustainable development, addressing the challenge of achieving high levels of economic growth without causing environmental damage and social upheaval. It underlines the vital role of the government in creating appropriate regulations for natural resource management and emphasizes the necessity of community participation in resource management.

South Jakarta, Indonesia

Prof. Dr. Henny Warsilah, DEA
Prof. Dr. Agus Heri Purnomo

Contents

About the Editors

Henny Warsilah is a research professor in development sociology (urban-rural) at the Center for Society and Culture Research, National Research and Innovation Agency (PMB BRIN). She graduated and obtained her PhD from Sorbonne University, Paris, France in 1995. Her research focuses on urban development, social transformation, resilient cities, and social conflict. Her latest publications include a book chapter, 'Resilience of Coastal City Facing Climate Amendment: Benoa and North Jakarta Bay Reclamation Cases' (Springer Nature, 2020), an edited volume on climate change, community response, and resilience titled *Building Resilient Cities in Coastal Urban Areas* (Elsevier, 2023), and a volume, *Assembling Nusantara: Friction, Mimicry and Resonance of the Development New Capital Nusantara*, (Springer Nature, Singapore, 2023).

Agus Heri Purnomo is a research professor at the Research Center for Society and Culture at the National Research and Innovation Agency. His research interests focus on the socio-economic aspects of marine and fisheries. He has also led two national research institutions, in socio-economics and in fish processing/technology.

List of Figures

List of Tables

Part I
Introduction

1 The Socio-Ecological Crisis in Southeast Asia and Indonesia from a Sustainable Development Perspective

Henny Warsilah

Abstract

This edited volume analyzes the issue of natural resource capitalization carried out in forest areas, along the coast, on peatlands, farmland, and within community customary forests, for the benefit of the mining and tin industries, plantations, and the food industry. Societal crises emerging from the capitalization of natural resources include social exclusion and marginalization, poverty, food and clean water shortages, floods, and other natural disasters. The aim of this book is to examine various socio-ecological crises in different natural resource areas. Its novelty lies in presenting an interdisciplinary and multi-perspective analysis rooted in a sustainable development approach. Data were collected qualitatively and analyzed descriptively.

Keywords

Natural resources · Socio-ecological crisis · Capitalization · Sustainable development

1.1 Wealth and Exploitation of Natural Resources in Indonesia

Indonesia, like other Southeast Asian countries, is fortunate to possess abundant natural resources which, along with human capital, constitute the most crucial components for national development. Countries such as Singapore, Japan, and Taiwan possess substantial potential natural resources compared to Indonesia, and their success in development is attributed to their well-developed human resources.

In order to benefit the entire population of the region, these natural resources must be exploited to their fullest potential, yet research shows that they have not been optimally employed for the welfare of the people. Additionally, Indonesia continues to export a significant amount of raw natural resources, leading to lower prices.

To ensure the availability of raw materials from natural resources, their utilization should be guided by the principles of sustainable development. Sustainable development involves meeting present needs without compromising the ability of future generations to meet their own needs. Over the past few decades, the exploitation of natural resources in Indonesia has indeed contributed to significant benefits for the country's economy. However, excessive exploitation has resulted in

H. Warsilah (✉)
Research Centre for Society and Culture - National Research and Innovation Agency, Jakarta, Indonesia
e-mail: hennywarsilah@gmail.com

H. Warsilah and A. H. Purnomo (eds.), *Approaching Natural Resources for Sustainable Development in Indonesia*, Sustainable Development Goals Series, https://doi.org/10.1007/978-981-96-6913-4_1

environmental and social damage, especially in regions rich in natural resources, such as the Kalimantan and Papua regions. Companies often fail to adhere to the principles of sustainable development or the promotion of community benefits when granting rights to manage natural resources, disregarding local communities and their environment. Unilateral control over natural resources without involving local communities in various regions has led to conflicts, both overt and latent, that have the potential to escalate at any moment.

In its national policy, the government has articulated a vision for the management of natural resources to ensure long-term benefits, aiming to achieve "fair, comprehensive, integrated, and environmentally-aware natural resource management for the welfare of society". The management process includes planning, implementation, monitoring, and evaluation, while addressing the interests of various sectors and administrative regions. It must also prioritize environmental sustainability, paying heed to the balance of ecosystems and the environment's carrying capacity.

1.2 Indonesia's Natural Resource Potential

The principles of natural resource management are as follows: (1) efficient resource management; (2) conservation of natural resources; (3) preservation of renewable natural resources; and (4) preservation of non-renewable natural resources (e.g., iron ore, nickel, tin, copper, gold, silver, and bauxite).

1.2.1 Petroleum

Indonesia's oil potential continues to decline due to continuous usage and the country has begun importing petroleum to meet domestic demand. Crude oil is an energy source for vehicles, factory machinery, and other purposes. The state electricity company (PLN) partially relies on petroleum for electricity generation. Petroleum resources are found both on land and beneath the ocean floor. Exploration and drilling for oil take place both on land and under the seabed (see Table 1.1).

Table 1.1 Oil mining on land and sea

Ground Drilling	Offshore Drilling	Processing Place
Aceh (Perlak)	Malaka Strait	Pangkalan Brandan
North Sumatra (Langkat)	Jawa Sea	Dumai (Riau)
South Sumatra (Sungai Gerong)	Cina Selatan Sea	Plaju (Jambi)
West Jawa (Cirebon)	Sunda Strait	Cilacap
Central Jawa (Cepu)	Makassar Strait	Balikpapan
East Kalimantan	Sulawesi Sea	–
South Kalimantan	Karimata Strait	–
Papua (Sorong, Biak)	Natuna Island	–

Source Abstracted from various sources and BPS data

1.2.2 Natural Gas

Liquefied natural gas, known as LPG (liquid petroleum gas) or LNG (liquid natural gas), is produced through natural gas liquefaction. This process also yields a type of light oil that serves as a raw material for industries like plastics and fertilizers. The primary source of natural gas lies in the Natuna Islands, with additional reserves located in Bontang (East Kalimantan) and Arun Nanggroe Aceh Darussalam.

1.2.3 Coal

Indonesia's first coal mining venture was initiated in Pengaron (East Kalimantan) in 1849, followed by Umbilin (West Sumatra) in 1892 and Bukit Asam (South Sumatra) in 1919. The Bukit Asam coal deposits are situated near the surface. While Bukit Asam produces high-quality aged coal, other regions of Indonesia yield lighter coal. Indonesia's coal production, which has increased by 273% over the past decade, ranks third globally.

1.2.4 Forest

Indonesia possesses one of the largest forest reserves globally, rivaled only by Brazil and Zaire. According to 2011 data from the Ministry of Forestry of the Republic of Indonesia, the country's forest coverage had reached 99.6 million hectares. However, regrettably there has been a notable decline in forested areas with the country's forests experiencing one of the highest rates of destruction globally. The annual deforestation rate reached 610,375.92 hectares, ranking the third highest in the world (Table 1.2).

Table 1.2 Forested area, selected countries

No	Country	Forest Area (Km^2)
1	Indonesia	99,600,000 Ha 99.600 km^2*
2	Russia	8,085,986 km^2
3	Brazil	4,714,920 km^2
4	Canada	3,101,340 km^2
5	America Serikat	3,034,070 km^2
6	China	2,054,056 km^2
7	Australia	1,632,912 km^2
8	Democratic Republic of Congo	1,329,707 km^2
9	Peru	685,536 km^2
10	India	677,598 km^2

Source Various sources and BPS
*Indonesia's forest area covers 99.6 million hectares or 50.9% of the total land area. 92.5% of the total forest area, equivalent to 88.4 million hectares, is within forested regions

1.2.5 Marine Resources

Two-thirds of Indonesia's territory consists of sea, leading to a significant potential for the country's marine wealth. According to a report by the Food and Agriculture Organization (FAO), Indonesia's sustainable potential for marine capture fisheries resources is estimated at around 6.5 million tons per year. Indonesia's captured fish production ranks third globally, following China and Peru. In farmed fish production, Indonesia secures the fourth position, trailing behind China, India, and Vietnam (FAO 2009).

Indonesia's marine wealth is also evident through the diversity of its marine biota. Indonesian sea is home to 8,500 species of fish, 555 species of seaweed, and 950 species of coral reef biota. This substantial biodiversity places Indonesia as the country with the largest marine biodiversity on a global scale.

1.3 Natural Resources Socio-Ecological Crisis

Among the various crises facing the world today is the social-ecological one. The ecological crisis has reached its final stage, seriously threatening the existence of planet Earth, its inhabitants, and all life on it. There are even threats that could lead to a gradual but certain end of life (Capra 2003, 1983). Upon closer examination, this crisis can be seen to be rooted in irresponsible human behavior toward the environment (Keraf 2014).

The definition of a socio-ecological crisis is a state of "accumulated suffering" where interconnected human and non-human elements serve as a condition for the expansion of capitalism. The concept of social ecology explains empirical cases that embroider the ruins of capitalism's praxis (Batubara et al. 2020). Among the causes of the socio-ecological crisis are uncontrolled logging and mining (Walhi 2021), as well as the rapid growth of the world's population and the implementation of the developmentalism ideology, which adheres to the belief that development is the ultimate goal and encompasses everything (Batubara et al. 2020). Batubara and Handriana (2021a; b) state that the term "socio-ecological crisis" in Indonesia has been extensively developed and utilized by non-governmental organizations to illustrate the interconnection between social and ecological issues rather than viewing them as separate entities. According to these authors, the socio-ecological crisis is not merely an "impact" but rather a "condition" of capitalism. In essence, the crisis becomes an inherent aspect of each round of capitalist development.

In the capitalist production process, commodities are crafted from raw materials extracted from fields, forests, or the Earth's depths. Similarly, machinery, gas, and oil, and even land allocated for constructing factory buildings or industrial plantation areas, serving as capitalist means of production, are none other than non-human entities. Capitalists spare no expense in accessing minerals, coal, oil, water, or land. These components arise from geological processes occurring over a specific timeframe. Capitalists procure these non-human entities at a low cost (Moore 2016), incorporating them into the production process, either directly or through multiple stages, until they become integral to the capitalist industry.

At the opposite end of the scale, individuals are deprived of their means of production (plantation land, agriculture), whether by the state or corporations. In some instances, this loss drives them to become wage laborers. Land grabbing in rural areas has compelled many to forfeit their means of production, which previously served as their economic foundation, often leading them to migration to industrial hubs, which they hope will fulfill their needs (Habibi 2021). The intense competition for jobs in the formal economic sector in urban regions frequently drives individuals to the informal economic sector, where they work as street vendors, hawkers, porters, and construction workers (Habibi 2021). Thus the loss of agricultural land or plantations burdens individuals with carrying the capacity, transitioning from being owners of means of production to becoming solely reliant on their labor power.

Capitalism is a production system centered around maximizing profits. The capitalist production system, as viewed by Karl Marx (2004), is characterized by capitalists (owners of capital) investing a certain amount of money to acquire workers and production facilities, enabling the production of commodities that are subsequently sold in the market. This selling process results in capital generating more money than initially invested. This increased nominal value of money is referred to as profit.

The capitalist production system operates in a continuous manner, not confined to a single cycle. According to Marx (2004), the system operates under the principle of capital accumulation, which compels capitalists to continually reinvest surplus profits into the production process. In essence, a portion of the total surplus profit obtained by the capitalist in the initial production period is reinvested into capital, initiating the subsequent production period. This accumulation of capital serves as the driving force behind the capitalist system. Examples of the natural resource political capitalism paradigm include the conservationist paradigm, the developmentalist paradigm, and the eco-populist paradigm (Witter and Bitmer 2005).

In Indonesia, the Mining Network NGO (JATAM) reports there are 3,092 unreclaimed mines throughout the country, causing extensive environmental damage. In coastal areas the socio-ecological crisis is associated with excessive sea reclamation efforts, leading to damage to marine and coastal ecosystems, as well as marginalization and exclusion of coastal communities. There are also efforts to destroy coral reefs and underwater ecosystems.

The socio-ecological crisis in Indonesia is largely driven by the implementation of mega projects, including:

1. *Sectoral Power*: continuing the land-grabbing regime (plantations and bases) and extractive industries (mining).
2. *The Master Plan for the Acceleration and Expansion of Indonesia's Economic Development (MP3EI)*, outlined in Presidential Regulation Number 32 of 2011, aims to transform Indonesia into a developed country by 2025. The plan envisions achieving a per capita income of US$14,250–15,500 and a Gross Regional Revenue Service (GDP) ranging from US$4.0–4.5 trillion. The MP3EI provides guidelines for ministries and regional governments to align their policies with Indonesia's economic development acceleration. The various programs outlined in the MP3EI are intended to advance Indonesia's vision of becoming the 10th largest country in the world by 2025, characterized by economic growth exceeding 6%.

3. *New MIFEE: Food Estate Program.* The food estate project, envisioned by President Joko Widodo to bring food diversity to the Indonesian nation, has encountered resistance from various environmental activists and has caused conflicts within civil society. This project represents the government's response to warnings from the UN Food and Agriculture Organization (FAO) regarding the risk of a food crisis due to the Covid-19 pandemic. The FAO highlighted that the pandemic may disrupt the food supply chain and lead to community malnutrition (FAO 2020).
4. *Green Development.* The tourism industry demands excessive sea reclamation, which damages marine and coastal ecosystems and results in the marginalization and exclusion of coastal communities, particularly fishermen residing in coastal areas.
5. *National Economic Recovery (PEN): Growth Regime and Economic Liberalization.* The global economic situation is becoming increasingly challenging due to the slowdown in production activities, causing millions of people to face the threat of lay-offs. The level of resistance by the International Monetary Fund (IMF) is expected to rise significantly, especially in developed countries. The combined threat of health and economic crises has also triggered turmoil in financial markets, with developing countries being particularly vulnerable.

Currently, various studies on development indicate that Indonesia's national development policies predominantly follow the paradigms of "developmentalism" and "conservationist developmentalism" (e.g., ecotourism). The dominance of sectoral power is evident, the land-grabbing regime (especially in the plantation sectors) and extractive industries (mining) being notable examples.

Reports from the National Commission on Human Rights highlight severe, systematic, and chronic human rights violations against Indigenous forest communities in the mining sector (gold, coal, nickel), plantations (palm, sugarcane, cocoa), and mining (conservation and forest protection). These violations involve various entities such as security police, the Indonesian National Armed Forces, companies (both state-owned and private), and sectoral regimes.

Despite this, many Indigenous forest communities (MHA) continue to uphold their ancestral (local) religions. The conflicts and appropriation of MHA living space in forest areas can be categorized as: (a) naturalization of land disposition (e.g., MIFFE, conservation in Lampung); (b) horizontal "bump" (e.g., Sumbawa, Maluku); (c) politics of territorial expansion (e.g., Aru Archipelago, Maluku); (d) politics of concessions and extraction (forestry, plantation, mining); and (e) politics of regional autonomy (Komnas 2013; Walhi 2001).

These cases demonstrate a common pattern among rural communities around the mining, marine, plantation, and forestry industries. The communities lack both protection of their rights and recognition of their living spaces, resulting in social exclusion, marginalization, criminalization, violence, terror, and violations of human rights, including gender justice. Many members of these communities find themselves trapped in post-relational poverty, marked by a decline in self-sufficiency and food security. This situation also leads to the erosion of traditional knowledge and local wisdom, as well as the loss of community identity, culture, language, and civilization.

In *The Three Ecologies*, Guattari (2008) asserts that the emergence of ecological issues with a sentiment will eventually threaten the survival of human life. This threat manifests as a decline in awareness of life, both individually and collectively. From an ecological perspective, there are at least three prerequisites for sustainable and inclusive development of coastal, rural, and forest areas:

(a). Every development activity must occur in an ecology-aligned location that meets biophysical requirements, following spatial planning for coastal, island, and ocean development areas.

(b). The importance of information on existing land use and waste disposal preparation,

including details on the type of waste entering coastal and marine areas.
(c). The quality level of coastal waters and air in rural areas, particularly those that can be rejuvenated, must not exceed the potential recovery capacity of the coastal area.

Furthermore, current ecological predicaments, according to Murray Bookchin (Batubara and Handriana 2021a, b: 33), cannot be dissociated from the Cartesian perspective, which frames social and non-human entities as binary entities. This hierarchical outlook positions humans at the center of the life network in the universe, granting them the power to control and dominate subordinate external entities (non-humans).

The government's choice of an extractive and exploitative economy of natural resources is purportedly influenced by several factors:

(1). The persistence over time of the residual consequences regime of the "growth and liberal economy".
(2). The triumph of the capitalist-neo-liberalistic character of the development paradigm school in natural resource political policies.
(3). A new wave of global capital space reorganization involving anti-socio-ecological justice in production, consumption, and distribution.
(4). The influence of the political-economic oligarchy and the ongoing corruption of natural resources within the state apparatus.
(5). The limited visibility and prominence of non-extractive and non-exploitative development alternatives for natural resources.

Signs of spatial reorganization and the extensive expansion of global capital are evident. Following the collapse of capitalism, the state becomes a market. Countries worldwide foster an array of desires, uniting based on the objective of economic dominance. It's as if they are addressing the crisis of capitalism through a "spacious temporal fix capital" (Harvey 2010).

While the socio-ecological crisis recognizes that environmental damage is intertwined with social activities, it does not imply that all humans share the same role and responsibility in triggering and addressing these issues. For instance, the depletion of groundwater wells in Semarang is not solely caused by domestic consumption. A significant factor contributing to the drying of shallow water wells owned by residents is the proliferation of the hotel industry in the province, with the consequent growing demand for water supply for hotel rooms (Batubara and Henny Warsilah 2010).

1.4 The Sustainable Development Perspective

A sustainable development paradigm represents a wise choice to prevent recurring socio-ecological crises in Indonesia. Sustainable development is a developmental process that optimizes the utilization of natural resources. Essential to this development is the capacity to satisfy the current generation's needs while safeguarding the health and well-being of future generations, rather than solely meeting immediate requirements. This means that economic progress must harness natural resources while enabling subsequent generations to perpetuate ongoing development (SMI 2022).

The UN's Sustainable Development Goals (SDGs) program has been operational in Indonesia since 2015, succeeding the Millennium Development Goals initiative. Among the goals aimed at achieving sustainable development are:

Poverty alleviation.

Eradicating hunger and achieving food security, improved nutrition, and promoting sustainable agriculture.

Fostering a healthier, more prosperous life that promotes well-being across all age groups.

Ensuring inclusive, quality education.

Advancing gender equality and women's empowerment.

Establishing robust infrastructure, encouraging innovation, and promoting sustainable industrialization, among others.

Promoting sustainable consumption and production patterns.

Taking significant measures concerning climate change and its impacts. (SMI 2022).

Sustainable development, when implemented effectively, is believed to enhance overall quality of life. Well-being constitutes one of the cornerstones of sustainable development. Other key concepts are the expansion of renewable resources, the preservation or enhancement of the quality of human life in the present and future, the judicious utilization of natural resources to minimize waste and environmental harm, and proficient natural resource management for future utilization.

Sustainable development principles apply to each sector as follows:

The economy: From an economic perspective, the principle of sustainable development aims to enhance worker skills, thereby increasing competitiveness. This, in turn, is expected to provide individuals with better job prospects and improved incomes. Moreover, it supports the development of fundamental infrastructure, including real estate and information systems.

Energy: Sustainable development necessitates more prudent energy consumption. Several methods can be adopted to conserve energy:

- Optimal utilization of renewable energy sources.
- Efficient management of available energy resources.
- Prioritizing the advancement of mass transportation.

Ecology: Ecological sustainability is the essence of this principle. To enhance environmental preservation, several measures can be undertaken, such as ensuring the existence of green spaces and fostering well-integrated transportation systems and buildings.

Social and culture: involvement or participation: Sustainable development mandates active participation from the broader community, with the government facilitating the process. Communities play a pivotal role in sustainable development initiatives. The government, acting as a community empowerment facilitator, must also be receptive to and accommodating of people's aspirations.

Equality or Equity: Equity is a fundamental goal of sustainable development. By upholding the principle of equality, this form of development seeks to alleviate economic disparities and ensure equal opportunities for all members of society.

1.5 Overview of the Book

The chapters in this book are not guided by a single overarching theoretical framework. Instead, they aim to depict various socio-ecological crises alongside some of the new trends discussed above. These discussions are rooted in natural resource policy, politics, law, socio-economic and cultural considerations, as well as environmental contexts.

The book aims to comprehend the socio-ecological processes and initiatives that have occurred, are ongoing, and will unfold. It seeks to deeply integrate socio-ecological conditions and impacts, shedding light on the significance of following these initiatives, particularly for vulnerable groups with limited influence, such as coastal communities, forest-edge communities, and inland communities.

The content of the book aligns with the four pillars of the SDGs for Indonesia:

(1). *Social development pillar*. Social development aims to ensure the equitable and fair fulfillment of fundamental human rights, contributing to the overall welfare enhancement of all individuals.

(2). *Economic development pillar.* Economic development encompasses achieving robust and sustainable economic growth through measures such as creating employment opportunities, fostering innovation, promoting inclusive industries, establishing adequate infrastructure, ensuring access to affordable clean energy, and nurturing collaborative partnerships.

(3). *Environmental development pillar.* Environmental development focuses on achieving effective natural resource management and a sustainable environment, serving as a foundation for the well-being of all life.

(4). *Political, legal, and governance development pillar.* The political, legal, and governance development aspect of SDGs involves establishing well-defined natural resource policies, enacting effective and transparent laws, and implementing accountable and participatory governance of natural resources. These efforts contribute to fostering stability, security, and a law-based state (Bappenas, SDG Goals).

The research for this book was conducted empirically and established a broad network encompassing what is understood as a socio-ecological framework. This approach acknowledges that many paradigms and dimensions within each of these terms are subjects of contention. Numerous social and ecological crises stem from the sustainable development paradigm, which is underpinned by three dimensions: political policy and natural resource management, socio-economic factors, and environmental management. In relation to each of these dimensions—political, economic, social, and environmental—some cases raise the question of whether it's truly possible to disentangle political, economic, and ecological factors from each other.

First, the legal, political, and policy dimension within natural resource management itself has given rise to the capitalization of natural resources. The enactment of Law Number 11 of 2020, known as the Omnibus Law, and the revision of the Minerals and Coal Law, have relaxed environmental protection measures (as discussed by Laely Nurhidayah et al. in Chapter 2). This situation is projected to lead to an average deforestation rate exceeding 5% per year, mining exploitation (particularly coal) surpassing 500 million tons per year, extensive infrastructure development under the concept of national strategic projects, and the expansion of oil palm plantations of up to 15 million hectares. Erwiza also highlighted that while the extractive tin industry has contributed significantly to state revenues, it has also caused environmental harm. Similarly, in Chapter 3, Firman Noor et al. emphasize the lack of adequate historical natural resource governance, asserting that Indonesia now requires a strong commitment to adopting a natural resource governance framework to ensure a sustainable environment as a foundation for its current status as a middle-income country.

Second, in Chapter 6, Maxensius et al. discuss the economic and social dimensions, emphasizing the necessity of establishing integration between economic, social, and environmental interests. However, the current challenge lies in determining the appropriate economic instruments and the necessary institutional arrangements to effectively sustain the market for environmental services. Henny Warsilah, on the other hand, highlights that the national economic interests related to food procurement through the food estate program do not necessarily stimulate the local economy. In fact, this program has led to a socio-ecological crisis, originating from deforestation to the introduction of food commodities plantation programs. Furthermore, Asmanah et al. (Chapter 7), Surati et al. (Chapter 11), Ekawati et al. (Chapter 10), Rusman et al. (Chapter 8), and Felani et al. (Chapter 5) emphasize that the values of ecological wisdom rooted in the local community's culture play a vital role in maintaining the equilibrium of natural resources, particularly among communities residing near forests. These communities uphold unwritten rules that govern the sustainable utilization of forest resources, preserving a tradition that has been passed down through generations.

Lastly, within the environmental dimension, Agus Heri, Zainal Arifin, Ary Wahyono, Kazumi, Henny Warsilah, Choerunisa, and Ismail point out that the socio-economic pressures on nature are substantial. The delicate balance between humans and nature is disrupted when local governments initiate welfare improvement programs, such as tourism development, that subsequently impact animal habitats and other marine environments. Robert Siburian and Dicky Rachmawan in Chapter 14 underscore the escalating global deforestation rates attributed to human activities, particularly in regions managed by Indigenous peoples. Indonesia, with the largest population of Indigenous peoples in Southeast Asia, has also been affected by this trend.

The book also presents chapters from Japan and Malaysia as a comparison for Indonesia, discussing how a socio-ecological crisis due to exposure to natural resources can affect the lives of marine biota, shellfish and birds, and showing the importance of sustainable natural resource management.

References

Bappenas, SDG Goals, https://sdgs.bappenas.go.id/.

Batubara and Warsilah, H (2010) Maleh Dadi Segoro: Social Ecological Crisis in the Coastal Areas of Demak and Semarang. Lintas Nalar Publisher. Yogyakarta

Batubara B, dan Handriana E (2021a) Urbanisasi Sebagai pabrik Krisis Sosial-Ekologis: Berdialektika dengan Prisma 1971–2021, Majalah PRISMA. https://www.prismajurnal.com/article/urbanisasi-sebagai-pabrik-krisis-sosialekologis-berdialektika-dengan-prisma-1971-2021/

Batubara B, Handriana E (2021b) From socio-ecological crisis to social ecology: the case of Suburbia Semarang. Prism 40(3):30–46

Batubara B, Warsilah H, Wagner I, Salam S, Semarang-Demak KP (2020) Maleh dadi Segoro: socio-ecological crisis in the Semarang-Demak Coastal Area. Cross Reason, Bantul

Capra F (1983) The turning point: science, society and rising culture. Flamengo, London

Capra F (2003) The hidden connections. Flamengo, London

FAO (2009) Fisheries or food security guidelines. Food and Agriculture Organization. https://openknowledge.fao.org

FAO (2020) The "State of the World's Forests" or "State of Food and Agriculture" series. Open Knowledge FAO. https://openknowledge.fao.org

Gallent PK (2010) Rural Sociol 75(1):28–57

Guattari F (2008) The three ecologies. The Athlone Press, London and New Brunswick, NJ

Habibi M (2021) The rate of capital and the dynamics of urban informal economy class. J Agrar Change 21(2). https://www.researchgate.net/publication/351825354_Masters_of_the_countryside_and_their_enemies_Class_dynamics_of_agrarian_change_in_rural_Java

Harvey D (2010) New imperialism: the genealogy and logic of contemporary capitalism. The Resist Book

Keraf SA (2014) Filsafat Lingkungan Hidup, Alam sebagai Sebuah. Sistem Kehidupan (bersama Fritjop Capra). Kanisius, Yogyakarta

Komnas (2013) Annual Report of the National Commission on Human Rights 2013. https://komnasham.go.id

Martinuzen J (1997) State, society and market: a guide to competition theories of development. Zed Books Ltd., London

Marx K (2004) Capital: a critique of political economy. Book 1. Hasta Mitra, Jakarta

Moore JW (2016) The rise of cheap nature. In Moore JW (ed) Antropocene or Capitalocene?: Nature, history, and the crisis of capitalism. PM Press, Oakland

Rachman NF, Yanuardy D (2014) Above the socio-ecological crisis how does the MP3EI megaproject work? In: Rachman NF, Yanuardy D (eds) MP3EI: master plan for the acceleration and expansion of Indonesia's socio-ecological crisis. Homeland Beta, Bantul

SMI (2022) Sustainable development: definition, concept and goals. https://ptsmi.co.id/pembelian-berkeuntungan.

Walhi (2001) Exclusion power package: collaboration of corporate oligarchy, state legitimacy and military/security authorities. https://koran.tempo.co/read/opini/457387/kuasaeksdinding-dan-krisis-agraria)

Walhi (2021) Socio-ecological crisis and capitalism. https://walhi-jogja.or.id/index.php/2021/11/10/krisis-social-ekologis-dankapitalism/#:~:text=Pada%20dasarnya%2C%20concept%20crisis%20social,Bossman%20

Witter and, Britner (2005) The role of molybdenum in agricultural plant production. Ann Bot 96(5):745–754. https://doi.org/10.1093/aob/mci226

Part II

Natural Resource Governance from a Political, Policy and Legal Perspective

Environmental and Societal Challenges in Indonesia: An Evaluation of Laws, Policies, and Natural Resources Governance

2

Laely Nurhidayah, Shawkat Alam, and Agus Suntoro

Abstract

Indonesia is rich in natural resources, but its keen focus on economic advancement has triggered environmental and societal issues, including climate change, biodiversity decline, habitat loss, flooding, and Indigenous land conflicts. In accordance with its aim of achieving the Indonesia 2045 vision, the government has introduced policies to attract businesses and investors, notably enacting the Omnibus Law (Law Number 11/2020) and the revised Mineral and Coal Mining Law, which have relaxed environmental protections. Consequently, there are concerns over escalating deforestation, intensification of mining, infrastructure projects under the national strategic initiative, and the potential spread of oil palm plantations to cover 15 million hectares. This chapter examines the laws, governance, and policies driving this development trajectory, emphasizing the government's mining policies and their social and environmental ramifications. A qualitative method is used, encompassing policy analysis, literature scrutiny, and insights from East Kalimantan mining case studies, to underscore the pressing need for reform in natural resource management.

Keywords

Natural resources · Governance · Indonesia · Laws and policies · Environmental issues · Ecological and social challenges · Mining policies

2.1 Introduction

Indonesia boasts a wealth of natural resources, including forests, minerals, peatlands, and fertile soils that sustain cash crops such as coffee, palm oil, cocoa, fruits, spices, and sugar. The mining sector is one of the country's largest sectors. Indonesia's mineral exports primarily comprise natural gas, crude petroleum, coal, nickel, bauxite, gold, tin, and copper (Elisha Sawe 2018). However, despite this natural abundance, 11.2% of Indonesia's population remains below the poverty line, experiencing environmental degradation due to the exploitation of these resources (Elisha Sawe 2018). Coal mining in East Kalimantan has led to adverse environmental and social consequences, including flooding, water

L. Nurhidayah (✉)
Research Center for Law - National Research and Innovation Agency, Jakarta, Indonesia
e-mail: lael003@brin.go.id

S. Alam · A. Suntoro
Macquire University, Sydney, Australia

H. Warsilah and A. H. Purnomo (eds.), *Approaching Natural Resources for Sustainable Development in Indonesia*, Sustainable Development Goals Series, https://doi.org/10.1007/978-981-96-6913-4_2

shortages, pollution, deforestation, biodiversity loss, land tenure disputes with locals, and tragic incidents of children drowning in abandoned coal mine pits (Yovanda 2019).

Despite the ongoing environmental degradation, Indonesia has not pivoted towards a more environmentally friendly approach to natural resource management. To realize its vision for 2045, the government has introduced various proactive and adaptive policies to entice businesses and investment. The recent introduction of Law Number 11/2020, also referred to as the Omnibus Law, along with revisions to the Mineral and Coal Mining Law, has relaxed environmental safeguards. Consequently, it is anticipated that there will be a rise in deforestation rates, averaging over 5% annually, mining activities (especially coal) surpassing 500 million tons annually, extensive infrastructure projects under the national strategic project framework, and the expansion of palm oil plantations covering up to 15 million hectares. The global community, acknowledging these concerns, has shifted towards a comprehensive approach emphasizing sustainability and sustainable development. This principle, originating from the 1972 United Nations Conference on the Human Environment, was highlighted in the *Our Common Future* report. The 2030 Agenda for Sustainable Development, adopted by UN Member States in 2015, provides a detailed plan for enhancing the well-being of people and the environment. In essence, sustainable development seeks to meet today's needs without compromising the ability of future generations to meet theirs (Brundtland 1987).

This chapter analyzes the laws, policies, and governance of natural resources, centering on the current development paradigm. In particular, it examines the government's stance on mining and its subsequent social and environmental implications. Utilizing a qualitative methodology, which encompasses legal and policy analysis, a review of existing literature, and empirical investigations via case studies from mining activities in East Kalimantan, the research underscores the urgency of overhauling the legal framework and governance of natural resources in Indonesia.

2.2 Methodology

This chapter employs a qualitative approach, encompassing literature reviews, policy analysis, and empirical research on mining effects in East Kalimantan. It focuses on the Omnibus Law on Job Creation No. 11/2020 and the Revised Mineral and Coal Mining Law No. 3/2020. Fieldwork in East Kalimantan involving interviews with local communities, officials, NGOs, and scholars was crucial for data collection. Direct observations were made in villages including Kampung Jawa, Sanga-Sanga Dalam, Sumbersari, Kutai Kartanegara, and East Kalimantan.

2.3 Theoretical Framework

This chapter is grounded in the concepts of environmental justice and sustainable development.

A contemporary development in environmental law (Eghosa 2023), *environmental justice* asserts that everyone deserves an equal right to a healthy environment. It mandates the equal distribution of environmental harms and benefits among all social groups (McLaren 2003). Core components of this framework include the equal distribution of environmental risks and rewards, inclusive participation in decision-making, access to legal remedies, the recognition of Indigenous and local knowledge, respect for cultural diversity, and ensuring that communities thrive. Central to this concept is the precautionary principle, which emphasizes caution in the face of scientific uncertainty about the environmental consequences of specific actions (White 2013). Entities must prove their operations would not harm health or impact minority groups, making environmental impact assessments essential for permits. Oil, timber, and minerals extraction can often degrade ecosystems and impact local cultures (White 2013).

Anchored in equity, *sustainable development* promotes balanced progress across its three dimensions: social, economic, and environmental. Defined in the Brundtland Report as meeting current needs without compromising future generations, sustainable development seeks a balance:

addressing ecological concerns without disrupting existing economic power structures (Sachs 1993). In mining, this means using sustainable methods to protect the environment while maintaining economic benefits. The social aspect of sustainable development encompasses a range of notions like justice, equitable resource distribution, enhancing quality of life, and values like freedom. Community involvement is key, meaning stakeholders, such as those impacted by mining, should participate in decision-making. Jabareen (2008) accentuates the importance of participation and empowerment within the sustainable development framework.

2.4 Findings and Analysis

2.4.1 Acceleration of Environmental Damage Following the Implementation of the Omnibus Law on Job Creation

On November 2, 2020, President Joko Widodo enacted Law Number 11 of 2020, titled Job Creation. This law is primarily aimed at streamlining business permits, covering environmental licenses, location permits, spatial planning, and implementing a risk-based approach to the permit system (Coordinating Ministry for Economic Affairs 2020). However, the Constitutional Court deemed the Job Creation Law unconstitutional in Decision Number 91/PUU-XVIII/2020, and urged the government to amend it. The government responded by introducing Government Regulation Number 2 of 2022 about Job Creation, bypassing the court's directive. This regulation was later converted into Law Number 6 of 2023 following an agreement between the government and the legislature on March 31 2023 (Wildan 2023).

Lembaga Bantuan Hukum (LBH) Jakarta (the Indonesian Legal Aid Institute) criticized the Job Creation Law for issues in the legal system, alleging its drafting was opaque and hasty, and the content potentially harmful to the public, workers, and the environment (Helmi et al. 2021). Similarly, the National Human Rights Commission pinpointed two concerns: the non-transparent drafting and discussion processes, and problematic clauses relating to environmental protection, such as changes to licensing mechanisms and corporate sanctions (Agus Suntoro et al. 2020). The Indonesia Center for Environmental Law (ICEL) voiced concerns over the law centralizing environmental management and protection, and the shift towards monetary penalties instead of criminal sanctions for environmental offenses (Indonesia Center for Environmental Law 2020).

2.4.1.1 Centralization of Authority

Law Number 22 of 1999, which focuses on decentralization, encompasses regional autonomy, administrative measures, and fiscal aspects (Jati 2012). The decentralization of the environmental sector is underscored in Law Number 23 of 2014, which delegates 11 responsibilities to regional governments. These include planning, strategic environmental studies, pollution control, and biodiversity, among others (Indonesian Government 2014). The regional government's role in the environment was further strengthened by Law Number 32 of 2009, granting them autonomy in tasks like planning, issuing permits, and impact management (Government and Parliament 2009). However, the introduction of the Job Creation Law has centralized authority within the central government, curbing the regional government's role in environmental management. Notably, Article 402A and Article 164 of this law emphasize that several authorities previously vested in ministers or regional entities now fall under the president's jurisdiction (Firmansyah et al. 2020).

A significant change is the central government's exclusive control over policies regarding environmental impact analysis and environmental management and monitoring efforts, as detailed in Article 63(e) of the revised Environmental Management and Protection Law. This move aims to streamline business permits and investments, even though the associated risks primarily concern regional authorities (Sianturi 2023).

The Job Creation Law also impacts the role of the Provincial Regency/City Environmental

Services. While previously there was a delineated authority structure between central and regional governments, centralization has now constrained regional governments in overseeing company regulations (Rimbawan et al. 2020). The centralization manifests in two main ways: the exclusive control over essential environmental permits by the central government; and the diminished regional authority to manage environmental and social challenges, especially those arising from centrally issued permits (Nugroho and Syahruddin 2021).

2.4.1.2 Changes in Environmental Permits

The Job Creation Law introduces a significant shift by replacing Indonesia's licensing system with a focus on simplifying business licensing. In the environmental realm, this translates to transitioning from environmental permits to environmental approvals, as outlined in the Job Creation Law Article 13 paragraph (2) revision of the Environment Law. Environmental approval can take the form of either an Environmental Feasibility Decree or a declaration of environmental management capacity by a business entity. Holders of this approval are obligated to allocate an environmental guarantee fund to a government-owned bank (Government and Parliament 2020).

Previously, an environmental feasibility decision was a prerequisite to obtaining an environmental permit, which served as a foundational tool for businesses to demonstrate their environmental responsibility to the government (Nugroho and Syahruddin 2021). This approval means the government either endorses or denies business operations based on environmental factors. The Job Creation Law's primary intent is to streamline environmental regulations for easier business operations, moving from a permit-focused system to a risk-based one (Zein and Nurasikin 2022). The essence of the traditional environmental licensing system, including the AMDAL, was to ensure businesses operate sustainably without harming the environment, aligning with sustainable development principles and intergenerational justice (Mahardika 2021).

Switching to environmental approvals will affect the government's ability to monitor and oversee environmental activities. Previously, businesses needed to meet specific licensing requirements before commencing activities. If any business changes resulted in environmental impacts, revising the environmental permit was mandatory. Under the new law, companies are not required to report any changes in their operations, even if they influence the environment, once they have received initial approval from the central authorities (Amri 2022). The Secretary of the Environment Service in Cirebon Regency, West Java Province, highlighted that the Job Creation Law diminished local governments' environmental oversight. With the central government now handling AMDAL approvals and curbing local governments' monitoring powers, environmental reports go directly to the central government, which may not always ensure on-the-ground checks (Fitroh 2023).

2.4.1.3 Restricting Access to Justice

In the context of sustainable development and environmental justice, the government holds the responsibility to safeguard, improve, and restore environmental quality. A key regulatory tool for this is the environmental permit issued by the government through State Administrative Decrees (Adharani 2016). These permits represent government decisions made under its public authority. Through these permits, a person or business entity is either granted permission or denied the right to undertake specific business activities. The issuance of permits typically rests with administrative officials tasked with serving the public (Devi et al. 2014). Within administrative law, the environmental permits given to companies fall under the decree (*beschikking*) category, which can be challenged in the State Administrative Court. Notably, such challenges are usually initiated when a party feels adversely affected by a definitive government act, leading to legal consequences for an individual or entity (Heriyanto 2018).

Permits, seen as a manifestation of *beschikking*, are recognized as State Administrative Decisions

(*Keputusan Tata Usaha Negara*/KTUN) and are legally binding and actionable. However, with the Job Creation Law transitioning from a permit-based system to environmental approvals, there is an inherent shift in their legal interpretation – from being an actionable state decision to merely a status label. Juridically, this transition means that these approvals can no longer serve as grounds for legal challenges in the State Administrative Court (Nugroho and Syahruddin 2021). Additionally, the Job Creation Law undermines citizens' access to justice by removing their right to contest the AMDAL for high-risk business endeavors. The law intentionally omits clauses related to objection mechanisms in the Environmental Management Law, specifically Articles 29, 30, and 31, which pertain to the AMDAL Assessment Commission. These mechanisms were pivotal for ensuring transparency, precision, and strategic use of the AMDAL as a measure against environmental degradation rather than just a bureaucratic requirement (Wayan et al. 2023).

Given these modifications – (a) the redefinition of environmental permits as non-actionable environmental agreements and (b) the removal of the AMDAL Evaluation Commission – the Job Creation Law appears to diminish essential democratic rights, namely access to justice. Communities are left without recourse to contest adverse environmental.

2.4.2 Law no. 3/2020 on the Amendment to Law no. 4 of 2009 on Minerals and Coal Mining and Its Impact on Social and Environmental Well-Being

Indonesia's Law No. 3/2020 was introduced as an amendment to the previous law, No. 4/2009, concerning mineral and coal mining. Its objectives were clear and commendable on the surface: enhancing mining management, advocating national interests, emphasizing eco-friendly practices, and promoting legal clarity to facilitate investments. Yet, despite these noble intentions, the law faced a significant backlash, primarily from non-governmental organizations (NGOs) who argue that it lacks adequate environmental safeguards and fails to prioritize the well-being of the broader community.

2.4.2.1 Lack of Participation and the Centralization of Power

The enactment of Law No. 3/2020 has profoundly transformed the regulatory landscape of coal and mineral mining in Indonesia, primarily emphasizing centralization. With the new provisions of Article 4(2) granting mining jurisdiction solely to the central government, local communities find themselves marginalized, losing their earlier clout on mining issues that directly concern them.

While Law No. 3/2020 acknowledges the significance of public engagement, Article 10(2) outlines an idealistic vision: decisions regarding mining zones should be characterized by transparency, accountability, and inclusivity. In theory, this should mean gathering insights from directly affected communities, joining forces with relevant government bodies, and addressing wider socio-environmental and human rights implications. However, the experience on the ground tells a contrasting story. For affected localities, this centralized framework has become an obstacle. When their environment faces threats from mining activities, their concerns often fall on deaf ears. Their natural advocates, the local governments, have been stripped of their traditional role in supervising and intervening in mining activities by this law.

Concerns regarding this centralization extend beyond just local communities. Franky Butarbutar, in his in-depth analysis for a judicial review of Law No 30/2020, highlighted the dangers inherent in this shift of power. He contended that concentrating power not only minimizes local governments' contributions but also distances them from the communities they serve. This could pose significant challenges to authentic public participation in mining-related dialogues. Meanwhile, Gusti Agung Made Wardana introduced a constitutional perspective. He opined that such centralization starkly contravenes the principle of subsidiarity, effectively sidelining the constitutional rights of local communities and placing them at an unwarranted disadvantage.

2.4.2.2 Criminalization

The recent addition of Article 162 in Law No 30/2020 has raised alarms about possibly criminalizing individuals who protest against mining operations. This article states that anyone who hinders or interrupts approved mining activities could face up to a year in jail or a hefty fine of 100 million rupiah. The potential consequences of this regulation have alarmed many stakeholders.

Highlighting these concerns, Gusti Agung Made Wardana, a scholar from the University of Gajah Mada, discussed the potential misuse of Article 162 during a judicial review of Law No 30/2020 at the Constitutional Court. Wardana pointed out that this article could be used to target human rights activists, particularly those affected by mining activities. He further noted that it could serve as a means of legal intimidation, stifling opposition (MKRI 2022). This controversial article appears to favor corporate interests over those of local communities. Conflicts between mining companies and residents have been on the rise. For example, in Kampung Jawa, East Kalimantan, residents have been protesting against a mining company, citing environmental damage from coal mining. Similarly, in Sumbersari village, East Kalimantan, the local community has raised concerns about mining pollution affecting their fisheries and farmlands.

2.4.2.3 Mine Land Reclamation and Rehabilitation: A Lackluster Implementation

Compounding these concerns is the lack of accountability observed in post-mining activities. Article 96 (b) of the same law mandates coal mining enterprises to undertake land reclamation and post-mining rehabilitation. Yet, glaring lapses persist in its enforcement.

A 2020 report by JATAM revealed that a staggering 3,092 coal mines remained unrehabilitated, a glaring oversight in corporate responsibility (Wibisono 2022). WALHI further contends that the present legislation offers loopholes for these companies. Specifically, they can potentially extend their mining operations twice over, each time for a decade, even if they default on their land reclamation duties. In essence, Law No. 30/2020 appears to compromise community rights and environmental protections, hidden behind regulatory pretense. The risk of Article 162 being used to stifle legitimate protests is alarming, signaling challenges for activists and communities. The inadequate attention to post-mining land rehabilitation underscores the need for stricter enforcement. Striking a balance between industrial growth, community rights, and environmental safeguarding is vital, placing a responsibility on policymakers to tackle these concerns.

2.4.3 Case Study: Unauthorized Coal Mining Exacerbates Environmental Harm

East Kalimantan Province has long been regarded as the epicenter of coal mining in Indonesia, with its vast resources playing a pivotal role in fueling the region's economic growth. In 2021, data reveals an impressive economic expansion in East Kalimantan, registering a growth rate of 2.48%. A deeper analysis of these figures shows that coal mining is the primary driver, contributing a whopping 45.94% to this growth. This dominant role of coal mining is further accentuated when compared to other significant sectors like manufacturing, which contributes 17.49%, and the construction sector, which chips in with 8.7%.

The immense contribution of coal mining to the economy is not just a factor of the vast reserves; it is also about the scale and reach of coal production and distribution in the region. Figures from the Department of Energy and Mineral Resources underscore this dominance, citing that in the said year, coal production soared to 221,851,935 metric tons. To put this in perspective, the coal produced could fill over 88 Olympic-sized swimming pools every day for a year.

Interestingly, the majority of this production is destined for international shores. A staggering 144,106,065 metric tons found its way to global markets, emphasizing East Kalimantan's role as a significant coal exporter. In contrast, domestic consumption, although substantial, was more restrained, accounting for 77,745,870 metric tons. This disparity between exports and domestic consumption underscores the global demand for East Kalimantan's coal, and how intrinsic the industry is to both local and international economies (East Kalimantan Government 2023).

While these numbers are awe-inspiring, they come with their own set of challenges and implications, especially when it comes to land allocation for mining activities. The Anti-Mining Advocacy Network, commonly known by its Indonesian acronym JATAM, has been closely monitoring this situation. Their records indicate that a vast stretch of land, approximately 5.2 million hectares, has been earmarked as coal mining zones.

The administration and governance of these mining areas fall under both regional and central authorities. The regional government, which includes the Governor, Mayor, and Regent, has been active in granting mining permissions. They have issued 1,404 Mining Business Permits (locally known as *Ijin Usaha Pertambangan* or IUP). On the other hand, the central government has taken responsibility for authorizing 30 Coal Mining Business Work Agreements, referred to in the local parlance as Coal Mining Concession Work Agreements (*Perjanjian Karya Pengusahaan Pertambangan Batu Bara*/PKP2B). This data is corroborated by the findings of Affandi (2022), highlighting the intricate web of permissions and agreements that facilitate the vast coal mining operations in the region.

Despite the issuance of thousands of mining business permits, illegal coal mining activities continue to flourish in East Kalimantan. The depth and extent of these illicit practices were shockingly illuminated by the revelations of Ismail Bolong, a member of the East Kalimantan Regional Police. Bolong's confession divulged his covert role as an illegal coal collector in the Marang Kayu District, Kutai Kartanegara (Kukar) Regency. More disturbingly, his admissions about establishing a secure supply chain by bribing high-ranking police officials have unearthed a deeply entrenched nexus between the illicit coal mining industry and those expected to enforce the law (Vicki 2022).

The proliferation of these unauthorized coal mining operations in East Kalimantan gained momentum post 2014. This surge can be attributed to the moratorium policy imposed that year, which halted the issuance of new mining business permits. On paper, the policy, rooted in the stipulations of East Kalimantan Province Regional Regulation Number 8 of 2013, aimed to ensure the appropriate implementation of reclamation and post-mining activities. However, in practice, the outcome was quite the opposite. With the Moratorium on Mining Business Permits (IUP) in place and assurances from the Governor, Mayors, and Regents across East Kalimantan, one would expect a decline in environmental degradation. Sadly, that has not been the case.

Instead of curbing the harmful effects, the region has witnessed sustained environmental harm stemming from both licensed and illicit coal mining operations. Many companies, both legitimate and underground, have displayed gross negligence in managing the fallout of their mining practices. Abandoned mine pits, and reminders of unsustainable mining operations, litter the landscape. These hazardous pits have become inadvertent death traps, claiming around 40 lives, a grievous testament to the mining industry's apathy. Moreover, the audacity of some mining entities sees them establishing operations perilously close to residential zones, endangering local residents and their way of life.

Conceptualized as a financial commitment, the Reclamation Guarantee Fund (*Dana Jaminan Reklamasi*/JAMREK) mandates mining companies to set aside resources to ensure that mined lands are restored to a viable state after mining activities cease. This initiative is overseen and regulated by the regional government. Regrettably, many mining enterprises default on this obligation, further straining the region's fragile ecosystem. More concerning is the regional government's inertia. Despite such glaring violations, there is a noticeable absence of stringent administrative and criminal legal sanctions against errant companies, highlighting a concerning lack of accountability and enforcement (Hilmawan 2016).

JATAM's intensive investigation paints a concerning picture of the state of coal mining in East Kalimantan province, particularly within the Kutai Kartanegara Regency. Out of the multiple locations flagged by JATAM for illegal coal mining, a shocking majority were situated within this very province. Even more alarmingly, of the 11 locations reported by JATAM to the relevant authorities and police, only a mere three garnered any response. And in those cases, only the lower-ranking individuals on site were held accountable. The proximity of one of these unauthorized mining sites to a police station in Kutai Kartanegara – a stone's throw away – further raises questions about the inefficacy or possible complicity of the local law enforcement. The deeper JATAM delved, the murkier the waters became. High-ranking military personnel linked with supporting these illicit activities, especially in places like Sumbersari and Loa Kulu in Kutai Kartanegara, faced no legal repercussions. In a more harrowing incident in Rempana village, East Kutai Regency, an individual who dared to voice opposition to these illegal operations was attacked with a weapon (statement given to JATAM, interview in Samarinda City, East Kalimantan, June 6, 2023).

The narrative takes a darker turn with insights provided by the Indonesian Forum for the Environment, or Wahana Lingkungan Hidup Indonesia (WALHI). As part of its efforts to represent the environmental interests of East Kalimantan, WALHI uncovered the heart-breaking reality of two villages, one of which was Kertabuana village in Kutai Kertanegara Regency, literally disappearing off the map due to aggressive coal mining. This upheaval from coal mining is not equally distributed; while migrant communities and corporations enjoy the spoils, the local inhabitants are left grappling with the dire consequences. Instead of being beneficiaries of the land's wealth, the local people are relegated to low-level roles such as daily wage laborers or coal mine guards, locally termed *Wakar*. The once self-sustaining communities are now at the mercy of external forces. Without the privilege of priority rights, they face the environmental aftermath and health hazards associated with coal mining. A telling instance is the report from a resident of Sebulu sub-district, Kutai Kartengara, who developed respiratory issues due to coal dust inhalation.

When it comes to illegal mining, the disparity is even starker. While local communities might receive some monetary compensation, the real control and operations are overwhelmingly in the hands of migrant workers (interview with WALHI Kalimantan Timur, Samarinda, East Kalimantan province, June 6, 2023). The onward passage of this illegally extracted coal, whether it is being funneled to legitimate permit-holding companies or directly shipped via pontoons to coal-dependent firms, is an area demanding further investigation (interview with WALHI Kalimantan Timur, Samarinda, East Kalimantan province, June 6, 2023). Both licensed and illegal coal mining threaten food security and economic resources for the residents of Sumbersari village, Kutai Kartanegara. Over the years, this village has gained prominence as a primary food hub, supplying an array of produce ranging from fresh vegetables to diverse horticultural products, and a thriving fisheries sector. It is interesting to note that much of the agricultural success of Sumbersari can be attributed to the industrious migrants from Java Island who have been nurturing this land since the 1980s.

In a recent interview, the Sumbersari village chief explained the gravity of the situation. The village's primary concern is safeguarding its vital water sources. Springs, which play a pivotal role in the village's agricultural and fisheries sectors, are under constant threat from upstream coal mining activities. Protecting these springs is not just about preserving nature; it is a desperate fight to safeguard the livelihoods of countless individuals who depend on them. As the village chief poignantly put it, the springs are the lifeblood of Sumbersari, fueling its agricultural prowess and, by extension, its economic vitality (interview in Kutai Kartanegara, East Kalimantan, June 9, 2023).

Haris Retno Susmiyati from Mulawarman University, East Kalimantan, who is also an active member of the Indonesian Caucus for Academic Freedom (KIKA), offers a broader perspective on the coal mining situation. While the glaring issues associated with illegal mining, from both legal and environmental viewpoints, cannot be denied, it is important to realize that even sanctioned coal mining is far from benign. Whether it is illicit miners operating in the shadows or officially recognized corporations wielding permits like IUPs or PKP2B, the environmental toll is palpable. The sheer vastness of the mining area, spanning an astounding 5.2 million hectares, exacerbates the ecological impact. Such an expansive scale of operations, as highlighted by Sucahyo (2022), is a recipe for significant environmental degradation, endangering not only the natural ecosystem but also the livelihoods of communities like Sumbersari.

Highlighting the complexities introduced by the centralized approach, the Supreme Audit Agency (BPK) has also weighed in. They examined the reconfiguration of permit management and the changing dynamics of authority in coal mining, as well as investigating the strategies employed to counter illegal mining activities.

According to the regional government, their ability to respond to these challenges has been restricted. Their primary course of action has been either to relay legal matters to the Joint Law Enforcement Team (GAKUM) or to directly involve the police. This shift in responsibility is a reflection of the broader reality that all permits and enforcement mandates now lie predominantly with the central government (interview with Muhamad Hamidin from the East Kalimantan Provincial Environmental Service, Samarinda, June 8, 2023). The overarching sentiment suggests that while centralization might streamline processes at a national level, it can also inadvertently neglect the unique challenges and nuances of individual regions like East Kalimantan.

2.5 Recommendation and Conclusion

2.5.1 Recommendation

Addressing these challenges requires a thorough overhaul of Indonesia's resource management strategy, striking a balance between attracting investment and ensuring environmental and social sustainability. Unchecked illegal mining activities, made more rampant due to weak enforcement and oversight, further exacerbate the problem. Mines left abandoned without proper rehabilitation not only ruin the landscape but also present risks to communities nearby.

A key move toward sustainable mining lies in refining the environmental impact assessment (EIA) process. Criticized for their exclusivity, EIAs should involve local communities, giving them insights into potential environmental hazards and risk mitigation plans. Ultimately, for a sustainable future, Indonesia must respect and uphold the three pillars: economic growth, environmental conservation, and social equity. As mining operations encroach on community territories, it is vital to respect their rights and traditions. Indonesia's development approach should be driven by commitment to social equity and environmental justice, ensuring a balanced and prosperous future for all.

A series of recommendations to the government to change, revoke, and revise regulations and policies in the mining sector, in addition to providing economic benefits, must remain true to the principles of sustainability and ecology. Regulations that need to be revoked include the Job Creation Law Number 6 of 2023 relating to the paradigmatic change from licensing to environmental approval concerning AMDAL, UKL, and UPL. Another recommendation is that the regional government be urgently reinstated as a supervisory institution in the environmental sector, with appropriate authority; given that there are areas for improvement in governance in terms of transparency, the accountability and supervision mechanisms of the regional government need to be strengthened. A third recommendation is to revise Law No. 3/2020 on the Amendment to Law No. 4 of 2009 on Minerals and Coal Mining to make the process participatory in the preparation of its norms, especially those related to community rights, and to revoke articles that have the potential to criminalize communities who object to mining activities.

2.5.2 Conclusion

Indonesia, endowed with vast natural resources, has been a prime spot for investment, especially in the mining sector. The country's push to boost economic growth has led to the enactment of the Omnibus Law and the revised Mineral and Coal Mining Law. However, these new regulations have created significant environmental and social issues.

A notable concern arising from these laws is the centralized approach to mining governance. This system may be intended to reduce corruption, but it effectively silences local communities, depriving them of their say in matters directly affecting them. This is especially troubling when local communities face negative repercussions from mining activities but find themselves without the legal means to address these problems.

Environmentally, the implications are vast. Predictions indicate a rapid rise in deforestation, potentially over 5% annually. This destruction impacts not only the climate but also the biodiversity and the Indigenous people dependent on these forests. The coal mining industry is set for massive growth, with forecasts predicting coal extraction of over 500 million tons annually. While this might benefit the economy, the environmental toll is substantial.

Moreover, with the national strategic project initiative propelling infrastructure development, and the expected expansion of oil palm estates covering an anticipated 15 million hectares, concerns are arising about land rights, habitat loss, and increased carbon emissions. These legal changes, designed to enhance investments, might inadvertently sacrifice environment, social wellbeing, and justice, thus conflicting with the tenets of sustainable development.

References

Affandi, AA (2022) Dari Wisata Juang Menuju Kota Limbang Tambang. Samarinda

Agus S et al (2020) Position paper on the draft job creation law (Omnibus Law) in a human rights perspective. Jakarta

Adharani, NA, Yulinda (2016) Permit function in environmental pollution control (case study: lawsuit for issuance of permit for liquid waste disposal in the Cikijing River). Jurnal Hukum Lingkungan 3(1):108–132

Amri FHA (2022) Environmental permits for business activities after the stipulation of Law Number 11 of 2020 concerning job creation. Bina Hukum Lingkungan 6 (3):438–452. https://doi.org/10.24970/bhl.v6i3.265

Brundtland GH (1987) Our common future: report of the world commission on environment and development. Geneva, UN-Dokument A/42/427

Coordinating Ministry for Economic Affairs (2020) Job Creation Omnibus Law. Jakarta

Dahlia KD, Alvi S, Syamsul A, Pandastaren T (2014) Environmental permits about environmental administration enforcement based on Law No 32 of 2009 concerning environmental protection and management. Jurnal USU Law II No 1(1):124–138

East Kalimantan Government (2023) Development of the East Kalimantan Energy Sector. Samarinda

Ekhator E (2023) Environmental justice in the global south: overview of some contemporary developments. In: A paper presented at the Human Rights Centre (HRC) Guest Speaker Series: Sustainability, Environment, Law, and Rights. Organized by the Human Rights Centre and University of Essex Law School, 2023, Available at SSRN: https://ssrn.com/abstract=4487210 or https://doi.org/10.2139/ssrn.4487210

Elisha S, Benjamin (2018) What are the major natural resources of Indonesia? Worldatlas. https://www.worldatlas.com/articles/what-are-the-major-natural-resources-of-indonesia.html

Firmansyah AT, Ema SS, Fenia AA (2020) The loss of democracy and regional autonomy through the corporatocracy of the omnibus law bill. Widya Yuridika 3(2):131. https://doi.org/10.31328/wy.v3i2.1492

Fitroh (2023) Interview with the Secretary of the Environment Service, Cirebon Regency, West Java Province

Government and Parliament (2020) Law No. 11 of 2020 concerning job creation. Indonesia

Helmi H, Fitria F, Retno K (2021) Use of the omnibus law in regulatory reform in the environmental sector in Indonesia. Masalah-Masalah Hukum 50(1):24–35. https://doi.org/10.14710/mmh.50.1.2021.24-35

Heriyanto B (2018) Absolute competence of state administrative courts based on the Paradigm of Law No. 30 of 2014 concerning government administration. Pakuan Law Rev 4(1):75–90. https://doi.org/10.33751/.v4i1.784

Hilmawan A (2016) Coal mining operations and its impact on sectoral and regional area: evidence of East Kalimantan, Indonesia. J Indon Appl Econ 6(1):22–42

Indonesia Center for Environmental Law (2020) Statement of Draft Job Creation Law. Jakarta

Indonesian Government (2014) Law Number 23 of 2014 Concerning Local Government

Jabareen Y (2008) A new conceptual framework for sustainable development. Environ Dev Sustain 10:179–192

Jati WR (2012) Inconsistency of the regional autonomy paradigm in Indonesia: the dilemma of centralization or decentralization. Jurnal Konstitusi 9(4):744–733

McLaren D (2003) Environmental space, equity and the ecological Debt. In: Agyeman J, Bullard RD, Evans B (eds) Just sustainabilities: development in an unequal world. Earthscan, UK

Mahardika AG (2021) Implications of the omnibus law on the constitutional right to a healthy environment. Jurnal Konstitusi 18(1):195–218. https://doi.org/10.31078/jk1819

Mahkamah Konsitusi Republic Indonesia (MKRI) (2022) Expert: mineral and coal mining law criminalizes human rights activists. https://en.mkri.id/news/details/2022-01-19/Expert:%20Mineral%20and%20Coal%20Mining%20Law%20Criminalizes%20Human%20Rights%20Activists

Nugroho W, Erwin S (2021). The legal politics of the job creation bill in the environment and forestry sector (a critical review). Jurnal Hukum Pembangunan 51(3):637–658. https://doi.org/10.21143/jhp.vol51.no3.3127

Rimbawan AY, Izziyana WV (2020) Omnibus law and its impact on agriculture and the environment. RECHTMATIG: Jurnal Hukum Tata Negara 6(2):19–28

Sachs W (1993) Global ecology and the shadow of development. In: Sachs W (ed) Global ecology. A new arena of political conflict. Zed Books, London, pp. 3–20

Sianturi A (2023) Degradation of regional government authority in the management and protection of environmental law after the omnibus law. J Acad Literat Rev 2(8):693–701

Sucahyo N (2022) Legal or illegal, coal mining destroys East Kalimantan. VOAI. 2022. https://www.voaindonesia.com/a/legal-atau-ilegal-tambang-batu-bara-menghancurkan-kaltim-/6464290.html

Vicki DP (2022) Problematics in the case of bribery of state officials in the case of illegal mining (Ismail Bolong). Jurnal Multidisiplin Madani 2(12):4414–4429. https://doi.org/10.55927/mudima.v2i12.2053

Wayan Sridiani N, Dwi RKSL, Ni NER (2023) Submission of Amdal Lawsuit based on omnibus law after constitutional court decision Number 91/PUU-XVIII/2020. Journal Evidence Of Law 2(1):67–74. https://doi.org/10.59066/jel.v2i1.239

White R (2013) Resource extraction leaves something behind: environmental justice and mining. Int J Crime Justice Soc Democr 2(1):50–64. https://doi.org/10.5204/ijcjsd.v2i1.90

Wibisono G (2022) Ketika reklamasi pasaca tambang di kaltim di abaikan perusahaan, https://kaltim.antaranews.com/berita/149693/ketika-reklamasi-pascatambang-di-kaltim-diabaikan-perusahaan

Wildan M (2023) Government Regulation instead of Law on Job Creation Becomes Law, Effective March 31, 2023. DDTC News, 2023

Yovanda (2019) Drowning deaths at disused mines in Indonesia renew calls for action, mangabey, https://news.mongabay.com/2019/07/drowning-deaths-at-disused-mines-in-indonesia-renew-calls-for-action/

Zein YA, Nurasikin (2022) Environmental spatial planning in the perspective of Law 11/2020 concerning job creation. IOP Conf Ser: Earth Environ Sci 1083(1):1–9. https://doi.org/10.1088/1755-1315/1083/1/012038

3 Natural Resources Governance in indonesia from a Political Perspective: The Marine, Fisheries, and Forestry Cases

Firman Noor, Meidi Kosandi, Agus R. Rahman, R. R. Emilia Yustiningrum, and Anta Maulana Nasution

Abstract

While Indonesia has abundant natural resources, until the current socio-ecological crisis it has been ignoring the environmental degradation consequences of full exploitation of its natural resources. There has been no suitable natural resources governance in the past, and Indonesia needs a strong commitment to adopting a natural resources governance framework in order to guarantee a sustainable environment as the basis for sustainable development, given Indonesia's current status as a high-middle-income country. This chapter aims to evaluate some of Indonesia's natural resources governance initiatives over the period 2004–2022. This evaluation is based on a qualitative framework in which data have been collected from secondary sources. Some stakeholders, especially the national government and civil society, still have low ideological bases and are the most practical in applying good governance in some natural resource sectors. Indonesia remains oligarchic in character, but it is time to formulate a multi-level natural resources governance framework based on the interests of the people. This framework should consist of norms, institutions, structures, and processes that determine how power over and responsibility for natural resources are regulated, exercised, and distributed, how decisions are taken, and how all the people can participate in and benefit from the management of natural resources.

Keywords

Socio-ecological · Natural resources · Governance · Sustainable · Development · Multi-level

3.1 Introduction

Governance principles formed part of the road to democracy of the new regime in Indonesia that followed the fall of President Suharto in May 1998. One such principle is the governance of natural resources, which involves the many complexities in both span and scope of extraction activities. Failure to organize governance and democracy represents a continuing crisis for Indonesia's Reformation regime.

F. Noor (✉) · A. R. Rahman · R. R. Emilia Yustiningrum · A. M. Nasution
Research Center for Politics - National Research and Innovation Agency, Jakarta, Indonesia

M. Kosandi
Department of Politics, Faculty of Social and Political Science, University of Indonesia, Jakarta, Indonesia

H. Warsilah and A. H. Purnomo (eds.), *Approaching Natural Resources for Sustainable Development in Indonesia*, Sustainable Development Goals Series, https://doi.org/10.1007/978-981-96-6913-4_3

From a political standpoint, the complexities of implementing government programs and policies is inevitably associated with the characteristics of the political system and the strategic position of natural resources as a source of revenue for the government as well as a source of political financing. In political terms, the challenge is how to restructure power in the new context of reformation without falling into the trap of oligarchy. Oligarchy is unavoidable, as can be seen both in developed and developing countries (Chesney 2018; Glassman 2021; Winters 2011), and it can occur both at national and local levels (Dewi 2022; Robison and Hadiz 2004; Winters 2014).

During the New Order era, it was widely understood that the country's economic prosperity was heavily reliant on the oil industry in Riau Province, which at the time could generate up to 1 million barrels of oil per day. Interestingly enough, Riau was then one of Indonesia's poorest provinces (Harmadi, Adji, and Mahdi 2019). Since society was not gaining anything worthwhile from its natural resources, it is hardly surprising that certain individuals desired Riau to be free of Indonesia.

Natural resources reflect state–civil society relations, and are ideally positioned to benefit the entire community (Dubbink 2013). Nonetheless, a system of political leadership and power distribution that has a special regard for the interests of particular individuals can have severe effects on both the survival of the ecosystem and human life (Buchanan 1954). Natural resources will inevitably be utilized as a source of power due to their strategic location. From this perspective, the right to ownership of natural resources is directly linked to the management of political or power interests, whether in terms of capturing or retaining power. In reality, those who own or control natural resources frequently use them as political tools. If necessary, a group in control of natural resources may utilize it to finance political ploys, including avoiding established political norms. In Indonesia, control of natural resources has the potential to be utilized to compensate for expensive political costs.

After the New Order, Indonesia's democratic life expanded once more with much experimentation in terms of culture, political processes, and institutions. In the early stages of reform, efforts could be seen to steer Indonesia back towards democracy, with alterations to the constitution, and laws that were more constitutional and aspirational becoming more effective. Both the right to a free press and the ability to create political parties were thus enforced. The establishment of professionalism within the military and the discharge of various political convicts has occurred in parallel with the enforcement of civilian supremacy. There has also been decentralization of power and autonomy, which has led to the frequent holding of regional elections.

However, democracy is still in its early stages in Indonesia. Numerous studies indicate that democracy is currently undergoing stagnation or retreat (Kurlantzick 2013; Mietzner 2012; Power and Warburton 2020; Winters 2014), and several studies suggest a situation of "delusional democracy" or post-democracy. The low degree of community involvement is currently perceived as the fundamental issue with Indonesia's attempt to implement democracy. A small group of elites control significant areas of decision-making that affect the interests of society. The interests of the elite and the oligarchs in their immediate environment appear to have taken over democracy in this situation. Democracy as a whole is still perceived as just being practiced at the procedural level; it has not been accomplished in its entirety.

The power of the oligarchy in the New Order has not diminished (Robison and Hadiz 2004); on the contrary, it has persisted and even gained strength in many areas of life. The current privileged and oligarch-dominated political structure eventually affects how natural resources are managed. With decentralization and the adoption of regional autonomy policies, political actors have enlarged their domains of influence locally and the power of the oligarchy has grown. This issue affects not only how natural resources are managed in an extractive and elitist manner, but also the quality of management results, political impacts that tend to be asymmetrical or progressively elitist, and the potential for continued maintenance of natural resources, which is more concerning than improving. In addition to outlining various repercussions and potential remedies that might be put into practice both now and in the future, this chapter will demonstrate the relationship between political circumstances, resource management, and diverse effects.

As an archipelagic state, Indonesia is famous for its natural endowment of rich and diverse natural resources from oil, gas, and minerals to forest and maritime resources (Table 3.1). The intensive exploitation of these natural resources resulting in degradation of the environment and the associated socio-spatial and socio-ecological crises began when Suharto took power in 1967 and continued until his fall in 1998. The Reformation era brought a new government which improved the governance of natural resource management based on decentralization. There are reciprocal relations between the governance of natural resources and the socio-ecological system dynamics (Febriamansyah, Yonariza, Ullah, and Shivakoti 2016) The three principles of decentralization, sustainability, and equity cover three types of relations between domestic and international, state and society, and center and periphery (Piattoni 2010).

3.2 Theoretical Background

The impetus provided by the Reformation regime in Indonesia offered the opportunity for the country to implement innovative developmental programs and projects guaranteeing sustainability and equity. A long-term sustainable development path was designed that embraced a sustainable environment ensuring natural resource governance, sustainability, and equity (Resosudarmo 2005).

This new approach concentrates on the twin aspects of governance, which is characterized by complicated relations of domestic versus international on one side, and state and society on the other (Piattoni 2010). While it does not equate to center and periphery, it reflects the basic relation between the national government and regional and local government. It means balanced centrality of the national government toward regional and local government. When adequate and suitable political processes are in place, emerging leaders will operate based on democratic transactions. On the other hand, there will be an oligarchy that replaces democratic leaders.

Oligarchy is defined as political power that is controlled by economic actors who use political instruments, mainly political parties, as their political vehicles so they tend towards anti-democracy (Bodislav 2012). Indonesia's domestic oligarchy is based on conglomerates in the ownership of the third generation of the founding families (Shimato 2022). The emergence of oligarchy has a destructive effect on equity for people, civil society, and the nation as a whole. The management of natural resources in the country may be trapped in oligarchic mechanisms that have degraded the environment and do not effectively improve sustainability and equity.

3.3 Methodology

This chapter uses a qualitative approach, collecting data through an analysis of secondary sources, to discuss Indonesia's natural resources governance from the perspective of politics, particularly during the administrations of Presidents Susilo Bambang Yudhoyono (SBY) and Joko Widodo. Qualitative analysis was undertaken on written documents such as books, journals, speeches by presidents and ministers, official government statements, national regulations, regional and international policies, national and local newspapers, treaties, and transcripts of interviews. Questions are asked about the unit of analysis and any subunits (Rowley 2002, p. 19), so the limitations that define the unit of analysis and subunits (Gerring 2004, p. 344) are significant when it comes to searching for evidence (Rowley 2002, p. 19). In this chapter, the unit of analysis is the Indonesian government.

Table 3.1 Natural Resources by Region and GDP, 2022

No	Sector	Region	GDP (Rupiah)	Location
1	Agriculture, [a] Foresty[b] Fishery	0 Forest area planning process 11[3]	2.4 Billion[1]	Not available
2	Mining	10[4]	2.4 Billion[1]	Provincial

Note a, b, and c are combined into one sector
Sources 1. Official Statistics News, 6 February 2023; 2. PPID Ministry of Environment and Forestry, "Minister of Environment and Forestry: Forest Area Boundaries Completed This Year", 30 January 2023; 3. Regulation of the Minister of Maritime Affairs and Fisheries of the Republic of Indonesia Number 22 of 2021; 4. Dwi Setyo Irawanto, et al., Whitepaper: 10 Mining Areas in Indonesia (Tangerang: Datanesia, 2022)

3.4 Result and Discussion

3.4.1 Normative Aspects: Review of Regulations

The collapse of the New Order regime did not cause the latest oligarchy to weaken (Robison and Hadiz 2007). Rather, they gradually reduced their domestic investments and successfully moved abroad, exploiting their ownership advantages (Lecraw 1993). Under the New Order regime, illegality dominated in the extractive industries in the main three natural resource sectors, mining, logging, and fishing (McCarthy 2011). These illegal activities were responsible for acute degradation of the environment (Robbi, Ismail, and Hoetoro 2020) and socio-spatial inequalities in urban areas (Mela and Toldo 2019). At the outset of the Reformation era, this old oligarchy attempted to adjust to the new President SBY, while at the same time, Reformation regimes sought to overcome illegal activities through natural resource control and governance (McCarthy 2011).

During President SBY's era, 2004–2014, the government passed 16 laws, including the Act on Minerals and Coal Mining. In general, these sought to orient normative aspects of natural resources to domestic interests, including better environment management and support for civil society, while not ignoring international cooperation. During the adjustment process, oligarchic actors eagerly attempted to reorganize the oligarchy, directly creating a number of new political parties: the Democrat Party (Demokrat) in 2001, the Party of Hanura (Hanura) in 2006, the Party of Great Indonesia's Movement (Gerindra) in 2008, the Democratic National Party (NasDem) in 2011, and the United Indonesia Party (Perindo) in 2014. In post- New Order Indonesia, they have a significant vote in the National Parliament elections, in which only the parties further are competent to propose presidential and vice-presidential candidates. Several political dynasties emerged (Aspinall and As'ad 2016; Editors 2017; Mukti and Rodiyah 2020) which ignored environmental concerns.

President Joko Widodo, who did not have a political party when he assumed office, failed to address the environmental problems left by the previous regime such as deforestation (Williams 2023), natural damage due to mining (Tegnan, Karjoko, Barkhuizen, and Bajrektarevic 2021), urban environmental problems (Dethier 2017), and damage to the marine environment (Waddell 2009). He launched the Omnibus Law, which was seen as a political breakthrough that assisted him to reach his policy goals without considering the democratic spirit (Hamid 2020).

The Ministry of Marine Affairs and Fisheries was relatively new compared to other ministries dealing with natural resources, such as agriculture, energy, and forestry. Previously, the fisheries sector was under the Ministry of Agriculture, and in 2000, a separate department/ministry was formed. For approximately 14 years (2000–2014), the marine and fisheries sector was not subject to political calculation. Even though it had become a separate department/ministry, in the early days it was not a priority in budgeting terms compared with the department/ministry of agriculture, which received three times more budget even though the non-tax state income (*Pendapatan Negara Bukan Pajak - PNBP)* is under the marine and fisheries sector (Satria 2015).

Indonesia did not have an outline or national marine and fisheries development policy until 2014. This changed after Widodo became President of the Republic of Indonesia in 2014. He asserted his maritime vision with the Global Maritime Fulcrum (GMF) slogan in 2014. Then in 2017, GMF was implemented into Presidential Regulation Number 16/2017 concerning Indonesian Ocean Policy (Marzuki 2017). GMF is defined as "A vision of Indonesia to become a maritime country that is sovereign, advanced, independent, strong, and able to make a positive contribution to regional and world security and peace following national interests".

The Indonesian Ocean Policy is a guideline for synergizing maritime-related policies and programs in various institutions, ministries, and local governments in Indonesia to realize Indonesia as a global maritime fulcrum (Adi 2018; Muhibat 2017). Presidential Regulation No. 16/2017 explained that there are seven pillars of Indonesia's maritime policy: (1) marine resource management and human resource development, (2) defense, security, law enforcement and safety at sea, (3) marine governance and institutions, (4) the economy, infrastructure, and increasing welfare, (5) management of marine space and protection of the marine environment, (6) maritime culture, and (7) maritime diplomacy. In 2022, President Joko Widodo issued Presidential Regulation No. 34/2022 concerning the Indonesian Ocean Policy action plan, which is a continuation of the policy that was issued in 2017. The Indonesian Ocean Policy action plan is set for five years from 2021 to 2025.

3.4.2 Implementation Aspects: Oligarchic Tendencies

Fisheries governance seeks to ensure the sustainability of fish stocks and fishing livelihoods. According to Spijkers et al. (2022), there are two aspects to fisheries governance, especially in relation to the sustainability of fish stocks: strict regulations on access to utilization of fishery resources; and collecting data and information to monitor stocks of fish resources.

Indonesian fisheries governance changed with the appointment of Susi Pudjiastuti as Minister of Marine Affairs and Fisheries (2014–2019) (Fabinyi and Barclay 2022). Pudjiastuti established that around 4000 large fishing vessels were owned by between eight and 20 entrepreneurs (Yadika 2019). Fisheries governance was dominated by giant corporations prioritizing profits over sustainability, and many criminal acts continued to be committed by large corporations, causing a loss of income for the state (TI 2022).

Pudjiastuti reformed fisheries governance by using strict regulations such as eradicating illegal, unreported, unregulated fishing (IUUF) and a moratorium on ex-foreign fishing vessels. Many comfortable parties were disturbed by these strict rules and decisive leadership, and a coalition was formed to oppose and overthrow Pudjiastuti, serving to confirm the existence of a strong oligarchic system shaping Indonesia's political economy (Scarpello 2020).

The oligarchy of fishery resources takes the form of the control by a handful of elites maintaining wealth (Nasution 2020a, b, July 29). It begins with the granting of permission or access to exploit the resources. One example is the case of lobster seed exports. During Pudjiastuti's term in office she issued ministerial decree No. 56/2016, prohibiting the export of lobster seeds, but in 2020 when she was replaced by Edhy Prabowo, the export of lobster seeds was permitted through ministerial decree no. 12/2020. This policy did not last long, however, with problems ranging from the monopoly on utilization and inappropriate permits, to corruption involving Prabowo (Nasution 2020a, b, December 4).

Indonesia, like coastal states around the world, is facing declining fish stocks and overfishing (Ayunda, Sapota, and Pawelec 2018; Heazle and Butcher 2007). Overfishing occurs due to the ineffectiveness of a country's regulations (Halim et al. 2020). Capture fisheries regulations are very closely related to licensing; the existence of strict regulations on permits will minimize the occurrence of overfishing and fishery resource oligarchy. The strict regulations issued by Minister Pudjiastuti have succeeded in reducing illegal fishing activities by foreign fishing vessels and fighting against the oligarchs who have been exploiting Indonesia's fishery resources massively without considering the sustainability of their resources (Cabral et al. 2018; Scarpello 2020). What Pudjiastuti did had a significant impact on the sustainability of fisheries resources and increased catches of small-scale fisheries (Khan, Gray, Mill, and Polunin 2018).

The forestry sector in Indonesia has been growing significantly since 2005 with approximately a 3% annual growth rate. Including both timber and forest products (such as pulp and paper), the share of GDP in the non-oil and gas sector in 2021 was 6%, or USD12.2 billion in total (BPS 2022). According to the report by the National Bureau of Statistics (2022), the industry attracted over 300 big companies and over 1000 small and medium enterprises, and involved over 1.5 million workers. This is the good news. The bad news is that natural resource management in the forestry sector in Indonesia during Yudhoyono and Widodo's administrations cannot be analyzed separately from other sectors such as agriculture, plantation, and mining. Forest land management and conservation are often subordinated to other land-use interests by the private sector, with government permission. Government regulation on forestry has not been reformed for many years, including under Yudhoyono's and Widodo's leadership. Indonesia has been managing its forests under Law No. 5 of 1990 on Forest Conservation and Law No. 41 of 1999 on Forestry.

These laws regulate the classification of forest conservation and land use throughout Indonesia. Land-use transfer from conservation forest to plantation or mining has also been enabled by various laws, such as the Foreign Investment Law 1967, Agrarian Law 1960, Local Autonomy Law 1999 and 2004, Mining and Coal Law 2009 and 2020, and Job Creation Law/Omnibus Law, 2020. It was not until 2021 that the government enacted Government Regulation No. 23 providing detailed regulation of the Omnibus Law 2020 specifically for the forestry sector.

The seeds of oligarchy in the forestry sector were sown under Suharto's New Order patrimonialism, and consolidated during the Yudhoyono and Widodo administrations. Timber and oil were the most important exports through which Suharto maintained economic growth, using them to fund infrastructure development throughout the country. During Yudhoyono's and Widodo's leadership, the decentralization of authority to local government exacerbated forest industry production permits and land transfer from conservation to industry.

Government Regulation No. 23 represents progress in natural resource governance in Indonesia. It comprehensively tackles conservation and forest industry issues that have been subject to scattered regulation under various laws in Indonesia since the New Order. For example, the government is required to formulate forestry planning, preserve watersheds, replant the equivalent of forests that were turned over to industry, promote social forestry (intercropping), and prepare food security programs (modern cattle farming inside the forest). However, implementation has not yet been seen in the few years since the introduction of the regulation, and big business interests in the forest industry have continued to grow. Despite the anti-mafia movement in forestry under Yudhoyono and the pro-poor policy during Widodo's administration, oligarchs in the forestry sector have been able to capture regulations in their favor.

In the mining sector, unlike forestry, there have been major changes in government regulation in both Yudhoyono's and Widodo's administrations.

Despite the stated objectives of the governance reform with the formulation of the new regulations (Law No. 4/2009, and Law No. 3/2020), which was to reform the flawed regime, however, in practice their implementation of the regulations has led to the flourishing and consolidation of oligarchy.

The 2009 law was intended to improve the income proportion for Indonesian people through an increase in the government's capital ownership share, while regulating the environmental impacts. Under the previous law of 1967, foreign investors were required to pay royalties to the central government. However, these were very low, ranging from 1 to 3% of the total export value, the income was sent to the central government, with no share for the local government or local people, and the environmental damage was severe.

The new law on minerals, mining, and coal (also called UU Minerba) required any foreign investment in mining to be matched up to 51% by a national share, either from the central government, state-owned enterprises (SOEs), local government, local SOEs, or private companies. The investment partnership took the form of stock sharing instead of contracts between government and foreign companies. So, the revenue sharing was no longer a few percent of royalty payment, but up to 51% dividend.

The law also required companies to provide an environmental impact analysis in order to obtain or extend a government extraction permits. The decentralizing of permits was intended to tighten control over environmental protection, with the hope that the local government would prioritize environmental protection over generating government income through uncontrolled permit issuance. These expectations, however, were not met in practice. Local governments issued thousands of permits for mining that caused massive deforestation. Although progress was made towards increasing the government share in multinational investments, the growth of local oligarchs, corruption, and collusion worsened environmental degradation during Yudhoyono's administration.

Widodo's administration's arguments for the new mining law in 2020 were to further increase Indonesia's share of the mining revenue, to strengthen control over local government permit issuance leading to excessive mining, and to reduce corruption and collusion. In the new law, the central government recentralized the authority for mining permit issuance, removed environmental impact analysis as a prerequisite for permit extension, legalized sanctions for obstructions to mining operations, and stipulated stimulus for downstream mining products by 0% royalty or royalty-free.

The implication of the new law for natural resources management is threefold: (1) massive exploitation and poor conservation; (2) stronger state control and suppression of opposition; and (3) consolidating oligarchy and marginalizing society. The exploitation of mines, especially coal and nickel, has caused deforestation and land grabbing in agriculture, and reportedly many fishermen have lost their livelihoods (Walhi 2022). Oligarchs who do business in mining and coal tend to participate in local electoral politics, especially by assisting certain candidates to ensure the outcome of the elections. In Indonesia, this is called *ijon politik*, which translates as "debt bondage politics". Thus the oligarchs establish their power in the local electorates and the elected governments.

3.5 Conclusion and Recommendation

3.5.1 Conclusion

The reform era presented a significant opportunity for the development of diverse regulations favorable for the preservation and utilization of natural resources to serve the overall welfare of the state and society. In a country that needs strong commitment and secure governance to guarantee a sustainable environment, this development provides hope and opportunity. In these new political circumstances, a range of regulations and statutes started to evolve, prioritizing domestic interests above international ones, providing more safeguards for the long-term preservation of natural resources, and enhancing employment opportunities for Indonesia's citizens.

Ultimately, however, the current legislation appears to be ineffective in addressing natural resource management methods, which deviate from the expectations of multiple stakeholders and result in numerous detrimental consequences for both the government and society. Furthermore, alongside the ongoing extensive environmental degradation and conflicts, there has been a notable consolidation of the oligarchy's presence and influence within the realm of natural resource management in Indonesia. The significance of these circles is magnified by the emergence of political avenues for obtaining authority, along with the gradual erosion of civil society.

Given the prevailing political circumstances, it is quite probable that natural resources will encounter a significant challenge in the future, akin to a metaphorical time bomb. The depletion of natural resources will not only occur at an accelerated rate, but it will also have adverse effects on the environment, society, culture, and legal aspects of the local community. From a political standpoint, this phenomenon is likely to yield adverse consequences as it reinforces the presence of an oligarchic system, exacerbating the disparity in benefits received by the privileged few as opposed to the general population. Enduring challenges of inequity and unfairness are likely to result, particularly within the domain of natural resource governance. The presence of abundant natural resources, rather than fostering prosperity, can be perceived as a calamity that fundamentally opposes the essence of the Constitution and *Pancasila*.

3.5.2 Recommendation

At the regulatory level, it is imperative to assert governance principles and replace ineffective policies by reviewing and revising the regulations governing natural resource management. Regulations related to natural resources should prioritize sustainability over the excessive exploitation that disrupts the environmental balance. The greedy behaviors of oligarchs, coupled with a lack of environmental consideration, have exacerbated these situations, leading to socio-spatial crises and disordered natural resource governance. It is crucial to explicitly incorporate economic sanctions into all legislative outputs, aiming to penalize oligarchs who arbitrarily cause environmental damage.

The principles for establishing good-practice multi-level governance are legitimacy, transparency, accountability, inclusiveness, fairness, integration, capability, and adaptability. These principles provide normative guidance for maintaining the sustainability and equity of natural resource governance. The involvement and cooperation of stakeholders from government, civil society, academia, and the private sector at governance level are imperative. Historically, civil society has often been the victim of injustices in natural resource management, suffering both environmental and material inequities. Natural resource management must provide positive impacts on the environment and the well-being of society. Governance practices that sacrifice the well-being of civil society must be discontinued, as they offer only short-term gains and long-term detriments. Fundamentally, natural resource governance should aim to enhance the overall welfare of society.

References

Adi TR (2018) Maritime culture empowerment under Indonesian Ocean policy. J Ocean Culture 1:102–117. https://doi.org/10.33522/joc.2018.1.102

Aspinall E, As'ad MU (2016) Understanding family politics: successes and failures of political dynasties in regional Indonesia. South East Asia Res 24(3):420–435. https://doi.org/10.1177/0967828x16659571

Ayunda N, Sapota MR, Pawelec A (2018) The impact of small-scale fisheries activities toward fisheries sustainability in Indonesia. In: Zielinski T, Sagan I, Surosz W (eds) Interdisciplinary approaches for sustainable development goals: economic growth, social inclusion and environmental protection. Springer International Publishing, Singapore, pp 147–167

Bodislav D-A (2012) Oligarchy versus democracy and regulation versus deregulation under the globalization effect. Theor Appl Econ 19(5):33–46

BPS (2022) Indikator Ekonomi Januari 2022. Badan Pusat Statistik (Indonesian Government), Jakarta, Indonesia

Buchanan JM (1954) Individual choice in voting and the market. J Polit Econ 62(4):334–343. https://doi.org/10.1086/257538

Cabral RB, Mayorga J, Clemence M, Lynham J, Koeshendrajana S, Muawanah U, Ghofar A (2018) Rapid and lasting gains from solving illegal fishing. Nature Ecol Evol 2(4):650–658. https://doi.org/10.1038/s41559-018-0499-1

Chesney M (2018) A permanent crisis: the financial oligarchy's seizing of power and the failure of democracy. Palgrave Macmillan, Switzerland

Dethier J-J (2017) Trash, cities, and politics: urban environmental problems in Indonesia. Indonesia 103(1):73–90. https://doi.org/10.1353/ind.2017.0003

Dewi KH (2022) Women local leaders within local oligarchy networks in post-reformasi Indonesia: existing conditions and challenges. In: Dewi KH (ed) Gender and politics in post-reformasi Indonesia: women leaders within local oligarchy networks. Springer Nature, Singapore, pp 141–146

Dubbink W (2013) Assisting the invisible hand: contested relations between market, state and civil society. Springer Science & Business Media, The Netherlands

Editors CR (2017) Joseph P. Kennedy, Sr. The life and legacy of the patriarch behind America's most famous political dynasty. CreateSpace Independent Publishing Platform, South Carolina

Fabinyi M, Barclay K (2022) Fisheries governance. In: Fabinyi M, Barclay K (eds) Asia-Pacific fishing livelihoods. Springer International Publishing, Singapore, pp 65–90

Febriamansyah R, Yonariza Y, Ullah R, Shivakoti GP (2016) The reciprocal relationship between governance of natural resources and socio-ecological systems dynamics in West Sumatra Indonesia, vol 4. Elsevier Science, San Diego

Gerring J (2004) What is a case study and what is it good for? Am Polit Sci Rev 98(2):341–354. https://doi.org/10.1017/S0003055404001182

Glassman RM (2021) Can democracy survive in the 21st Century? Oligarchy, Tyranny, and Ochlocracy in the age of global capitalism. Springer International Publishing, Switzerland

Halim A, Loneragan NR, Wiryawan B, Fujita R, Adhuri DS, Hordyk AR, Sondita MFA (2020) Transforming traditional management into contemporary territorial-based fisheries management rights for small-scale fisheries in Indonesia. Mar Policy 116:103923. https://doi.org/10.1016/j.marpol.2020.103923

Hamid A (2020) Analysis of the importance of Omnibus Law Cipta Kerja in Indonesia. Int J Sci Res Manag 8(08):236–250

Harmadi SHB, Adji A, Mahdi S (2019) Regional inequality in Indonesia: pre and post regional autonomy analysis. In: Paper presented at the 15th Indonesian Regional Science Association (IRSA) International Conference, Banda Aceh, Indonesia

Heazle M, Butcher JG (2007) Fisheries depletion and the state in Indonesia: towards a regional regulatory regime. Mar Policy 31(3):276–286. https://doi.org/10.1016/j.marpol.2006.08.006

Khan AMA, Gray TS, Mill AC, Polunin NVC (2018) Impact of a fishing moratorium on a tuna pole-and-line fishery in eastern Indonesia. Mar Policy 94:143–149. https://doi.org/10.1016/j.marpol.2018.05.014

Kurlantzick J (2013) Democracy in retreat: the revolt of the middle class and the worldwide decline of representative government. Yale University Press, New Haven

Lecraw DJ (1993) Outward direct investment by Indonesian firms: motivation and effects. J Int Bus Stud 24(3):589–600. https://doi.org/10.1057/palgrave.jibs.8490247

Marzuki K (2017) Indonesia's National Sea Policy: concretising the global Maritime Fulcrum. RSIS Commentary (052)

McCarthy JF (2011) The limits of legality: state, governance and resource control in Indonesia. In: Aspinall E, van Klinken G (eds) The state and illegality in Indonesia. Brill, Leiden, pp 87–106

Mela A, Toldo A (2019) Socio-spatial inequalities in contemporary cities. Springer, Berlin

Mietzner M (2012) Indonesia's democratic stagnation: anti-reformist elites and resilient civil society. Democratization 19(2):209–229. https://doi.org/10.1080/13510347.2011.572620

Muhibat SF (2017) Indonesia's new ocean policy: analysing the external dimension. Mar Affairs: J Natl Mar Found India 13(2):50–61. https://doi.org/10.1080/09733159.2017.1410936

Mukti HH, Rodiyah R (2020) Dynasty politics in Indonesia: tradition or democracy? J Law Legal Reform 1(3):531–5381. https://doi.org/10.15294/jllr.v1i3.38090

Nasution AM (2020a) Dua kejanggalan aturan ekspor benur dan indikasi oligarki dalam kasus dugaan korupsi Edhy Prabowo. The Conversation. Retrieved from https://theconversation.com/dua-kejanggalan-aturan-ekspor-benur-dan-indikasi-oligarki-dalam-kasus-dugaan-korupsi-edhy-prabowo-151122

Nasution AM (2020b) Benih Oligarki dalam Pengelolaan Sumber Daya Kelautan. Detik.com. Retrieved from https://news.detik.com/kolom/d-5112530/benih-oligarki-dalam-pengelolaan-sumber-daya-kelautan

Piattoni S (2010) The theory of multi-level governance: conceptual, empirical, and normative challenges. Oxford University Press, Oxford

Power T, Warburton E (2020) Democracy in Indonesia: from stagnation to regression? ISEAS - Yusof Ishak Institute, Singapore

Resosudarmo BP (2005) The politics and economics of Indonesia's natural resources. Institute of Southeast Asian Studies (ISEAS), Singapore

Robbi I, Ismail M, Hoetoro A (2020) Environmental degradation in Indonesia 1969–2016. In: Paper presented at the 23rd Asian Forum of Business Education (AFBE 2019), Bali, Indonesia

Robison R, Hadiz V (2004) Reorganising power in Indonesia. The Politics of Oligarchy in an Age of Markets. Routledge, London

Robison R, Hadiz VR (2007) Reorganising power in Indonesia: the politics of Oligarchy in An Age of

Markets. Bull Indones Econ Stud 41(3):395–396. https://doi.org/10.1080/00074910500306619
Rowley J (2002) Using case studies in research. Manag Res News 25(1):16–27. https://doi.org/10.1108/01409170210782990
Satria A (2015) Politik Kelautan dan Perikanan. Yayasan Pustaka Obor Indonesia, Jakarta
Scarpello F (2020) Susi versus the rest: the political economy of the fisheries industry in Indonesia during Jokowi's first term. Asian J Polit Sci 28(2):122–141. https://doi.org/10.1080/02185377.2020.1774908
Shimato H (2022) Transitions in Indonesian conglomerates—with a focus on the third generation of the founding families. Mitsui & Co. Global Strategic Studies Institute Monthly Report, pp 1–9
Spijkers J, Mackay M, Turner J, McNeill A, Travaille K, Wilcox C (2022) Diversity of global fisheries governance: types and contexts. Fish Fish 24(1):111–125. https://doi.org/10.1111/faf.12713
Tegnan H, Karjoko L, Barkhuizen J, Bajrektarevic AH (2021) Mining corruption and environmental degradation in Indonesia: critical legal issues. Bestuur 9(2):90
TI (2022) Korupsi, Penghindaran Pajak, dan Manipulasi Data Perdagangan. Kertas Kebijakan Sektor Perikanan. Transparansi Internasional, Jakarta
Waddell S (2009) Rising to the challenge of providing legal protection for the Indonesian coastal and marine environment. In: Cribb R, Ford M (eds) Indonesia beyond the water's edge: managing an archipelagic state. Institute of Southeast Asian Studies (ISEAS), Singapore, pp 172–194
Walhi (2022) Tinjauan Lingkungan Hidup: Membangkang Konstitusi Mewariskan Krisis Antar Generasi. Jakarta: Wahana Lingkungan Hidup Indonesia (WALHI)
Williams DA (2023) The politics of deforestation and REDD+ in Indonesia: global climate change mitigation. Taylor & Francis, Abingdon
Winters JA (2011) Oligarchy. Cambridge University Press, New York
Winters JA (2014) Oligarchy and democracy in Indonesia. In: Ford M, Pepinsky TB (eds) Beyond Oligarchy: wealth, power, and contemporary Indonesian politics. Cornell University Press, Cornell, pp 11–33
Yadika B (2019) Menteri Susi: 20 Pengusaha Kuasai 4 Ribu Kapal di RI. Liputan6.com. Retrieved from https://www.liputan6.com/bisnis/read/4008199/menteri-susi-20-pengusaha-kuasai-4-ribu-kapal-di-ri

4 Prospects and Trends in Natural Resource Governance: A Case Study of Ex-Tin Reclamation Land in East Belitung, Indonesia

Uus Faizal Firdaussy, Tatang Rusata, and Hasbullah

Abstract

Belitung Timur, located on Sumatra Island, has been a significant center for tin production in Indonesia since the seventh century. However, as many tin mining operations have gradually ceased, former mining areas have become abandoned and unproductive. Reclamation efforts, mandated by government regulations, have often needed to catch up, achieving only a 25% success rate. This abandonment of reclaimed mining lands has led to the emergence of illegal mining activities, posing risks to the social, cultural, and economic well-being of the local community. This chapter uses a case study methodology to review and analyze the Trigona beekeeping initiative on reclaimed tin mining land in Belitung Timur Regency as an alternative for post-extractive tin mining industries. The study shows that cultivating Trigona bees on lands reclaimed from former tin mining operations offers a viable alternative and a promising model for the sustainable utilization of post-mining landscapes. It promotes both ecological restoration and economic resilience in East Belitung, enhancing land rehabilitation and providing a stable income, and it aligns well with market preferences for honey. However, challenges such as overpopulation and market fluctuations persist, necessitating improved governance and stakeholder cooperation.

Keywords

East Belitung · Governance · Sustainable development · Tin mine reclamation land · Trigona honeybee farming

4.1 Introduction

Tin is a staple commodity in the Bangka Belitung Archipelago province of Indonesia and the region has been famous for it since colonial times. It is estimated that tin mining began as early as the era of the Srivijaya Kingdom in the seventh century (Erman 2008; Swastiwi et al. 2017). The tin mining system known as *timah ladang* is conducted by excavating tin sand while residents clear the forest for agricultural fields. Until the eighteenth century, the Bangka Belitung community mined tin by sifting it during their farming

U. F. Firdaussy (✉) · T. Rusata
Research Centre for Society and Culture - National Research and Innovation Agency, Jakarta, Indonesia
e-mail: uusf001@brin.go.id

Hasbullah
Department of Agriculture, East Belitung Regency, Belitung, Indonesia

H. Warsilah and A. H. Purnomo (eds.), *Approaching Natural Resources for Sustainable Development in Indonesia*, Sustainable Development Goals Series, https://doi.org/10.1007/978-981-96-6913-4_4

seasons (Swastiwi et al. 2017). To this day, tin remains second only to agriculture for livelihoods in this area (Sulista et al. 2019).

Bangka Belitung province is not only the main contributor of tin production and reserve in Indonesia (see Fig. 4.2) but it also makes Indonesia the second largest tin producer in the world after China with a total annual production of approximately 74 million tons of tin ore (see Table 4.1 and Fig. 4.1) (National Minerals Information Center 2023).

Most onshore tin mining operations in the East Belitung district have ceased by now, leaving the former mining lands in a critical and neglected state. The Ministry of Environment and Forestry has reported that 275,500 hectares of land in the Bangka Belitung Archipelago province are in a critical condition due to tin ore mining. In the East Belitung district alone, 7337 hectares of land are deemed very critical, 48,074 hectares are critical, 187,099 hectares are somewhat critical, and 9950 hectares are potentially essential (Haryadi et al. 2018) (Fig. 4.2).

According to Haryadi et al. (2018), critical lands in East Belitung since the cessation of onshore mining operations fall into two categories. Some have been successfully reclaimed by companies, in compliance with the Ministry of Energy and Mineral Resources Regulation No. 34/2017 on mine land reclamation. Other areas remain neglected or have failed reclamation efforts. Haryadi et al. (2018) further note that the average success rate for reclamation is no higher than 25% of the government's annual target. This shortfall is compounded by insufficient efforts in the reclamation process.

Considering the issues outlined above, a study on the management of reclaimed mining land is urgently needed. However, the management of such land is complex. Paradoxically, countries abundant in natural resources often perform worse in governance compared to countries with fewer natural resources. This phenomenon is termed the 'resource curse' by social scientists (Auty 2002). Humphreys et al. (2007) assert that although these countries are rich in natural resources, extraction activities often hinder, rather than facilitate, sustainable development.

To address the issues of critical and neglected land resulting from tin mining operations in East Belitung, a local community initiative has emerged that focuses on the diversification of Trigona beekeeping *(Heterotrigona itama)* on reclaimed mining lands as an alternative to post-extraction activities. This chapter focuses on the impact of this initiative on post-tin mining lands in East Belitung. Using a sustainable development perspective, which emphasizes the equilibrium between the economic, social, and environmental needs of present and future generations, the study aims to shed light on the viability and benefits of this ecological venture.

4.2 Study Area

Our study explores the Trigona beekeeper experience in East Belitung Regency, an administrative district located on Belitung Island in the Bangka Belitung province (Fig. 4.3). The Trigona beekeeping initiative represents a crucial cultural element in both the historical and socio-cultural contexts of the East Belitung Island community, functioning as both a local product and a trade commodity. The complex relationship between humans and bees has significantly shaped the history and identity of the inhabitants of East Belitung Island, where Trigona honeybee production is widespread, spanning seven sub-districts within the East Belitung Regency: Manggar, Damar, Kelapa Kampit, Simpang Renggiang, Gantung, Simpang Pesa, and Dendang.

With a population of approximately 127,000 people (Pemerintah Kabupaten Belitung Timur 2021), Belitung Island's economy is primarily supported by agriculture and mining activities. The markedly low level of industrialization and sparse population contribute to making the East Belitung Regency one of the least polluted areas in terms of air quality, creating an ideal environment for apiculture.

Table 4.1 Indonesia tin production and forex reserve forecasting, 1994–2022

Year	Average Tin Price (USD/ metric tons)	Indonesia production (kilo/tons)	Indonesia reserve	China production	Forex reserve (million dollars)
1994	5.464	34.000	750.000	50.000	186
1995	6.214	38.400	750.000	52.000	239
1996	6.165	38.500	750.000	60.000	237
1997	5.647	47.000	750.000	65.000	265
1998	5.540	40.000	750.000	79.000	222
1999	5.404	48.000	800.000	62.000	259
2000	5.436	48.000	800.000	97.000	261
2001	4.484	51.000	800.000	79.000	229
2002	4.061	54.000	800.000	80.000	219
2003	4.895	70.000	800.000	50.000	343
2004	8.513	66.000	800.000	110.000	562
2005	7.380	80.000	800.000	120.000	590
2006	8.781	90.000	800.000	125.000	790
2007	14.537	102.000	800.000	135.000	1.483
2008	18.510	96.000	800.000	110.000	1.777
2009	13.574	55.000	800.000	115.000	747
2010	20.406	56.000	800.000	120.000	1.143
2011	26.054	42.000	800.000	120.000	1.094
2012	21.126	41.000	800.000	110.000	866
2013	22.283	95.200	800.000	110.000	2.121
2014	21.899	76.000	800.000	96.000	1.664
2015	16.067	52.000	800.000	110.000	835
2016	17.934	52.000	800.000	92.000	933
2017	20.061	83.000	800.000	92.000	1.665
2018	20.145	85.000	800.000	95.000	1.712
2019	18.661	77.500	800.000	84.500	1.446
2020	17.125	53.000	800.000	84.000	908
2021	32.384	70.000	800.000	90.000	2.267
2022	31.335	74.000	800.000	95.000	2.319

Source The World Bank, 2023; National Minerals Information Center, 2023

4.3 Methodology

This study employs a qualitative case study approach focused on the governance of reclaimed mining land between 2019 and 2023. Three critical questions were identified at the outset of the study: (1) How did the Trigona beekeeping initiative start on reclaimed tin mining land in East Belitung? (2) Can this initiative serve as a sustainable development policy alternative for post-extraction activities in the tin mining industry in East Belitung? (3) How is the governance of the Trigona beekeeping initiative supported by all stakeholders? These questions align with the study's overarching purpose, which is to describe how the Trigona beekeeping initiative functions on reclaimed tin mining land and to evaluate its potential as an alternative for post-extraction activities in the tin mining industry for the East Belitung community.

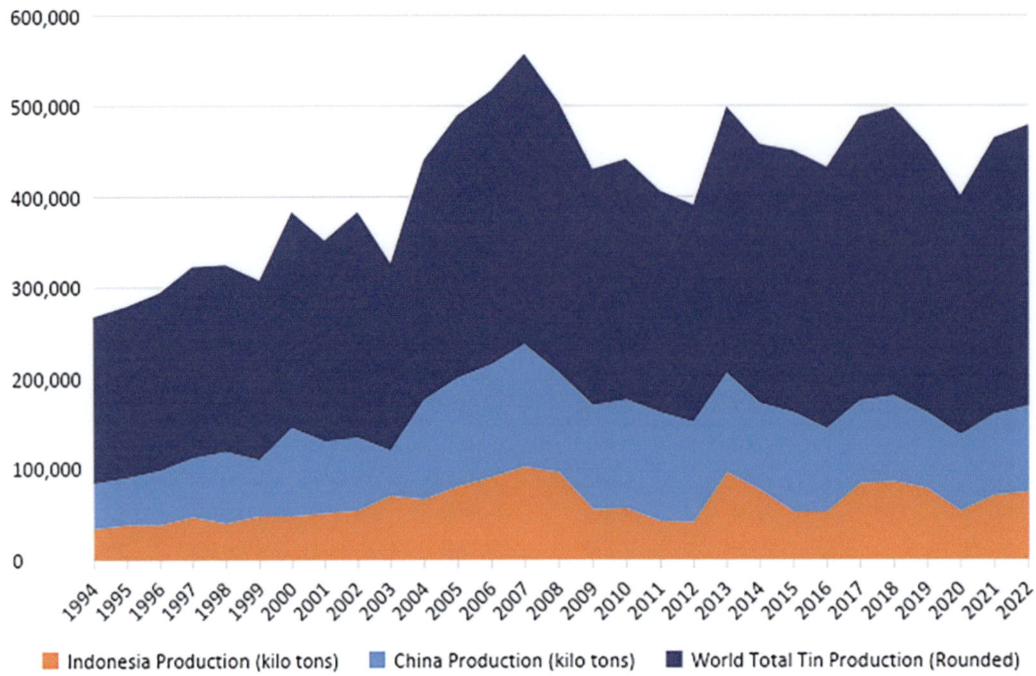

Fig. 4.1 Tin production comparison, 1994–2022. (*Source* (National Minerals Information Center, 2023))

The data in our study were collected from qualitative interviews with three key informants and through participant observation, a data collection method where the researcher engages in the daily activities and interactions of a specific group. This approach was chosen to help the researcher gain a more comprehensive cultural perspective on the subject of the study (Brennen 2021).

The informants for this research included a Trigona beekeeper (I3), key figures from the Raje Teran East Belitung Trigona beekeeping community (I1), and an East Belitung Regency government official (I2). We collaborated with Trigona beekeepers in East Belitung Regency whose livelihoods derive from beekeeping and who have the capacity to meet both national and international demand. We also worked with independent beekeepers, defined as those who primarily produce Trigona honey for the local market, personal consumption, and as gifts.

The primary focus of the analysis is on the beginning of the Trigona beekeeping initiative, its management, and how the actors are interacting with each other to develop this initiative. In particular, the study investigated the initiative's prospects for sustainable tin mining reclamation land governance.

4.4 Theoretical Framework

The concept of environmental governance is inherently dynamic and adapts to complex challenges (Mulyana et al. 2023; Neo and Chen 2010). Various theoretical frameworks, such as dynamic and adaptive governance, have been employed to explain this adaptive capacity (Folke et al. 2005). Governance is a facilitative condition for achieving the goals of sustainable development (Warsilah et al. 2017). Henny Warsilah (2017) asserts that resource management will advance towards sustainable natural resource management if development successfully integrates social, economic, and environmental aspects. She contends that management for sustainable development must ensure that all facets of development are conducted in a balanced manner and do not disproportionately favor one dimension over the others.

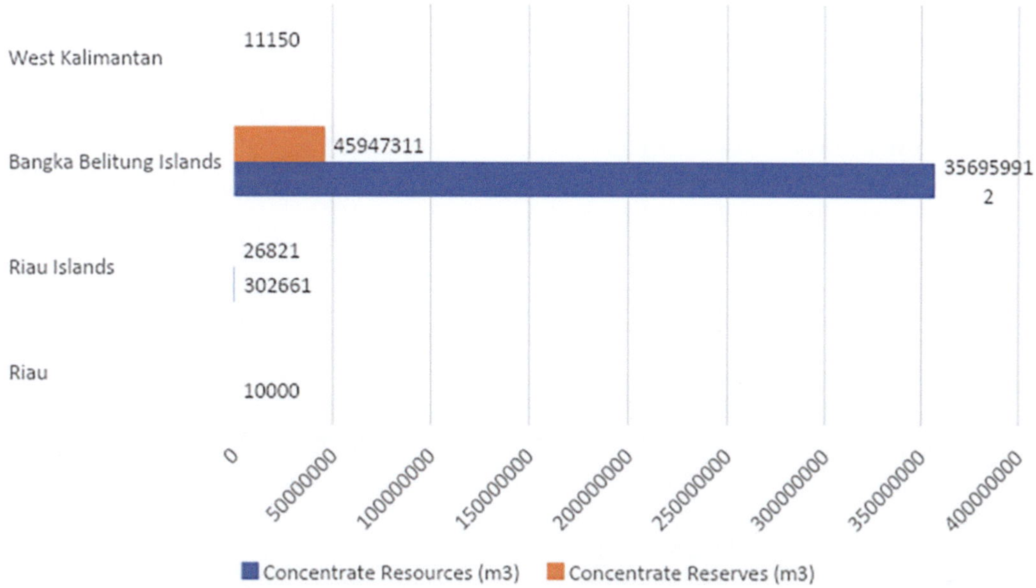

Fig. 4.2 Bangka Belitung contribution to national tin resources and reserves. (*Source* Vanessa et al., 2018)

Natural resource management is a complex phenomenon with paradoxical outcomes. Countries endowed with abundant natural resources often exhibit poorer management performance than their resource-scarce counterparts (Auty 2002). Humphreys et al. (2007) contend that paradoxically, resource extraction often acts more as an impediment than as an asset to sustainable development. In a practical context, Dwivedi (2001) observes differing viewpoints on environmental governance between developing and developed countries. In developing nations, the focus generally centers on protecting environmental resources both as a means of livelihood for impoverished communities and as a post-colonial development effort. Conversely, governance in developed nations tends to address more established issues (Dwivedi 2001). Moreover, in resource-dependent developing countries, any increase in per-capita income is often offset by the depletion of natural resources and environmental degradation (Nurrochmat et al. 2022). Developed nations, however, generally employ more structured and measurable approaches, incorporating innovations in extraction methods and environmental monitoring technologies.

Mining governance, as noted by Anggraini et al. (2019), is a double-edged sword. Traditional mining contributes positively to societal welfare but also generates negative impacts, such as health issues, road degradation, river pollution, soil erosion, noise pollution, and land conflicts. Effective governance thus necessitates government guidelines and oversight to manage these consequences (Anggraini et al. 2019). Auty (2002) notes that in terms of efficiency, the negative impacts arising from mining are often more damaging to non-mining sectors than to what are internationally considered competitive sectors.

Indonesia faces a monumental challenge in the realm of environmental management, especially when it comes to balancing development and sustainability agendas. Good governance is not a zero-sum game between the state and society; rather, it is the outcome of clean, democratic, and effective administration (Nurrochmat et al. 2022). Given the multi-scale complexity inherent in environmental management, significant capacity is needed to address these issues effectively (Lemos and Agrawal 2006). Effective governance practices must also take into account the role of local resource governance in fostering innovation

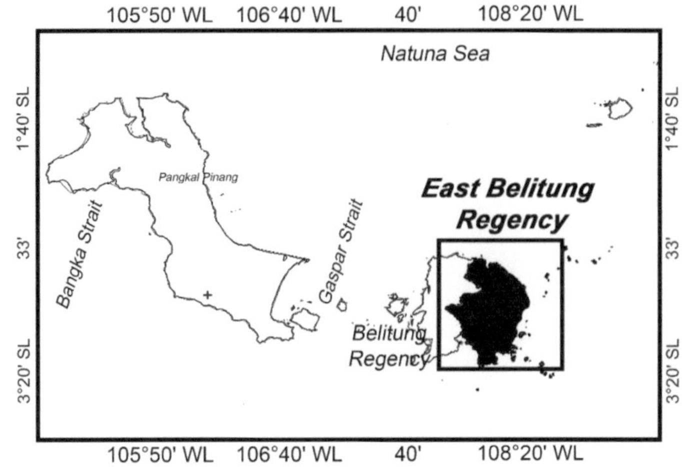

Fig. 4.3 Map of east belitung regency. (*Source* Perda Nomor 13 Tahun 2014 tentang RTRW Kabupaten Belitung Timur Tahun 2014–2034, 2014))

and enabling local adaptive strategies. Emphasis should be placed on the relationship between the characteristics of local governance and the interactions that influence adaptation to regional environmental conditions.

4.5 Result and Findings

4.5.1 The Trigona Beekeeping Initiative on Reclaimed Land from a Sustainable Development Perspective

From the perspective of sustainable development (Warsilah et al. 2017), several themes emerged from the data analyzed, spanning social, economic, and environmental dimensions. The social dimension encompasses the cultural aspects; the economic covers governance aspects; and finally, the environmental dimension of Trigona beekeeping encompasses the technical aspects of this activity and its impact on the environment.

4.5.1.1 The Environmental Dimension

Technical expertise plays a crucial role in this innovative approach. The technical knowledge was disseminated among the Trigona beekeeping community members through training sessions, particularly those led by I1, a pioneer of the KRTB community. Some farmers have also gained insights from YouTube content.

Trigona Sp. Bee Colony

The initial step in cultivating Trigona honey bees involved acquiring a Trigona bee colony (see Fig. 4.4b). The Trigona honey bee community obtains colonies through various methods. In the early stages of this initiative, colonies were primarily sourced by cutting down mature trees that already housed existing colonies. However, farmers are now capable of running their own breeding programs. This was corroborated by I3, who obtained their first colony from an existing one located on their reclaimed mining land. In addition to purchasing colonies, I3 has found them in diverse environments, including logs, underground locations, and even wells.

In Eastern Belitung, the cultivated colonies consist of two specific types of Trigona honey bees: Itama (*Heterotrigona itama*) (Fig. 4.4a) and Levi (*Tetragonula laeviceps*). However, only the Itama type is marketed commercially, while the Levi type is used for personal consumption by the farmers. The Itama colony is chosen for its higher production yield.

Location of Colony Hive

The colony should be placed in a location not exposed to direct sunlight and near a water source. This helps maintain the humidity and temperature inside the box at optimal levels. Cultivating

Fig. 4.4 **a** Trigona bee (*Heterotrigona itama*); **b** Trigona bee colony; **c** Trigona bee honey

Trigona bees on land that was previously used for tin mining offers advantages, such as the availability of water and feed to support colony growth. Land reclaimed from tin mining and replanted with acacia as a pioneer species proves highly suitable for honey bee cultivation. This is because acacia can flower year round, ensuring a consistent feed supply for the bees. Trigona bees also favor acacia because it produces resin, which the colony uses to make propolis for building their nests. However, according to I3, the downside of acacia is that it only tends to live for about six years.

The most important consideration when positioning a colony is the number of colonies in a specific area. This must be regulated to avoid overpopulation creating an imbalance between population size and the availability of feed in that area. While the ratio of land area or vegetation to the number of colonies has not been fully calculated, experience from informants suggests that one hectare of land with dense acacia can accommodate 100 colonies. However, if there is ample water and the acacia is also healthy, some areas can accommodate up to 300 colonies per hectare.

Product Characteristics

The characteristics of Trigona honey (Fig. 4.4c) vary due to differences in cultivation land and can be divided into two types. Honey cultivated in areas with multiflora vegetation tends to have a darker color, while honey produced in areas with monoflora or monoculture vegetation, such as on reclaimed lands, tends to be bright yellow. Interestingly, the characteristics of honey produced on reclaimed lands align well with the preferences of the majority of East Belitung Trigona honey consumers, who favor bright yellow honey. This has encouraged many honey bee farmers to relocate their colonies to reclaimed mining lands.

Honey Harvesting

Honey can be harvested once the honeycomb is full. The rate at which it fills up depends on both the size of the colony and the environmental carrying capacity. The chosen harvesting frequency varies based on these factors. For instance, the Raje Teran Community typically harvests every 45 days, while KRT harvests their honey once a month. The average honey production per colony is about 500 ml per harvest, with the highest recorded yield reaching 800 ml per harvest.

4.5.1.2 The Environmental Impact of Trigona Beekeeping

The environmental impact of cultivating Trigona honey bees on post-mining land is a compelling case study in sustainable land management. Trigona honey bee cultivation on reclaimed tin mining land serves as an ideal model for managing rehabilitated critical lands that previously held no economic value. Through Trigona honey bee cultivation, local communities can derive significant economic benefits without altering the landscape of the reclaimed land. This approach not only sustains the ecological integrity of the land but also transforms it into a productive resource.

Reclaimed land predominantly covered by *Acacia mangium* proves to be ideal for producing Trigona honey, which meets the high demand in major markets such as Malaysia. The honey produced is bright yellow and sweet, indicating that the reclamation methods employed are highly compatible with market preferences. Moreover, the year-round flowering of *Acacia mangium* supports continuous honey production, ensuring a stable and ongoing source of income for the communities involved. This demonstrates that post-reclamation landscapes can be effectively repurposed to support sustainable and economically beneficial activities, reinforcing the viability of integrating ecological restoration with community-based agricultural initiatives.

4.5.1.3 Organizational Aspect—The Social Dimensions

Actors

Trigona beekeepers in East Belitung include both those who are part of a community and those who are not. Beekeepers who choose to be part of a community often join the Indonesia National Beekeeping Community, known as ILMI, or the local Trigona beekeeping communities Komunitas Raje Teran Belitung and Komunitas Asoka Pelo.

Another key player in this ecosystem is the KRTB harvesting team. To ensure the quality of the produced honey, harvesting is carried out exclusively by this team. The Raje Teran Community has 10 such harvesting teams spread throughout East Belitung.

The number of independent or individual farmers is not currently well documented because they are not part of a community. However, they do place their colonies on reclaimed land with acacia vegetation which has not yet been designated protected forest and remains under the control of tin mining companies. Beekeeping activities on the company's reclaimed land do not present a problem for the companies, as these activities are considered non-damaging to the land.

Farmers from Simpang Pesak are ideal Trigona honey farmers. Their community can already balance the amount of vegetation or land with the maximum number of colonies to prevent overpopulation in any given area, ensuring optimal honey production. In contrast, farmers in Bira once experienced an overpopulation issue, with up to 4000 colonies in Bira alone. Despite the investment already made, these Bira farmers chose to stick with it. By 2023, only 1000 colonies remained on the land in Dusun Bira, with some colonies having been abandoned and others relocated to less crowded areas.

Community

The Trigona honeybee community, known as the Raje Teran Community or KRTB, was established to accommodate existing farmers who were facing marketing difficulties. Later on, the community became a facility for members to learn from each other by conducting short training sessions to share knowledge about Trigona honey bee cultivation. The community also holds meetings to improve relationships, although now this is more often done through a WhatsApp group.

Before the setback event of overproduction and imbalance of supply and demand occurred, the community had reached more than 200 members, although the membership declined as members felt honey bee farming no longer met their expectations. A community called Asoka Pelo has been established to address marketing issues with honey that do not comply with the Malaysian market. Asoka Pelo focuses on marketing honey that has a different color from the honey sent to Malaysia, aiming to market darker-colored honey so that all farmers can benefit.

Not all Trigona honey bee farmers are part of the community. According to I3, the community is needed because the productive life of fast-growing acacia trees in reclaimed former mining lands is only six years. Therefore, sustainable planting of acacia is necessary to ensure the continuity of honey bee cultivation, which according to I3, will be easier if done collectively within a community. I3 also stated that to secure long-term contracts with buyers, the quantity of honey to be produced must be guaranteed. To ensure a stable honey supply to meet the Malaysian contract, KRT built a storage warehouse equipped with freezers and dehumidifiers.

The Beginning of the Trigona Beekeeping Initiative

The cultivation of Trigona bees, locally known as *Madu Kelulut*, has a long history on Belitung Island. Rather than being managed on a commercial scale, Trigona honey bee cultivation is generally viewed as a supplementary activity or hobby, with honey production occurring on a small scale.

Over time, enthusiasts of Trigona honey have taken the initiative to form a community, Komunitas Raje Teran Belitung (KRTB), aimed at addressing the marketing challenges facing the Trigona honey industry in Eastern Belitung.

KRTB's primary task is to market the honey produced by its members, which is branded as Raja Teran Belitung honey or RTB honey. Starting with 15 people and a production rate of 15 L of honey per month, the community reached its peak in 2021, boasting 200 members and the ability to produce up to 4 tons of honey per month.

The initiative to start Trigona beekeeping on land reclaimed from tin mining began with KRTB recognizing the need for a showcase facility to demonstrate Trigona honey bee cultivation. This land would serve as a venue to present the potential of Trigona beekeeping to visitors and the community at large. Based on consultations between KRTB and the Kelubi Village government, it was determined that Trigona beekeeping is a non-destructive form of cultivation. The Kelubi Village government suggested a protected forest area, originally reclaimed from tin mining, for the community's use.

KRTB began by placing a few colonies as an experiment. These colonies turned out to yield high-quality honey that matched customer demand and demonstrated good productivity. From this point, KRTB farmers began to place their colonies in the reclaimed tin mining land in Dusun Bira, Kelubi Village. Following this increased activity, the community initiated a formal process to secure land management permissions and eventually obtained a social forestry license.

Recognition

The initiative for Trigona honeybee farming on former tin mining lands has received recognition from various sectors, both private and governmental. The community in East Belitung, including the mining community, has acknowledged the potential of Trigona honeybee farming. This form of farming requires relatively few specialized skills and can produce results within the first three months. This makes it a faster form of cultivation compared to others, such as seasonal plants that require specialized skills. Additional recognition has come from the East Belitung district government and the Bangka Belitung provincial government, which has helped the community obtain a social forestry license in the protected forest area of Dusun Bira, Kelubi Village, from the Ministry of the Environment. The Directorate General of Intellectual Property Rights, under the Ministry of Law and Human Rights of the Republic of Indonesia, has also granted a geographical indication for Teran honeybees as a unique species native to Belitung Island.

4.5.1.4 Economic Dimensions

The Land

The land reclaimed for Trigona honeybee farming is quite extensive in the East Belitung district. It spans about 220 hectares in Dusun Bira, Kelubi Village, which has been designated the Raje Teran Honeybee Village. Additionally, there are 12 hectares in Selingsing managed by the village government-owned enterprise (BUMDes) of Selingsing Village, another 12 hectares in Selingsing managed by the Raje Teran Community (KRT), and some land in Tebat Rasau managed by the Tebat Rasau Community. There are also individually owned parcels of land, such as one owned by I3.

Trigona Honey Market

The Trigona honey market in East Belitung is divided into three categories: the export market, the domestic market outside the island, and the local market within the island of Belitung. The current foreign market focus is Malaysia, where the Raje Teran Community has successfully secured a long-term contract for their Trigona honey, amounting to about one ton per month. Meanwhile, the domestic market for Raje Teran honey includes Jambi, Depok, Bandung, Bogor, and Bali.

Management

The community's role in managing Trigona honey bees focuses on marketing, capital, and education. Markets for the honey were originally secured by

I1, who is also a pioneer in the commercialization of Trigona honey. I1 successfully penetrated markets in Jambi, Depok, Bandung, and Bogor, and also secured a long-term contract to supply honey to a consumer in Malaysia at a rate of 1 ton per month.

In terms of capital, KRTB serves as a bridge between managers and capital owners. In 2019, during the early days of KRTB, Trigona beekeeping was not yet considered a promising business in East Belitung. At that time, I1 not only made initial investments in Trigona honeybee cultivation for his operations but also provided capital to community members to start their own Trigona honeybee cultivation through a partnership scheme. The balance between self-management and partnership is 50:50. This approach ensures that sales contracts are fulfilled. I1 emphasizes the importance of this initial investment to convince those around him that Trigona honeybee cultivation is a viable business.

Unlike KRT, I3, an individual Trigona honeybee cultivator, did not start with land already reclaimed by a tin mining company. Instead, I3 started from scratch, reclaiming land that was the remains of an old 19th-century tin mine. He began the reclamation process in 2013 and only became interested in Trigona honey bee cultivation in 2021. I3 initiated his Trigona beekeeping venture by collecting Trigona honeybee colonies found on his land—in old tree trunks, underground, and even in a well on his property.

Significance

This research also finds out how the initiative of cultivating Trigona honey bees affects the lives of beekeepers in East Belitung. According to I1, farmers have at least 10 colonies of Trigona honey bees, and at most, they have 300 colonies. Therefore, it can be assumed that farmers with 10 colonies will generate at least Rp. 600,000 per month. This is based on the assumption that harvesting occurs every 45 days and yields 10 L per harvest. With a price per liter of around Rp. 100,000, farmers can earn about Rp. 1,000,000 per harvest. Multiplying by 2 and then dividing by 3 months, they can earn Rp. 600,000 per month. According to I1, during the harvest season or the dry season, farmers with 10 colonies can earn up to Rp. 1,000,000 per month.

Successes and a Setback for Honey Cultivation

In the Trigona beekeeping initiative, farmers have not only achieved significant gains but also experienced setbacks. The setbacks occur because Trigona honeybee cultivation is not well managed. Beekeepers in East Belitung were late to realize that cultivation must balance the number of colonies with the vegetation available to support those colonies. Therefore, the number of colonies in a particular area has to be limited. If overpopulation occurs, then honey production will decrease drastically. This is what happened in East Belitung from the end of 2022 to early 2023. Due to an imbalance between the number of colonies and the availability of feed, honey production significantly decreased and failed to meet farmers' initial expectations.

In addition to the issue of overpopulation, the initiative in East Belitung also faces challenges related to pests and balancing supply and demand. For instance, monkeys fond of honey have proven very difficult to manage. On the supply and demand side, farmers become discouraged when their harvests are not absorbed by the existing market. The sudden increase in the number of farmers has made it impossible for the market to absorb the entire harvest.

Problems related to oversupply, pests, and overpopulation have led to a drastic reduction in community involvement in Trigona beekeeping. Some miners who had switched to bee farming returned to their old business as tin miners. This shift has also resulted in conflicts. Those who had previously tried bee farming and had grown accustomed to working on that land eventually began illegal mining there again, causing damage to the reclaimed land.

According to I1, I2, and I3, the shift occurred because the mining community is accustomed to seeing instant results. When the potential of Trigona honey bees became apparent, they flocked to it. However, when the results were not as expected, some even went as far as burning their colonies or leaving them in neglect.

4.6 Discussion

Tin mining in East Belitung Regency, whether legal or illegal, has contributed to the welfare of the local community. As mining has become the principal livelihood in addition to the agricultural sector, the community is finding it difficult to break free from their dependency on this extraction business. This poses a challenge bearing in mind the need for rehabilitation and sustainable natural resource management (Ibrahim et al. 2018; Sulista 2019). Tin miners in Bangka Belitung province exhibit a reluctance to discontinue mining operations, attributable to the inconsistency of political policies and the weakness of the legal culture (Prapti Rahayu et al. 2023).

The local government has sought to rehabilitate former tin mining land, introducing alternative economies in the hope that mining communities can transition to other industrial sectors. However, this effort has not fully succeeded in reducing mining activities in East Belitung. One of the challenges is that the alternative sectors offered, such as tourism and plantation industries, have not been sufficiently attractive to a community that is accustomed to mining (Sulista 2019). Both industries require specialized skills, especially in the field of plantations, which may be difficult for the local community to acquire.

The shift from tin mining to Trigona honey bee farming by some community members has gained attention. Those accustomed to instant results from mining—where they can earn income by the evening if they mine in the morning—find the switch difficult; however, some residents have succeeded in profiting from this type of farming, encouraging others to enter the same business. The disaster that resulted from oversupply and overpopulation, following which the number of Trigona honey bee farmers decreased dramatically, underscores the critical role of government in promoting or aiding market development. On the other hand, the community also has a part to play in managing quotas. The governance and business model of Trigona honey bees shows there is significant potential for managing abandoned lands based on local ecosystems, demonstrating that stakeholders can manage to prevent overproduction and maintain a balance between the environmental carrying capacity and the number of colonies. This issue is not only present in the honey industry but also in the floating cage aquaculture industry (Lusiastuti et al. 2020).

The residents' expectation of instant gratification makes it hard to come to terms with declining yields in Trigona honey beekeeping. Some individuals, frustrated by this, have resorted to burning their bee colonies, while others have simply neglected theirs. According to another source, the issue is not that the honey is not selling; rather, it is selling for less than they had anticipated. This has led some former bee farmers to revert to their previous occupation of tin mining, causing further conflicts in the process. Adding to these challenges are frequent criminal activities associated with Trigona honey bee farming, such as the theft of bee colonies, illegal harvesting of acacia wood, and the use of synthetic insecticides by neighboring farmers, underscoring the need for prudent and sustainable governance in natural resource management. Farmers who choose not to return to mining find themselves with limited options.

4.7 Conclusion and Recommendation

4.7.1 Conclusion

Governance of former tin mining areas, such as East Belitung, Indonesia, has experienced significant changes. Trends include sustainable governance, community participation, and economic diversification. Community involvement in decision-making regarding the use of reclaimed land is key in this effort. The initiative taken by the local community in East Belitung to introduce Trigona beekeeping is an effective way to rehabilitate former tin mining land. Apart from being an alternative source of income for local communities, it also reduces their dependence on extractive businesses that are vulnerable to environmental damage and global fluctuations. However, the success of this initiative also depends on wise management practices. Problems in the management of Trigona honey bee honey in East Belitung include overpopulation or excessive supply. It is important to consider good management, effective marketing, monitoring bee colonies, and preventing bee health problems as part of successful farming practices, which means that intervention from the government or other related parties is needed. To overcome overpopulation and oversupply, it is necessary to consider the balance of the ecosystem and a sustainable economy.

In conclusion, the social, economic, and environmental aspects of the Trigona beekeeping initiative using tin reclamation land has been shown to fulfill the role of a sustainable management model. This crucial model needs to be promoted and disseminated in other regions to achieve greater long-term benefits for local communities and the environment. The East Belitung initiative exemplifies the successful application of sustainable development theory, integrating both social, economic, and ecological benefits. It has not only assisted the restoration of degraded ecosystems but also offers new prospects for improving the local community's economy.

4.7.2 Recommendations

A number of recommendations need to be implemented to ensure the successful governance of the economy and environment on former tin mining lands, particularly in East Belitung. First, the government should facilitate new marketing channels to address the problem of excess honey supply. Second, the community should be implementing Trigona colony quotas to help maintain a balanced ecosystem and a sustainable economy, ensuring that Trigona honey production remains available throughout the year without experiencing a significant decline due to overpopulation. Third, the government must strive to prevent illegal miners from operating on former tin mining lands by tightening supervision of these areas. This supervision is essential to ensure that the land does not become an illegal mining site again. If this issue is left unchecked, not only will the model of Trigona honey production on the former mining land fail, but it will also trigger recurring environmental disasters and a larger social crisis. Furthermore, the government should also support research and development to provide scientific evidence for both the products and the business models. In this way, the Trigona beekeeping initiative can become a governance model for the sustainable use of post-mining landscapes. This will bring significant economic benefits to mining communities, reducing dependence on tin mining which is vulnerable to global price fluctuations, as well as providing a sustainable alternative income for local communities. With these steps, we can achieve a smoother transition from the mining sector to alternative economic sectors in East Belitung.

References

Anggraini D, Soemarno S, Riniwati H (2019) The effect of traditional gold mining toward the socio-economic and environmental aspect in Sepauk District Sintang Regency. Indones J Environ Sustain Dev 10(1):51–57. https://doi.org/10.21776/UB.JPAL.2019.010.01.09

Auty R (2002) Sustaining development in mineral economies: the resource curse thesis. Sustain Dev Mineral Econ. https://doi.org/10.4324/9780203422595

Brennen BS (2021) Doing qualitative research. In: Qualitative research methods for media studies, pp 16–30. https://doi.org/10.4324/9781003122388-2

Dwivedi R (2001) Environmental movements in the global south: outline of a critique of the 'livelihood' approach. Global Social Movements. https://doi.org/10.1057/9780230554443_11

Erman E (2008) Rethinking legal and illegal economy: a case study of Tin Mining in Bangka Island. Southeast Asia: History Culture 2008(37):91–111. https://doi.org/10.5512/SEA.2008.37_91

Folke C, Hahn T, Olsson P, Norberg J (2005) Adaptive governance of social-ecological systems. Annu Rev Environ Resour 30:441–473. https://doi.org/10.1146/annurev.energy.30.050504.144511

Haryadi D, Darwance D, Salfutra RD (2018) Implementasi Tanggungjawab Reklamasi Pertambangan Timah Di Pulau Belitung. Progresif: Jurnal Hukum 12(2):2083–2101. https://doi.org/10.33019/PROGRESIF.V12I2.972

Humphreys M, Sachs JD, Stiglitz JE (2007) Introduction: what is the problem with natural resource wealth? In: Escaping the resource curse. Columbia University Press

Ibrahim I, Haryadi D, Wahyudin N (2018) Knowledge of the context, behavior, and expectations of miners about the tin mining policies and practices in Bangka Belitung. Masyarakat, Kebudayaan Dan Politik. https://doi.org/10.20473/mkp.v31i42018.358-367

Lemos MC, Agrawal A (2006) Environmental governance. Annu Rev Environ Resour 31:297–325. https://doi.org/10.1146/ANNUREV.ENERGY.31.042605.135621

Lusiastuti AM, Taukhid MM, Prayitno SB, Sugiani D, Caruso D (2020) Building and improving the capacity of fish and environmental health management strategy in Indonesia. IOP Conf Ser: Earth Environ Sci. https://doi.org/10.1088/1755-1315/521/1/012016

Mulyana W, Prasojo E, Suganda E, Moersidik SS, Mulyana W, Suganda E, Prasojo E, Moersidik SS (2023) The conceptual models of dynamic governance toward sustainable urban water management in metropolitan area. Springer Nature Switzerland, pp 243–271

National Minerals Information Center (2023) Mineral Commodity Summaries 1996–2023. U.S. Geological Survey. https://www.usgs.gov/centers/national-minerals-information-center/mineral-commodity-summaries

Neo BS, Chen G (2010) Dynamic governance - embedding culture, capabilities and change in Singapore. In: Dynamic Governance—Embedding Culture, Capabilities and Change in Singapore. World Scientific Publishing

Nurrochmat DR, Sahide MAK, Fisher MR (2022) Making sustainable forest development work: formulating an idea for a more appropriate green policy paradigm. Front Environ Sci 10:783718. https://doi.org/10.3389/FENVS.2022.783718/BIBTEX

Pemerintah Kabupaten Belitung Timur (2021) Data Kependudukan. Portal Pemerintah Belitung Timur. https://portal.beltim.go.id/data-kependudukan/

Perda Nomor 13 Tahun 2014 Tentang RTRW Kabupaten Belitung Timur Tahun 2014–2034 (2014) https://jdih.belitungtimurkab.go.id/content/rencana-tata-ruang-wilayah-kabupaten-belitung-timur-tahun-2014-2034

Prapti Rahayu D, Faisal F, Darwance D, Jaka Ferdian K (2023) Environmental and social injustice: impact and sustainability of small scale tin mining under Indonesia's new mineral and coal regulation. Cepalo. https://doi.org/10.25041/cepalo.v7no2.3137

Sulista S (2019) The sustainability of livelihoods of villages around the Tin Mine in Bangka Islands. Jurnal Bina Praja. https://doi.org/10.21787/jbp.11.2019.43-52

Sulista B, Provinsi K, Bangka B, Jalan P, Belitung N, Perkantoran K, Kepulauan P, Belitung B (2019) Tambang Inkonvensional: Peran Masyarakat Dan Daya Tarik Ekonomi Bagi Penambang. Jurnal Teknologi Mineral Dan Batubara 15(1):63–75. https://doi.org/10.30556/JTMB.VOL15.NO1.2019.348

Swastiwi AW, Nugraha SA, Purnomo H (2017) Lintas Sejarah Perdagangan Timah Di Bangka Belitung Abad 19–20. In: Sobuwati D, Nuraini (eds) Balai Pelestarian Nilai Budaya Kepulauan Riau

The World Bank (2023) Commodity markets. In: World Bank Commodity Price Data (The Pink Sheet): annual prices, 1960 to present. https://doi.org/10.4324/9781315706863-15

Vanessa A, Tampubolon A, Haeriah A, Dedi A, Widi BN, Pardiarto B, Sayekti B, Sunuhadi DN, Mulki D, Putra E, Emowo Labaik G, W HMH, Eddy HR, Sinulingga IK, Muksin I, Mauritz J, Kisman Raja M, Zulfikar (2018) Indonesian Mineral Yearbook. Center for Mineral Coal and Geothermal Resources Geological Agency, Ministry of Energy and Mineral Resources

Warsilah H, Wardiat D, Fatimah N, Rochmawati, Nadila SM (2017) Pembangunan inklusif dan kebijakan sosial di Kota Solo, Jawa Tengah. Yayasan Pustaka Obor Indonesia

5 The Socio-Ecological Crisis of Abandoned Coal Mining Pits in East Kalimantan: Efforts to Fight Injustice

Mirza Satria Buana, Mochamad Felani Budi Hartanto, and Dian Andi Nur Aziz

Abstract

The presence of abandoned coal mining pits in East Kalimantan has emerged as a significant issue negatively affecting both the environment and the local community. From 2011 to 2021, a recorded total of 40 individuals, mostly children, lost their lives due to sinking into these abandoned pits. Beyond the loss of human life, environmental degradation, air pollution, and water contamination have occurred as a consequence of irresponsible mining practices and inadequate post-mining environmental reclamation. Against this backdrop, this chapter aims to answer the following questions. First, how does the legislation regulate the mechanism of post-mining reclamation, particularly in delivering remedies to victims and affected citizens? Second, what social and solidarity movements exist for victims and affected citizens to access legal and non-legal remedies? The novelty of this research lies in its examination of the resistance against mining corporations and the state displayed by victims and their families pursuing their rights to remedy in the areas affected by mining operations, based on the implementation of the United Nations Guiding Principles on Business and Human Rights (UNGPs) framework. Qualitative research methodology and descriptive analysis are employed, with a specific focus on the rights to remedy of victims in cases of human rights violations. The data collection process involves document analysis and case studies related to victims of former mining pits in Kalimantan. The data utilized in this study are derived from government reports, the National Commission on Human Rights, non-governmental organizations, legal documents, and relevant literature.

Keywords

Mining · Coal · Remedy · Human rights · UNGPs · Kalimantan

M. S. Buana (✉)
Faculty of Law, Lambung Magkurat University, Banjarmasin, Indonesia
e-mail: mirza.buana@ulm.ac.id

M. F. B. Hartanto · D. A. N. Aziz
Research Centre for Society and Culture - National Research and Innovation Agency, Jakarta, Indonesia
e-mail: moch063@brin.go.id

D. A. N. Aziz
e-mail: dian097@brin.go.id

5.1 Introduction

This chapter examines the social, cultural, and legal effects of the abandoned coal mining pits in East Kalimantan, a significant issue affecting both local communities and the surrounding

H. Warsilah and A. H. Purnomo (eds.), *Approaching Natural Resources for Sustainable Development in Indonesia*, Sustainable Development Goals Series, https://doi.org/10.1007/978-981-96-6913-4_5

environment. From 2011 to 2021, 40 individuals, most of them children, lost their lives due to sinking into these abandoned mining pits (JATAM 2020). Beyond the loss of human life, environmental degradation, air pollution, and water contamination have occurred as a consequence of inadequate post-mining environmental restoration efforts.

East Kalimantan Province is a province rich in natural resources. It is well known as a stronghold of extractive industries, particularly coal mining. In the period 2018–2022 there were approximately 1404 mining permits, and 30 mining indentures (Ministry of Energy and Mineral Resources 2023). In Kutai Kartanegara Regency, there are 264 coal mines. There are 39 companies operating in Samboja district, with a total land area of 382.63 ha. In Sepaku district, which is included in Nusantara Capital City, five active mining companies operate with a total land area of 30.40 ha. At least 19 abandoned pits have been identified (Mongabay 2022). It is reported that most of the victims of the abandoned pits did not receive a proper legal remedy. It can therefore be hypothesized that the government and companies ignore the constitutional rights of the victims and people living near the pits. In other words, government and non-state actors are involved in what can be termed human rights violations by omission (Gasser 2007).

There are several previous studies regarding the effects of extractive industry in the East Kalimantan context. Maimunah (2014) narrates the women's perspective of survival amidst extractive coal mining practices in East Kalimantan, particularly in its capital city, Samarinda. The study describes the negative effects on the environment and shrinking of living space caused by the coal mining boom, with business practices that are not accountable, and the ineffective control by central government of regional authorities that has resulted from the regional autonomy policies of 2001–2010 (Maimunah 2014). Furthermore, Buana (2017) explores the struggle of local and Indigenous peoples in the Bontang district to defend their religious-spiritual values against the modernization brought by multinational corporations (Buana 2017). The construction of Nusantara Capital City, the new capital city of Indonesia, which includes some regions of East Kalimantan, has caused many social, political, and legal issues, particularly in former mining areas. The development has caused Indigenous groups and local communities to lose their autonomy and land rights, as there is no statutory protection for affected citizens in the Nusantara Act (Buana et al. 2023). Previous studies have not specifically examined the victims' perspective on seeking remedy in the ex-mining areas. In the context of the present study, citizens is a more general term, encompassing local and migrant residents living near the mines. The study contributes to the existing understanding of development and investment practices, particularly where a clash between citizens' rights and development projects is concerned.

This chapter aims to answer the following questions. First, how does the legislation regulate the mechanism of post-mining reclamation, particularly in delivering remedies to victims and affected citizens? Second, what social and solidarity movements exist for victims and affected citizens to access legal and non-legal remedies? In answering these questions, this study employs the United Nations Guiding Principles on Business and Human Rights (UNGPs) framework as a 'theoretical touchstone' (United Nations 2011) and the UN Sustainable Development Goals (SDGs) to identify the sustainability aspects of the socio-ecological crisis in East Kalimantan. Civil and political rights are also at the center of the analysis, particularly the rights of citizens to access effective remedies, and freedom from fear (right of security) (International Covenant on Civil and Political Rights 1966). The cross-cutting analysis also concerns economic, social, and cultural rights, particularly regarding the rights of citizens to enjoy a humane and safe environment (Lillich 1986, p. 74). In their efforts to obtain remedies, affected citizens who live near the areas have many options for dispute settlement. A 'citizen-as-plaintiff' can therefore do 'forum shopping' to obtain the most favorable and beneficial outcome (Cornel Law School 2022).

This research addresses issues that are crucial for achieving SDGs, particularly Goal 6 (Clean Water and Sanitation), Goal 11 (Sustainable Cities and Communities), Goal 13 (Climate Action), Goal 15 (Life on Land), and Goal 16 (Peace, Justice, and Strong Institutions). It explores the socio-environmental challenges posed by abandoned mining sites, highlighting the importance of addressing water contamination caused by mining activities, the necessity of proper post-mining land reclamation to ensure that communities are not exposed to hazardous conditions, the need to mitigate the environmental damage caused by irresponsible mining practices, the importance of reclaiming and rehabilitating degraded land and preventing further biodiversity loss due to mining activities, and the importance of creating a just and equitable society where the rights and safety of all citizens are upheld, in alignment with the overarching goals of the SDGs.

5.2 Methodology

This chapter employs a socio-legal methodology to depict broad aspects of legal issues (Banakar and Travers 2005). Legal scholarship is at its core, but qualitative research methodology and descriptive analysis are employed, with a specific focus on the rights of victims to remedy in cases of human rights violations. The data collected includes case studies related to victims of abandoned pits in East Kalimantan. In analyzing the existing legislation and regulations in Indonesia, the study establishes that the current laws on mining, post-mining restoration, and access to remedy in Indonesia have many normative flaws. Due to the unwillingness of the government and non-state actors (companies) to settle disputes and to fulfill citizens' rights to an effective remedy, victims and affected citizens are engaging in 'forum shopping' for remedies through both legal mechanisms and social and solidarity movements.

5.3 Results and Findings

5.3.1 Context: East Kalimantan and Its Extractive Industry

Indonesia's coal reserves are the sixth largest in the world, constituting approximately 4% of the global total (Kementerian Energi dan Sumber Daya Mineral 2022). Indonesia has demonstrated the capacity to both satisfy domestic demand and fulfill export requirements. In the year 2022, national coal production reached 687 million tons, exceeding the government's target of 663 million tons (see Table 5.1). This substantial production volume enabled Indonesia to not only meet its domestic needs, which amounted to 193 million tons or 116% of the 166 million tons target but also to exhibit an extensive surplus in national coal production capacity. Consequently, Indonesia was able to satisfy export demands to the tune of 494 million tons (Kementerian Energi dan Sumber Daya Mineral 2023b).

Table 5.1 Coal production and export plan

Year	national coal production plan, million tons	National coal export plan, million tons
2022	663	497
2023	694.5	460

Source Ministry of Energy and Mineral Resources (2023)

Ongoing data indicates that as of September 11, 2023, production had reached 509.11 million tons, equivalent to 73.31% of the production target. Coal exports realized amounted to 236.15 million tons, or 51.34% of the planned export volume. For domestic consumption, the government has established a Domestic Market Obligation (DMO) of 176.80 million tons, of which 71.06 million tons, or 40.19%, have been fulfilled (MODI 2023a). As stated in the Decree of the Minister of Energy and Mineral Resources of the Republic of Indonesia Number 267.K.MB.01/MEM.B/2022, the DMO policy requires 25% of coal production to be allocated for domestic consumption, especially for electricity and industry sectors.

In 2022, non-tax state revenue (*Penerimaan Negara Bukan Pajak* or PNBP) derived from the energy and mineral resources sector (ESDM) reached Rp351 trillion, which amounted to 138% of the set target of Rp254 trillion. This figure significantly surpassed the entire 2021 revenue of Rp184 trillion (Kementerian Energi dan Sumber Daya Mineral 2023b). Revenue supplements from the energy and mineral resources sector (ESDM) have resulted in a notable elevation in the contribution of revenue-sharing funds to regional governments in 2022, reaching a total of 109.18 trillion rupiah. This amount exceeded the planned target by a significant margin, registering a surplus of 138.43% compared to the designated plan and substantially surpassing the 2021 actual realization of 44.34 trillion rupiah (MODI 2023b).

The Ministry of Energy and Mineral Resources of the Republic of Indonesia has reported that, as of 2021, Indonesia's coal reserves amounted to 38.84 billion tons (see Table 5.2), with the majority of these reserves situated in Kalimantan and Sumatra. Kalimantan accounts for a substantial portion of these reserves, containing 62.1% of the total potential coal reserves and being the largest coal resource region in Indonesia, with 88.31 billion tons of resources and 25.84 billion tons of reserves. Additionally, Sumatra possesses 55.08 billion tons of coal resources and 12.96 billion tons of coal reserves. Assuming an annual average coal production of 600 million tons, the estimated lifespan of these coal reserves stands at 65 years, assuming no new reserve discoveries (Kementerian Energi dan Sumber Daya Mineral 2023a).

Table 5.2 Indonesia's coal reserves

	Million tons
Coal reserves	38,840
Average annual production	600
Coal resources	143,700

Source Ministry of Energy and Mineral Resources, 2023

According to official data, as of September 2022, the Indonesian government has issued a total of 1626 Mining Business Licenses for Minerals and Coal, with 892 of these licenses specifically designated for coal mining (Kementerian Energi dan Sumber Daya Mineral 2022). Based on the latest data from the Central Statistics Agency, there were a total of 193 active coal mining companies operating in East Kalimantan (ANTARA News 2023), located in the middle north of the Indonesian territory (see Fig. 5.1). Nonetheless, it is crucial to note that these data sets still necessitate the synchronization of permit holder information. The primary issue in the management of coal mining businesses in East Kalimantan lies in the lack of harmonization among permit holder data, which poses a substantial risk of significant financial losses to the state (Lawi 2019). The research conducted by Transparency International underscores that the lack of transparency in granting Mining Business Licenses (IUP) within the mining sector, characterized by restricted access to data and information and minimal involvement of affected communities, gives rise to significant corruption risks during the authorization process (Transparency International 2017).

The lack of data clarity extends to the matter of post-mining reclamation. As highlighted in the report by the National Commission on Human Rights (Komnas HAM), former excavation sites resulting from coal mining operations are, notably, left without appropriate measures such as fencing or warning signage. These pits are situated close to residential areas (Komnas HAM 2016). The East Kalimantan Mining Advocacy Network (Jaringan Advokasi Tambang or JATAM), a civil society organization actively engaged in advocating mining-related issues, has documented the persistent menace of abandoned mining excavations in the province of East Kalimantan, with a total of 1735 disused pits in the region. Notably, within the municipality of Samarinda, 349 of these former pits remain unclosed, lacking any reclamation or restoration measures. The abandoned mining excavations in East Kalimantan have tragically resulted in a substantial loss of human

Fig. 5.1 Map of Kalimantan. (taken from https://geoportal.esdm.go.id/emo/)

life. According to a report from the year 2021, a total of 40 fatalities have been recorded due to the presence of these excavations, even within the proposed new national capital region (IKN). Among these cases, the city of Samarinda has been the most heavily affected, contributing to 23 out of the 40 reported fatalities. The Kutai Kartanegara (Kukar) region has also reported incidents, with 13 fatalities documented in the mining excavations of Kutai Barat. Additionally, Penajam Paser Utara and Paser have each reported one and two cases, respectively (CNN Indonesia 2023).

On a national scale, JATAM National (2021) has documented that between the years 2014 and 2020, a cumulative total of 168 fatalities have tragically occurred as a consequence of accidents involving open mining pits throughout Indonesia. Furthermore, a substantial risk continues to persist, with 3092 open pits situated near densely populated residential areas remaining, containing hazardous water and heavy metal (JATAM 2021b).

Based on the abovementioned conditions, it is pivotal to merge all business activities, especially extractive ones, into the legal framework of human rights. The human rights mechanism aims to make sure that development and business activities can be enjoyed by all citizens, in particular those who live near development projects and industries.

5.3.2 Human Rights Enter Business: Remedy as a Mechanism to Survive

International human rights provisions, both binding (covenants and conventions) and non-binding (declarations and principles) state that both state and non-state actors are the obligators of human rights. They are duty bearers who have obligations to respect, protect, and fulfill the human rights and fundamental freedoms of their citizens and non-citizens (Gasser 2007). Both state and non-state actors that fail to exercise, or ignore, the rights of citizens are considered violators of human rights, both in commission (active act) and omission (ignorance) (Gasser 2007). Non-state actors (business enterprises or companies), as specialized organs of society performing specialized functions, are required to obey all applicable existing laws and to respect human rights (United Nations 2011). Respect and obedience to the existing laws are the requirements of human rights-based development, in which human rights become a constitutive part of development (Uvin 2007, p. 601). And, as rights holders, citizens must have a legal constitutional guarantee that when their rights are breached, there will be appropriate, accessible, and effective remedies. The ICCPR stresses this right by saying: "Everyone shall have the right to recognition everywhere as a person before the law" (International Covenant on Civil and Political Rights 1966). In the context of disputes or grievances arising from development practices, the government should recognize the "right to survival" in the legal system of rights and remedies. It is a customary international *jus cogens* norm, elaborated out of the recognition of the consensus of non-binding instruments claiming the right to development (Acharya 2014, p. 360).

When a breach of rights happens, the state or, in its operative sense, the government, must take appropriate steps to ensure, through judicial, administrative, legislative, or other appropriate means, that when such abuses occur within their territory and/or jurisdiction those affected have access to effective remedies (United Nations 2011). This principle is manifest in the survival right to development, which focuses on the ability of communities to survive development processes. To create a conducive ability to survive, development projects should be surrounded by a set of legal mechanisms and provisions that can elevate and defend people's rights *vis-à-vis* government actions, forcing legislators to adopt rights-based policies, and in turn promoting and creating a rights-enabling economic environment (Sengupta 2008). Citizens need rights that are claimable through legal remedies (Pratiwi et al. 2022, p. 9). Those claimable rights should be of paramount importance in the provisions regarding development. This is the essence of the survival right to development.

This principle is also in line with the ICCPR provision on effective remedy, which states: "Everyone whose rights and freedoms as outlined in [the] Convention are violated shall have an effective remedy before a national authority notwithstanding that the violation has been committed by a person acting in an official capacity" (International Covenant on Civil and Political Rights 1966). Regarding this principle, the government is obligated to provide substantive remedies, either in judicial or non-judicial mechanisms, including administrative, legislative, or other mechanisms. Remedy can include broader aspects encompassing apologies, restitution, rehabilitation, financial or non-financial compensation, and punitive sanctions (whether criminal or administrative, such as fines), as well as the prevention of harm through, for example, injunctions or guarantees of non-repetition (Teitel 2002, p. 893).

In a procedural sense, the mechanism for remedy should be impartial, secure from judicial corruption, and free from discrimination and political or other influences that could impede or manipulate the outcome. This fair and impartial procedure is relevant to the ICCPR non-discrimination principle and the guarantee of equality before the law that states: "All persons are equal before the law and are entitled without any discrimination to the equal protection of the law. In this respect, the law shall prohibit any discrimination and guarantee to all persons equal and effective protection against discrimination on any ground…" (International Covenant on Civil and Political Rights 1966).

Procedures are pivotal because they can determine whether the outcome is meaningful or mere tokenism. The government must ensure the effectiveness of domestic judicial mechanisms when addressing business-related human rights violations, including considering ways to reduce legal, practical, and other relevant obstacles that could lead to a denial of access to remedy

(United Nations 2011). Types of obstacles can be defined as follows. The first kind can be pressures inflicted on citizens by other state agents and business actors. This condition happens due to an imbalance between the parties to business-related human rights claims. Realistically speaking, citizens are never on the same footing as business actors. Any acts of intimidation and pressure relate to freedom from fear and the right of citizens to obtain security, particularly within the judicial system (Spigelman 2010, p. 547). The ICCPR guarantees these rights and freedoms, stating: "(1) No one shall be subjected to arbitrary or unlawful interference with his privacy, family, home, or correspondence, nor to unlawful attacks on his honor and reputation; (2) Everyone has the right to the protection of the law against such interference or attacks" (International Covenant on Civil and Political Rights 1966).

The second internal obstacle is the absence of legal standing for claimants. Citizens such as Indigenous peoples and migrants cannot access judicial processes in a home state court due to their distinctive characteristics or lack of administrative documents, regardless of the merits of their claim. In other words, they are systematically excluded from legal protection. Third, the cost of bringing claims to court is prohibitive and cannot be waived or reduced to reasonable levels through government support. In some jurisdictions, civil proceedings are not free, with claimants responsible for unsecured costs including lawyers' fees. This barrier is structural because it is stated in the legal framework. And it is detrimental because it can directly breach citizens' rights to legal representation (Soy and Hing 2019, p. 7).

The fourth hurdle comes from both the government and business actors. The latter influences the former to legislate to regulate the legal responsibilities of corporate groups under domestic and civil laws, but the norms facilitate the avoidance of accountability. This is 'intended to fail' legal provision. Rather than providing legal certainty and access to justice for claimants, lawmakers create a 'lacuna' to prevent companies from accepting accountability and legal responsibility. At best, this constitutes legislative corruption. Finally, the fifth obstacle also comes also from law enforcers who are reluctant to investigate individual and business involvement in human rights violations. This practice of judicial corruption contradicts the ICCPR's principle, which guarantees: "All persons shall be equal before the courts and tribunals ... everyone shall be entitled to a fair and public hearing by a competent, independent, and impartial tribunal established by law" (International Covenant on Civil and Political Rights 1966).

As regards non-judicial remedy mechanisms, the government must also provide a complementary state-based system for remedy alongside the judicial mechanism. In order to ensure their effectiveness, non-judicial mechanisms should be: (1) *legitimate*, enabling trust from the stakeholder groups for whose use they are intended, and being accountable for the fair conduct of grievance processes; (2) *accessible*, being known to all stakeholder groups for whose use they are intended, and providing adequate assistance for those who may face particular barriers to access; (3) *predictable*, providing a clear and known procedure with an indicative time frame for each stage, and clarity on the types of process and outcome available and means of monitoring implementation; (4) *equitable*, seeking to ensure that aggrieved parties have reasonable access to sources of information, advice and expertise necessary to engage in a grievance process on fair, informed, and respectful terms; and (5) *transparent*, keeping parties to a grievance informed about its progress, and providing sufficient information about the mechanism's performance to build confidence in its effectiveness and meet any public interest at stake (United Nations 2011). These guiding principles will be tested in concrete cases in abandoned mining pits in East Kalimantan, Indonesia. However, the legal system and its regulatory mechanism should first be scrutinized in relation to human rights principles.

5.3.3 What the Laws Say

There are statutory obligations for companies concerning post-mining reclamation. The 2009 law stresses that post-mining activities, including reclamation, are integral parts of mining concession activities. Thus, the concession holders (companies) cannot disregard them legally. Generally speaking, post-mining activities should be undertaken in well-planned, systematic, and sustainable ways. Reclamation is a more specific effort to manage and restore the quality of the environment in and near the mining area. The two processes are inseparable. The government must control post-mining activities, including reclamation processes, starting from the very beginning of the concession contract (the planning stage) (Law No.4/2009 on Mineral and Coal Mining, Article 6). The Law emphasizes the administrative requirements for both the holders of ordinary (IUP) and special concession (IUPK) permits to provide guaranteed funds for post-mining reclamation activities. Without this requirement, the production-stage permit cannot be issued by the government. And, if they disobey the obligation, both central and local government officials (the Ministry, the Governor, the Regent/the Mayor) can assign a third party to carry out reclamation activities funded by the guarantee fund (Law No.4/2009 on Mineral and Coal Mining, Article 100). However, there is uncertainty about who the third party is, and how exactly the third party undertakes the reclamation processes. By failing to provide clear and enforceable laws in this regard, the government could be a violator of human rights by omission (Gasser 2007).

Furthermore, both the 2010 Government Regulation and the 2014 Ministerial Regulation emphasize preventive measures by obligating permit holders to stipulate the planning of reclamation and post-mining activities before approval of the permit by the government. The two stages of mining activities – exploration and production – must include reclamation and post-mining documents. The planning document should emphasize four activities: (1) reclamation at the remnants of mining projects (mining pits); (2) the maintenance of reclamation; (3) community development; and (4) controlling mechanisms (GR No.78/2010 on Reclamation and Post-Mining Activities, Article 10; The Energy and Mineral Ministerial Regulation No.17/2014 on Reclamation and Post-Mining Activities, Article 12). In preparing the planning document, companies must consult with government institutions and communities living near the mining project (GR No.78/2010, Article 11). However, the regulation does not elaborate on how public participation should be achieved. In other words, there is no meaningful participation in these preventive measures (Soy and Hing 2019). Despite there being stage-by-stage evaluations of permits (both exploration and production) (GR No.78/2010, Article 22), there is no citizen control mechanism to control and evaluate the practice of mining projects and companies' post-mining activities, including reclamation. Citizens lack rights to obtain information, consult, and provide clearance/consent for nearby mining projects that affect their livelihoods.

The regulations also re-emphasize the obligation of permit holders to provide the guarantee fund in the form of term deposits (GR No.78/2010, Article 29; The Energy and Mineral Ministerial Regulation No.17/2014, Article 12). They also stress that if the guarantee fund is insufficient to exercise reclamation and post-mining activities, the permit holders (companies) must add the fund shortage (GR No.78/2010, Article 34). In case of failure to accomplish all the obligations, the permit holders are subject to administrative sanctions, such as written warning, temporary stoppage of mining activities, and/or permit revocation (GR No.78/2010, Article 50). However, the word 'and/or' in this norm is problematic, because it means that the government can either choose one of several alternative sanctions or exercise all sanctions consecutively. This norm creates legal uncertainty and ineffective implementation which could lead to a violation of human rights (Lillich 1986).

In 2020, the government revised the Mining Law by adding repressive norms alongside administrative measures. Criminal sanctions (maximum of a five-year prison term and a fine of Rp. 100 billion) were imposed for both permit holders who disregard the reclamation procedures and those who fail to provide the guarantee fund (Law No.3/2020 on the 1st Revision on Law No. 4/

2009 on Mineral and Coal Mining, Article 103). Before the enactment of the 2020 law, the government could only exercise administrative requirements (preventive measures) for permit concession holders. Concession permits issued from 2020 onward will be covered by both administrative and criminal (repressive measures) sanctions. By having both sanctions, the government aims to obtain the full cooperation of permit holders to exercise post-mining activities and reclamation. However, the nature of the legal system is prospective, meaning it can only be exercised going forward. The 2020 Mining Law cannot prosecute retrospectively other past crimes. This creates a 'lacuna' or legal vacuum for this specific context of East Kalimantan's mining pits.

The rights of citizens to obtain effective remedies are not regulated by all the legislation and regulations on mining in general and reclamation/post-mining activities in particulars. In other words, there is a lack of statutory protection for citizens who live nearby and are directly affected by mining projects. The legislation and regulations merely focus on the company's obligation, but disregard the citizens' rights to obtain effective remedy. The absence of mechanisms is a violation of human rights by omission (Gasser 2007). The absence of statutory protection for citizens is proof that the government does not fully understand the inseparable relationship between human rights and business.

Despite the absence of legal protection, citizens who live nearby and are directly affected by mining projects still have access to justice through judicial mechanisms. The 2019 Regulation of the Supreme Court provides opportunities for citizens to challenge government conduct and/or its absence through citizen lawsuit and public-interest tort lawsuit mechanisms (*onrechtmatige overheidsdaad*) (The Supreme Court Regulation No.2/2019). The legal foundations of this lawsuit are human rights law (The 1945 Constitution, Article 28; Law No.39/1999 on Human Rights, Article 28), environmental law (Law No.32/2009 on Preservation and Protection of Environment, Article 92), and judiciary law (Law No.48/2009 on Judiciary, Article 5, 10). The basis of the lawsuits is government actions that can cause the form of legal actions (*rechtshandeling*) and real actions (*feitelijke handeling*). Real actions are actions that are not aimed at a legal effect but can cause legal consequences. In the East Kalimantan context, citizens of Samarinda City utilized this lawsuit, and the court granted their petitions. However, although legal access is available, the implementation of the Court's decision remains burdensome, as the government was reluctant to cease its policy. This 'legal blockage' forces citizens to go 'forum shopping' to another non-judicial mechanism.

5.3.4 Citizens 'Forum Shopping' to Obtain Remedy

The concept of 'forum shopping' is taken from Kebeet von Benda-Beckmann's article on dispute settlement in Minangkabau society, which describes disputants who have a choice between different institutions and base their choices on what they hope the outcome of the dispute will be (Benda-Beckmann 1981). This concept is relevant to citizens' efforts to obtain rights and remedy beyond the unsatisfactory legal 'forum'.

As mentioned previously, irresponsible coal mining extraction activities in the coal-rich East Kalimantan province have led to many problems, not merely environmental degradation, but also social problems. This situation is exacerbated by lax law enforcement, reflected in the location of open coal mines in extreme proximity to residential areas. It contradicts the provisions outlined in Minister of the Environment Regulation Number 4 of 2012 concerning Environmentally Friendly Indicators for Open-Pit Coal Mining Enterprises and/or Activities. The regulation stated that the edge of the pit should be at least 500 m from a residential area, yet the fact is that many pits are located less than 500 m: in Bengkuring, an area of Samarinda City, there is a pit located less than 100 m from a residential neighborhood (see Fig. 5.2).

The absence of post-mining reclamation of the former coal mining pits has led to water pooling in these voids, forming what appear to be huge lakes. Water contamination in abandoned mines

Fig. 5.2 Satellite Image of Bengkuring Area, Samarinda. (taken from Google Maps with digital ruler used to measure the distance)

potentially produces acid mine drainage, which is susceptible to pollution and can have a negative ecological impact (Amy et al. 2023). Hazardous pits located close to residential areas lack any barriers or warning signs, making it easy for children from the neighborhood to approach, play around, and even swim in them. The first incident of children drowning in an abandoned mining pit occurred on July 13, 2011, when three children regarded as good swimmers were playing around the abandoned mining pit in Sambutan Subdistrict, Samarinda City, went swimming in it, and tragically lost their lives. As of now, there is no legal clarity for the victims' families. There is no criminal justice process, and the company that owns the abandoned pit has only provided "charitable aid" amounting to IDR 15,000,000 (about US$1000) for the victims' families (Komnas HAM 2016). In all, 40 individuals are recorded as having drowned from 2011 to 2021, 33 of them children (JATAM 2021a).

The imbalance of power between the victims and the corporation results in intimidation and pressure applied by the latter on the former. Ignorance of the issue on the part of government and law enforcement, and lack of attempt at reclamation, legal actions against irresponsible companies, or proper legal remedy for the victims, led the victims to actively resist injustice and seek remedy. Through so-called forum shopping they are choosing alternative strategies to obtain remedies.

A petition was created in 2015 by one of the victim's mothers, Rahmawati, who lost her son in an abandoned mining pit in Samarinda. The petition urges the government to shut down and penalize the company responsible for the abandoned mine that resulted in the children's deaths. The petition has garnered support with over 10,000 signatures and was submitted to the Minister of Environment and Forestry as well as the Mayor of Samarinda. Following the submission of the petition, Rahmawati was granted an audience with the Minister of Environment and Forestry.

The day after the petition was submitted, victims' families, accompanied by JATAM, an NGO focusing on human rights and environmental advocacy, filed a complaint with the National Commission on Human Rights (Komnas HAM) regarding their right to justice and remedies. This complaint resulted in the establishment of an investigation team by Komnas HAM for alleged human rights violations related to post-mining pits in East Kalimantan. The team issued two recommendations to the East Kalimantan provincial police chief, the Governor, and the Ministry of Environment and Forestry: to carry out more detailed investigations and to request additional information from the environmental expert. The team also believes that the sentence received by the perpetrator was too low and neither provided a deterrent effect nor fulfilled a sense of justice (Komnas HAM 2016). However, of the 17 companies owning pits that must be held responsible, only four were given light sanctions by the Ministry of Environment and Forest in the form of a ban on activities for only 160 days (JATAM 2019).

JATAM and 23 victims' mothers held a solidarity action on Mother's Day (December 22, 2015), displaying photos and victims' memorabilia installations to gather community support and sympathy regarding the case. This solidarity action encouraged the Ministry of Environment and Forestry to hold a mining work meeting in East Kalimantan in June 2016 and resulted in an integrity pact signed by 125 mining companies throughout East Kalimantan. This integrity pact contains the mining companies' commitment to safeguard and ensure that their ex-mines do not claim any more victims (JATAM 2019). Aside from the integrity pact, the actions also encouraged the Provincial Council of East Kalimantan to form a special committee on reclamation and investigation of ex-mine cases (*Pansus*) in 2016, which issued recommendations in 2017, 2019, and 2023. The *Pansus* recommends stakeholders take action against 21 illegal mining business permits that are still operating and asks the government to transparently open the disbursement of mine reclamation guarantee funds (DPRD Provinsi Kalimantan Timur 2023). At the time of writing, the victims and their families still feel that they have not received an effective remedy from the corporations and government. The injustice experienced by the victims and their families can lead to forum shopping and other self-interested strategies (Overdevest and Zeitlin 2014). The variety of pathways and mechanisms adopted by victims and their families to seek justice and remedies unintendedly shows the unwillingness of the government and non-state actors (corporates) to settle the conflicts and to fulfill citizens' rights to an effective remedy.

5.4 Conclusion and Recommendation

5.4.1 Conclusion

The loss of 40 lives in abandoned coal mines that has resulted from irresponsible and inadequate post-mining environmental restoration efforts illustrates the socio-ecological crisis in East Kalimantan. The fact that there are hundreds of abandoned pits without any reclamation and restoration efforts in East Kalimantan raises criticism of the control and management of mining reclamation guarantee funds that mining companies are required to deposit before carrying out exploitation. The East Kalimantan case also shows that the existing legislation and regulations in Indonesia regarding mining, post-mining restoration, and access to remedy have failed to provide legal protection, let alone legal remedy, for affected citizens. According to the norms stated in human rights instruments, it can be concluded that as the state is ignoring the rights of citizens, both state and non-state actors are considered violators of human rights, both in commission (active act) and omission (ignorance). Non-state actors (corporates) have failed to obey all applicable existing laws of mining, post-mining restoration, and access to remedy, nor have they respected the human rights of communities. The lack of effective reclamation efforts leads to water contamination, unsafe living conditions, and environmental degradation, an injustice that leads victims and communities, including NGOs, to seek justice and remedies through forum shopping. The various pathways and mechanisms taken by victims and their families to seek justice and remedies unintendedly show the unwillingness of the government and non-state actors (corporates) to settle the conflicts and to fulfill citizens' rights to an effective remedy, contravening the SDG objectives.

5.4.2 Recommendation

The government must identify and compel companies to close pits located near residential areas, evaluate and revoke their permits as appropriate, and undertake environmental recovery and remediation, including psychological support for the victims' families. Law enforcement authorities must continue the criminal law enforcement process for the deaths of the 40 victims and prosecute those most responsible, rather than framing them as purely accidental tragedies. In its quest to achieve the SDGs, the government should strengthen legislative and regulatory frameworks for mining, promote sustainable industrial practices, enhance environmental protection and restoration efforts, improve community engagement and participation, ensure access to justice and remedies, and foster sustainable economic development and alternative livelihoods for affected communities.

References

Acharya, U (2014) The Future of human development: the right to survive as a fundamental element of the right to development. Denver J Int Law Policy 42(3)

Amy SW, Dani U, Muslim D, Nasruddin D, Nuryahya H, Nurhasan R, Agustin DS, Nu'man H (2023) The forming of acid mine drainage based on characteristics of coal mining, East Kalimantan, Indonesia. J Ecol Eng 24(7):301–310

ANTARA News (2023) Jumlah Inspektur Tambang di Kaltim harusnya ditambah. ANTARA News Kalimantan Timur. https://kaltim.antaranews.com/berita/177834/jumlah-inspektur-tambang-di-kaltim-harusnya-ditambah

Banakar R, Travers M (Eds) (2005) Theory and method in socio-legal research. Hart Publishing

Benda-Beckmann KV (1981) Forum shopping and shopping forums: dispute processing in a Minangkabau village in West Sumatra. J Legal Plural Unoff Law 13(19):117–159. https://doi.org/10.1080/07329113.1981.10756260

Buana MS (2017) Can human rights and indigenous spirituality prevail over state-corporatism? A narrative of ecological and cultural rights violation from East Kalimantan, Indonesia. J Southeast Asian Human Rights. https://doi.org/10.19184/jseahr.v1i1.5282

Buana MS, Listiningrum P, Nugroho PA, Marbun AAY (2023) The Nusantara Capital city project: why development and human rights do not always mix. 16(1):185–223. https://doi.org/10.1515/ldr-2022-0063

CNN Indonesia (2023) Lubang Bekas Tambang Kaltim: 40 Tewas sejak 2011, Didominasi Anak. nasional. https://www.cnnindonesia.com/nasional/20220204192046-12-755269/lubang-bekas-tambang-kaltim-40-tewas-sejak-2011-didominasi-anak

Cornel Law School (2022) forum shopping. https://www.law.cornell.edu/wex/forum_shopping

DPRD Provinsi Kalimantan Timur (2023) Laporkan Hasil Kerja, Begini Rekomendasi Pansus Investigasi Pertambangan. https://dprd.kaltimprov.go.id/post/laporkan-hasil-kerja-begini-rekomendasi-pansus-investigasi-pertambangan

Gasser U (2007) Responsibility for human rights violations, acts or omissions, within the "Sphere of Influence" of Companies. Berkman Center Research Publication No. 2007-12. https://doi.org/10.2139/ssrn.1077649

International Covenant on Civil and Political Rights (1966).

JATAM (2019) Lubang Tambang Maut_Bagaimana Lubang Tambang Membunuh Generasi Muda Kaltim. JATAM

JATAM (2020) 40 Hari Tanpa Penjelasan Pemerintah dan Kepolisian, JATAM Kaltim & LBH Samarinda Lapor dan Tagih Penyelesaian Kasus Anak Tewas di Lubang Tambang ⋆ JATAM. In JATAM

JATAM (2021a) Mining pit takes its 40th victim during the 2021 Climate Change Conference (COP26). https://www.jatam.org/en/mining-pit-takes-its-40th-victim-during-the-2021-climate-change-conference-cop26/

JATAM (2021b) Ditengah pembahasan konferensi Iklim COP 26, Jatuh lagi Korban Lubang Tambang ke 40 ⋆ JATAM. JATAM

Kementerian Energi dan Sumber Daya Mineral (2022) Keputusan Menteri Energi dan Sumber Daya Mineral Republik Indonesia Nomor 301.K/MB.01/MEM.B/2022 Tentang Rencana Pengelolaan Mineral dan Batubara Nasional Tahun 2022–2027

Kementerian Energi dan Sumber Daya Mineral (2023a) Cadangan Batubara Masih 38,84 Miliar Ton, Teknologi Bersih Pengelolaannya Terus Didorong. ESDM. https://www.esdm.go.id/id/media-center/arsip-berita/cadangan-batubara-masih-3884-miliar-ton-teknologi-bersih-pengelolaannya-terus-didorong

Kementerian Energi dan Sumber Daya Mineral (2023b) Siaran Pers Kementerian Energi dan Sumber Daya Mineral Republik Indonesia Nomor: 026.Pers/04/SJI/2023 Tanggal: 30 Januari 2023

Komnas HAM (2016) Pelanggaran Hak Asasi Manusia dalam Kasus Eks Lubang Tambang Batubara di Kalimantan Timur

Lawi GFK (2019) Pemerintah Diminta Tuntaskan Data Izin Usaha Pertambangan di Kaltim. Bisnis.com

Lillich RB (1986) Civil rights. In: Human Rights in International Law: Legal and Policy Issues. Oxford University Press

Maimunah S (2014) Mencari Makroman di Tanah Pinjaman: Perempuan Makroman di Tengah Perubahan Agraria dan Perjuangan Komunitas Menghadapi Pengerukan Batubara - Sajogyo Institute (No. 10, 2014)

Ministry of Energy and Mineral Resources (2023) ESDM One Map - Exploring Energy and Mineral Resources of Indonesia. https://geoportal.esdm.go.id

MODI (2023a) Realisasi Produksi & Penjualan Batubara. https://modi.esdm.go.id/produksi-batubara

MODI (2023b) Kontribusi Dana Bagi Hasil (DBH) Bagi Pemda. https://modi.esdm.go.id/dana-bagi-hasil

Mongabay (2022) Mine pits expose the holes in Indonesia's plan to relocate its capital. Mongabay Environmental News. https://news.mongabay.com/2022/04/mine-pits-expose-the-holes-in-indonesias-plan-to-relocate-its-capital

Overdevest C, Zeitlin J (2014) Assembling an experimentalist regime: transnational governance interactions in the forest sector. Regul Govern 8(1):22–48. https://doi.org/10.1111/j.1748-5991.2012.01133.x

Pratiwi CS, Listiningrum P, Anwary MAZ Al (2022) Critiques on contemporary discourse of international human rights law: a global south perspective. Human Rights in the Global South (HRGS), 1(1):1–12. https://doi.org/10.56784/hrgs.v1i1.3

Sengupta, A (2008) Elements of a theory of the right to development. In: Arguments for a Better World: Essays in Honor of Amartya Sen: Volume I: Ethics, Welfare, and Measurement. Oxford University Press

Soy K, Hing V (2019) Legal framework of the right to legal representation. In: Upholding the Right to Lawyer. Swisspeace

Spigelman J (2010) The forgotten freedom: freedom from fear. Int Compar Law Q 59(3):543–570. https://doi.org/10.1017/S0020589310000229

Teitel R (2002) Transitional justice in a new era. Fordham Int Law J 26(4):893

Transparency International (2017) Penilaian Risiko Korupsi dalam Pemberian Izin Usaha Pertambangan di Indonesia

United Nations (2011) Guiding Principles on Business and Human Rights. United Nations. https://www.ohchr.org/sites/default/files/documents/publications/guidingprinciplesbusinesshr_en.pdf

Uvin P (2007) From the right to development to the rights-based approach: how "human rights" entered development. Dev Pract 17(4/5):597–606

Part III

Natural Resources Governance from an Economic and Socio-Cultural Perspective

Developing an Inclusive Carbon Market in the Forestry Sector: The Role of Institutional Setting and Governance

6

Maxensius Tri Sambodo, Mesnan Silalahi, Alan Ray Farandy, and Syarif Hidayat

Abstract

Forest resources have not been able to provide for the welfare of their surrounding communities, with the jargon of sustainable forests and prosperous communities still far from a reality. The carbon market in the forestry sector will provide opportunities, but how should institutional and governance arrangements support the carbon market to work more inclusively and sustainably? This is important because the carbon market needs to balance economic, social, and environmental interests—or a triple win. This study uses a mixed-methods approach. A quantitative approach is used to find hidden information related to forest carbon in Indonesia, while a qualitative approach is based on documents and newspaper reports. This study draws two conclusions. First, the internal problems of the forestry sector remain quite complex, and this could hinder accelerated involvement in the carbon market. Second, in addition to accommodating the global norm for adopting carbon trading, it is important to facilitate smallholders and Indigenous peoples playing an active role in the carbon market. Both can be well developed if symmetrical steps are taken to improve institutions and governance by integrating into the global carbon market.

Keywords

Carbon market · Forestry · Institution · Governance · Inclusive

6.1 Introduction

A study conducted by the Department of Civil Engineering at the University of Hong Kong and the Southern University of Science and Technology found that tropical forest carbon stores released in 2001–2005 were 0.97 billion tons of CO_2 equivalent per year, increasing to 1.99 billion tons in 2015–2019 (Susetyo 2023). The study stated that changing the status of forests to non-forests contributed 48% of the release of carbon emissions into greenhouse gases and was followed by transport (21%), fires (12%), factory waste (11%), agriculture (5%), and industry (3%). The highest contribution of the forestry sector will certainly threaten the achievement of sustain-

M. T. Sambodo (✉) · M. Silalahi · A. R. Farandy
Research Center for Behavioral and Circular Economics - National Research and Innovation National Research and Innovation Agency, Jakarta, Indonesia
e-mail: maxe001@brin.go.id

S. Hidayat
Research Center for Cooperatives, Corporations, and People's Economy- National Research and Innovation Agency, Jakarta, Indonesia

H. Warsilah and A. H. Purnomo (eds.), *Approaching Natural Resources for Sustainable Development in Indonesia*, Sustainable Development Goals Series, https://doi.org/10.1007/978-981-96-6913-4_6

able livelihoods for the surrounding community. Reducing the area of forest will not only increase greenhouse gas (GHG) emissions but also reduce the carrying capacity of ecosystems and will become a threat to the sustainability of human life and other living things.

Indonesia has changed its GHG reduction target as stated in the Enhanced Nationally Determined Contribution. In the Enhanced Nationally Determined Contribution (NDC) document, which was approved by the government in September 2022, Indonesia will reduce its emissions target by 31.89% through its own efforts (CM1) and by 43.2% with efforts in collaboration with international agencies (CM2). The emission reduction targets from the forestry sector (FOLU) in both CM1 and CM2 scenarios are higher than those of other sectors. The burden of reducing emissions from the FOLU sector certainly needs to be supported by adequate resources and capabilities. This NDC update is designed in line with the Long-Term Low Carbon and Climate Resilience Strategy (LTS-LCCR) 2050. An important part of achieving the NDC is the issuance of Presidential Regulation No. 98 of 2021 concerning the Implementation of the Economic Value of Carbon to Achieve NDC and Control of Greenhouse Gas Emissions in National Development. Apart from carbon pricing, this regulation also addresses carbon trading, carbon taxes, and results-based payments.

The Indonesian REDD + (Reduction Emissions from Deforestation and Forest Degradation, Role of Conservation, Sustainable Management of Forest and Enhancement of Forest Carbon Sink) National Strategy 2021–2030 document (MEF 2022) states that the forestry and land sector is estimated to contribute to a reduction of 17.4–25.5% in emissions from the NDC target by carrying out a series of strategies and activities to achieve the FOLU Net Sink Target in 2030. The framework for achieving REDD + requires the support of four interconnected elements: architecture; tools for implementation (means of implementation/MoI); institutions, regulations, and systems; and implementation.

Indonesia has a target to rehabilitate 12 million hectares of degraded land and the REDD + scheme will be an important instrument to make this happen. The REDD + framework will be closely linked to COP decisions, the Warsaw Framework, and the Paris Agreement. The implementation strategy focuses on strengthening REDD + architecture and institutions, managing REDD + implementation, and, in a paradigm shift, involving stakeholders and sharing benefits. Concerning REDD +, Indonesia has twice submitted FREL (Forest Reference Emissions Level). FOLU Net Sink Target 2030 (− 140 $MtCO_2$ emission level) will be achieved through a series of policies, namely: reducing emissions from deforestation and forest degradation; increasing carbon capture capacity for natural forests; increasing carbon capture from soil systems; reducing emissions from forest fires and peat decomposition; and law enforcement.

Applying the paradigm of sustainable development in the forestry sector is a challenge for Indonesia. As stated by the Ministry of Environment and Forestry, the total area of forest areas in Indonesia will reach 125.76 million hectares in 2022 (Widi 2023a), but around 36.7% of 25,863 villages located in forest areas are in the poor category (Susetyo 2022). The potential of forest resources has not been able to improve community welfare, as indicated by the Village Development Index (Indeks Pembangunan Desa / IPD), a framework designed to facilitate attainment of the basic services necessary to fulfill minimum service standards at the village level. The level of village development directly correlates with the degree of establishment, hence enhancing the village's competitiveness (Harmadi et al. 2020). Villages are classified as established, developing, or disadvantaged. Table 6.1 shows the Village Development Index for forest villages. We define forest villages as villages that are located around or in the forest area. In 2018 there were 75,436 villages 2748 villages inside forests, then 17,933 villages around forest areas. Thus about 27.4% of the villages in Indonesia are inside and around forest areas, and the majority of those villages are in the developing stage.

Table 6.1 Village development index of forest villages, 2018

	Inside forest area (%)	Around forest area (%)	Inside and around forest area (%)
Disadvantaged Villages	9.05	28.74	37.79
Developing Villages	4.17	55.46	59.63
Established Villages	0.07	2.51	2.58
Total	13.29	86.71	100.00

Source Author's calculation

As Table 6.1 shows, efforts to optimize the use of forest resources to increase income, employment opportunities, social justice and maintain sustainability remain a challenge. Forestry resources have not been able to increase development either within or around the forest. More serious efforts are needed to encourage more development villages to become established villages. Increasing the added value of forestry resources needs attention. The economic potential of carbon from forests needs to be increasingly utilized, including for other environmental services.

Carbon market opportunities, both mandatory and voluntary, need to be utilized to maintain forest sustainability and community welfare. Many business actors, including Indigenous peoples, have been involved in the carbon market through voluntary emissions trading systems (ETS), verified carbon standards (VCS), gold standards, Plan Vivo, and climate, community, and biodiversity standards (CCBS). The most important aspect of involvement in the voluntary market is Article 6 of the French Agreement (Witoelar 2023) on corresponding adjustments (CA). CA aims to ensure that the two countries involved in carbon trading do not both count the same emission reductions and the carbon credits will remain with Indonesia. However, to ensure that CA is implemented, infrastructure support related to institutional capacity and technical capabilities is needed to support this mechanism (Witoelar 2023).

This chapter combines concepts of sustainable livelihoods, institutions, and proper governance to analyze how the carbon market can be one of the economic instruments for improving the welfare of communities that have been given responsibility for managing forests. The current challenge is how society can adapt to a new norm of forest carbon, and what preparations need to be undertaken so that the community can play an increasingly active role in the carbon market. The chapter focuses on the forestry sector in the village context because this sector not only bears the brunt of the emission reduction target stated in Indonesia's NDC, but also enjoys a great opportunity to overcome the problems of poverty, inequality, and unemployment. The chapter show how sustainable development principles can be integrated into the forestry sector (SDG 15), promoting inclusive economic growth (SDG 1, SDG 8, SDG 10), enhancing institutional capacities (SDG 16), and fostering strong partnerships (SDG 17) for effective climate action (SDG 13).

6.2 Methods

This study uses mixed methods. A qualitative approach involved studying documents and referring to reports from mainstream newspapers. For the quantitative approach, diverse forest carbon literature related to sustainable livelihoods, poverty, local champions, market, Indonesia, community, and social governance was collected from the Scopus© database. Keywords with the mentioned intensity range (ranging) from high to low and the network of linkages between keywords are illustrated in Table 6.1. Content analysis was performed on these articles using the author–index keywords. Co-word analysis was applied using VOSViewer to acquire a view of the global research in selected fields, with some details provided in the context of Indonesia (Table 6.2) (Callon et al. 1986). Co-word analysis explores the interaction between keywords across a population of texts and such co-occurrences are depicted in graph form. Qualitative meta-analysis was performed to obtain a clear and comprehensive picture of findings regarding a specific phenomenon by analyzing text in a corpus of documents investigated in some studies (Timulak 2014). A qualitative meta-analysis can produce stronger and more valid conclusions by combining the findings of different qualitative studies. Relevant research was synthesized to construct the narratives and connections between the keywords, thus highlighting interesting and important issues.

Table 6.2 Search strings used for the co-word analysis

Scope	The search string
Indonesian context of research related to sustainable livelihood, poverty, local champion, market, community, and social governance	TITLE-ABS-KEY("forest carbon" AND ("sustainable livelihood*" OR poverty OR "local champion" OR market OR Indonesia OR community OR social OR governance)) AND (LIMIT-TO(EXACT KEYWORD, "Indonesia"))

Table 6.3 Area and distribution of indigenous forest indicator areas* to June 3, 2022

Province	Area (thousand Ha)
Kalimantan Utara	402.5
Sulawesi Selatan	128.9
Sulawesi Tengah	121.4
Aceh	112.6
Kalimantan Barat	108.1
Maluku Utara	58.2
Maluku	49.6
Bengkulu	19.6
Papua	18.8
Riau	18.7
Kalimantan Timur	17.0
Sulawesi Barat	10.4
Banten	9.2
Sumatera Barat	5.0
Jambi	3.7
Kalimantan Tengah	3.4
Papua Barat	2.6
Jawa Barat	0.6
Bali	0.5

Note * Customary Forest indicator area: area within a state forest that has not yet received a legal product in the form of a regional regulation, but has been confirmed as customary forest by the Regent or Mayor
Source Mustika (2022)

6.3 Theoretical Framework

The concept of sustainable development as stated in the 1987 document *Our Common Future* is that current development should not sacrifice the ability of future generations to meet their needs. The document also mentions two important things. First, needs are strongly related to what the poor need, and should be sensitive to priority scales. Second, fulfillment of needs can be limited by conditions of technology and social organization in the present and the future. Still, within the framework of sustainable development, the sustainable livelihood approach (SLA) has been a topic of research since the early 1990s. Natarajan et al. (2022) try to sharpen this approach by inserting 'livelihoods in flux'. They argue that strategies will always face conditions of uncertainty, both in the short term, for example changes in commodity prices, pest attacks, floods, or even disease outbreaks/pandemics, and in the long term, for example, structural changes in the economy, an aging population, and climate change.

The initial stage of SLA analysis was developed by the UK's Department for International Development (DFID) through the Economic and Social Research Council (ESRC) and the Natural Resources Institute (NRI), which was established in 1997. At this early stage, studies were also assisted by the Institute of Development Studies (IDS), which identified five types of asset: human, natural, financial, physical, and social. Natarajan et al. (2022) reorganized the five asset components into three parts: financial and physical, power relations, and climate and context or environmental relations. We argue that forest carbon will relate to these assets and the process of connecting the 'asset' with the forest carbon norm requires good institutions and proper governance in the carbon market (Fig. 6.1).

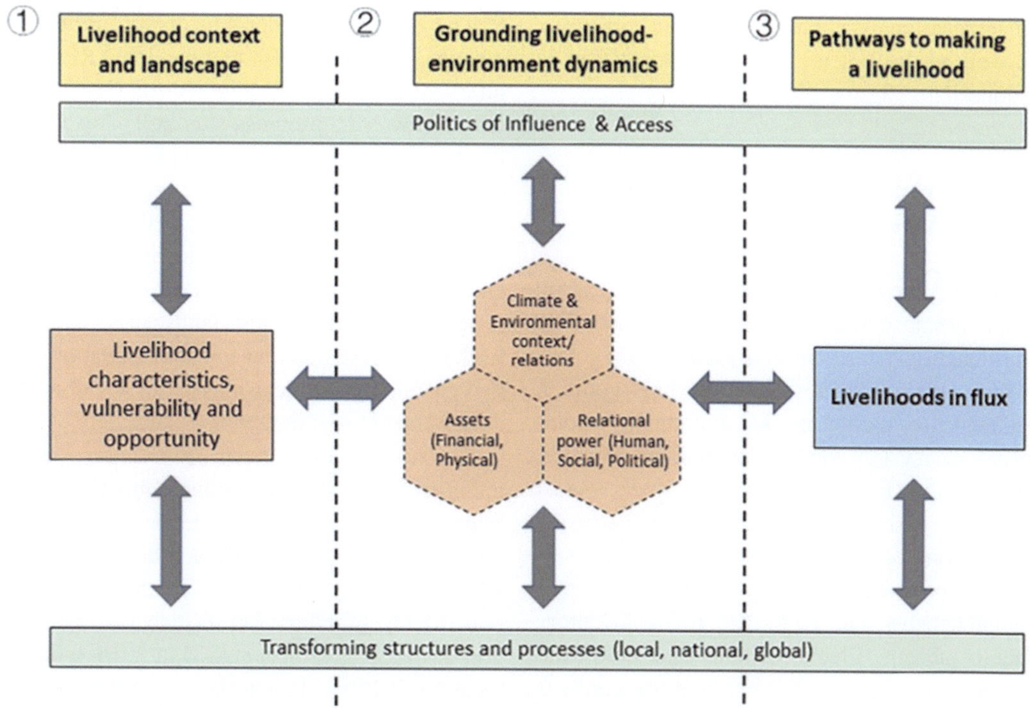

Fig. 6.1 Sustainable livelihoods framework for the twenty-first century (*Source* Natarajan et al. (2022))

According to North (1994), in analyzing economic performance over time, two assumptions are not well understood, namely the importance of institutions and of time (North 1994). Institutions shape the incentive structure of a society, and political and economic institutions are all important determinants of economic performance. Thus, finding an institutional model that can tightly knit together the social system will be an important key to realizing sustainable long-term economic growth. Considering the idea of how economic growth can be achieved optimally, Rodrik (2003) concluded that institutional quality is the key. It is important to build institutions that can protect property rights, manage conflict, enforce laws and regulations, and align economic incentives with social benefits and costs, which are the foundations for long-term growth. Rodrik also said that to achieve high and sustainable economic growth in the long term, stronger institutions are needed. Ensuring that such an institutional setting is on the right track requires both good and proper governance (Hidayat 2016). Referring to the framework of a new-generational way of thinking about (of) governance, Hidayat (2016) offered the idea of an alternative concept that he termed "proper governance". This concept rests on four main principles: development, democracy, social inclusivity, and local context. This means that both the state and society's roles must be considered equally important. Meanwhile, the existence of governance itself should be viewed as a tool to achieve four main objectives: acceleration of economic development, a democratic life, respect for the rights of every citizen, and the creation of social inclusiveness (North 1994; Rodrik 2003; Hidayat 2016; Hidayat and Indrayani 2019).

6.4 Results and Discussion

The analysis in this section is divided into two parts. The first is a survey of several case studies from the literature of carbon market implementation. The second part discusses institutional and governance analysis based on current practices, as well as the potential for implementation in Indonesia.

6.4.1 Implementation of Carbon Market

Figure 6.2 shows terms relating to forest carbon in Indonesia, such as REDD + , governmentality, additionality, deforestation, environmental governance, agency, political ecology, land conflict, peatland fire, carbon market, community monitoring, ecosystem service, carbon governance, conservation biopolitics, carbon stock, forest inventory, carbon offset, data complementarity, and carbon accounting. Some of the main issues related to forest inventory in Indonesia are: comprehensive and up-to-date forest inventory data is absent; illegal logging and deforestation remain critical problems; land-use conflicts persist, often driven by competing interests such as agriculture, mining, and infrastructure development; forest fires, often linked to land clearing and peatland degradation, cause haze and air quality problems in the region; enforcement of forest protection laws is inconsistent, and monitoring efforts face challenges; integrating biodiversity data into forest inventory remains an ongoing challenge; ensuring Indigenous communities' participation and recognizing their rights remains an issue; transparency in forest inventory data and making it accessible to the public, researchers, and stakeholders remain a concern; and inventory efforts need to be enhanced through technology (remote sensing and geographic information system or GIS).

Thus, there are still many aspects needing attention so that the Indonesian forestry sector can be well connected with the minimum level of governance of the global carbon market. The following four experiences can provide information about Indonesia's exploitation of cooperation opportunities or carbon markets.

First, the Ministry of Environment and Forestry (MoEF) of the Republic of Indonesia selected Kalimantan as a pilot area for trialing a national REDD + program under the Forest Carbon Partnership Facility (FCPF), and this local REDD + intervention was shown to reduce GHG emissions. However, progress was still slower than the local action plan target, under which GHG emissions are expected to be reduced by 22.38% from the historical baseline for 2020 (Kiswanto et al. 2021). The local government should accelerate strategies for reducing GHG emissions by restoring degraded forest landscapes, avoiding further deforestation and forest degradation, and making appropriate decisions for socio-economic development programs.

Second, several projects from Indonesia are included in the carbon trading scheme through Plan Vivo, such as Nanga Lauk (the Rimbak Pakai Pengidup), Kapuas Hulu, West Kalimantan, which helps the Nanga Lauk community maintain 1450 ha of village forest (Plan Vivo Foundation 2019). This project started in 2016, with the local NGO PRCF Indonesia as coordinator, and involved 226 village households. The report states that with the Plan Vivo certificate, the potential for carbon reduction that can be achieved is 1308 tons of CO_2 per year from the Nanga Lauk Village Forest (LTS International Ltd 2019). At the same time as a Plan Vivo certificate is being obtained, Lestari Capital is also providing a new financing scheme through the Sustainable Commodities Conservation Mechanism (SCCM). With support from Cargill (Hariyanto 2018), protected forest areas can be expanded. Economic opportunities that can be developed by local communities include honey, bamboo, rattan, and ecotourism (Capital 2023).

Third, the Bujang Raba project aims to maintain biodiversity in Sumatra's Barisan Hills through the REDD + program. The project, involving five local community groups, started in 2013, with KKI Warsi as coordinator. To date 125,820 Plan Vivo certificates (PVCs) have been issued to 934 households participating in this project, with 101,640 PVCs approved for the future. More than 5339 Ha of primary forest are to be protected, with a net carbon benefit of 40,000 tons of CO_2 per year (Plan Vivo Foundation 2023a). Another REDD project is Durian

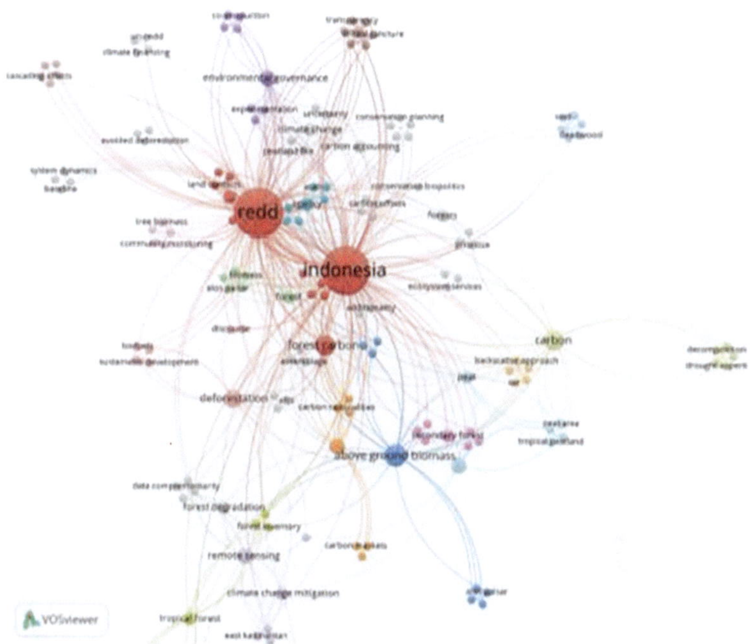

Fig. 6.2 Indonesian context of research related to sustainable livelihood, poverty, local champion, market, community, and social governance

Rambun, aimed at protecting the forest village of Rio Kemunyang in Jambi. The coordinator of this activity is Fauna and Flora Indonesia, involving 78 households, with 16,100 PVCs issued to date. The total village forest is 3616 Ha, with 2516 Ha of protected area and 1100 rehabilitation areas. The potential carbon gain is 6618 tons of CO_2 per year (Plan Vivo Foundation 2023b). The Bujang Raba project generated Rp. 350 million in income for the forest management community from the sale of PVCs in 2019, increasing to Rp. 1 billion in 2020 (Kompas 2020). These funds are certainly very beneficial for the community to improve their standard of living. Likewise, it is open to the green finance companies such as Lestari Capital, to also enter business development activities while preserving forests.

Thus, there are at least three general patterns that can provide lessons for strengthening the forest-based carbon trading ecosystem in Indonesia. First, at the site level, support from local NGOs is needed to map the carbon potential that can be created; and these NGOs will become the agencies that deliver the work plan towards a carbon financing scheme. Second, there is the 'addition' side in the sense of expansion of areas that can be protected, or economic activities that can be developed. All this certainly complements and strengthens existing carbon financing schemes. These additional activities can also reduce risks to achieving sustainable livelihoods. Third, the work of preparing quality proposals requires the support of good resources and capabilities. Thus, multi-stakeholder collaboration, both government (state actor) and non-government (non-state actor) becomes important. It is interesting to note from the experience of Plan Vivo that there is a lot of academic support behind the effort to build an integrated system.

There are at least two major challenges that need to be solved so that carbon trading can provide maximum benefits for the people of Indonesia, as mandated in Article 33 of the 1945 Constitution. First, how to accelerate or prioritize this program in areas that still have well-maintained forest areas, but where welfare conditions are still relatively poor. Figure 6.3 shows how many villages are underdeveloped in the Papua region. On the other hand, in 2020, the area covered by forest land in the provinces of West Papua and Papua is approximately 25.29 million hectares and 8.96 million hectares respectively, or 81.7% and 93% of the land, respectively (BPS 2023). These figures are the highest nationally. Papua's very high forest cover needs to be maintained and developed, especially considering the challenges of its social conditions. According to the 2018 village potential data (*Podes*), in Papua province, there are 5552 village-level areas (5456 villages and 96 sub-districts); of the villages, 10 are new independent ones, 693 are developing and 4753 underdeveloped (Papua 2022a). This situation is in line with the poverty profile in Papua. In March 2022, the percentage of poor people in rural areas reached 35.39%, while in urban areas it was 5.02% (Papua 2022b).

In the book *Otonomi Khusus dan Kutukan Sumber Daya Alam Papua* [*Special Autonomy and the Curse of Papua's Natural Resources*], published by INDEF and Greenpeace (Indef 2022), the three provinces with the highest natural resource curse in Indonesia are named as East Kalimantan, Papua, and West Papua. Extractive policies can be seen in the increasing rate of deforestation in the last two decades, which has potentially resulted in the loss of 7.5 million hectares of natural forest in Papua. The book shows the importance of correcting the current development model by giving Indigenous communities a more substantive role. This is in line with the principles of free, prior, and informed consent (FPIC) when making decisions that impact people's lives. The community must be actively involved in formulating development plans. Equally important is improving governmental governance to enhance quality, effectiveness, and accountability, reduce corruption, and maintain the sustainability of natural resources.

Second, even though carbon trading will provide benefits, new problems can arise, particularly in relation to the recognition of Indigenous peoples. It is very possible that with the large economic value of carbon, forests will be controlled by big investors, potentially marginalizing and displacing Indigenous communities (National Coordinator of the Mining Advocacy Network/Jatam, Melky Nahar) (Kompas 2023). This occurs because carbon trading is predominantly conducted by industry. The public does not yet fully understand the carbon trading scheme or its benefits. Carbon trading industries do not normally portray themselves as environmentally friendly, potentially masking greenwashing behavior. The most important consideration is recognizing the rights of Indigenous peoples and local communities and safeguarding the traditional practices they use to protect forests.

Therefore, it seems very reasonable that the implementation of FOLU Net Sink in 2030 must be accompanied by the establishment of customary forests. As the Executive Director of the Legal and Community Association (Huma), Agung Wibowo (Mustika 2022), states, 'there is a motivation for proposing customary forests and the reasons are more motivated by economic interests than the protection of forest ecosystems and customary order'. Huma also states that 'every request from the community regarding forest management, such as social forestry or customary forests, are still hampered by legality issues'. This is because legality must be obtained through formal legal mechanisms such as regional regulations, even though the majority of Indigenous or local communities use informal legal mechanisms to resolve every problem. Public relations records show that as of 2019, there have been 297 regional legal products recognizing Indigenous communities, with a potential area of 453,832 hectares of customary territory whose status has not been clarified. The customary forests quoted in FoLU Net Sink 2030 only cover an area of 79,000 Ha in 89 customary communities. According to the recommendations from the customary forest coordination meeting, which has been a long process, a wider area of customary forest should have been stated.

6.4.2 Strengthening the Inclusive Forestry Sector Carbon Market Policy Ecosystem

The Plan Vivo Foundation, a registered charity in Scotland originally founded to help small farmers in Chiapas, Mexico, to plant trees, is now an important player in carbon credits on a global level. It began work in 1994 through a collaborative project with the University of Edinburgh, El Colegio de la Frontera Sur, and other

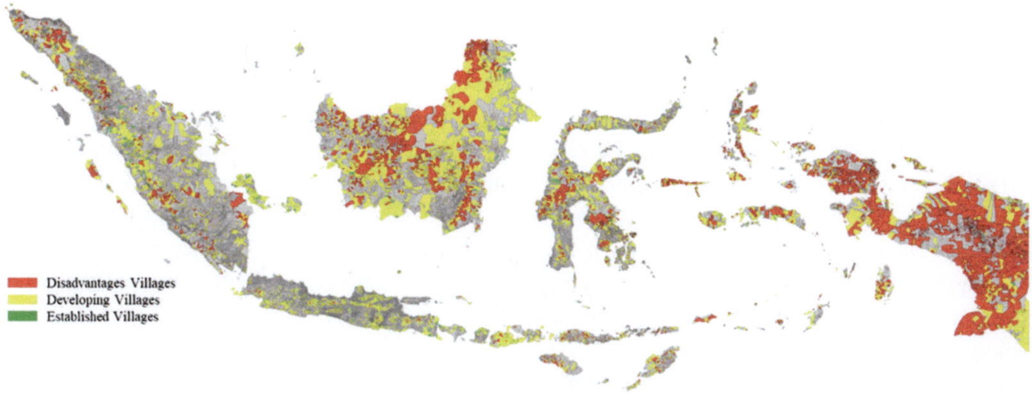

Fig. 6.3 Map of village development index of forest villages, 2018

local partners. The Plan Vivo carbon crediting system has great potential to overcome emissions from the forestry sector at a small scale; the institution conducts audits and validations. Importantly, this scheme establishes a connection between emission reduction efforts and improving the quality of life of communities living around the forest. Efforts to empower the community are ongoing, and a voluntary carbon market for payment for ecosystem services is being developed. Many PVCs have been awarded to communities working to protect the surrounding environmental services. The surrounding community benefits from non-timber forest products, namely carbon. Plan Vivo is managed independently by the non-profit BioClimate Research and Development. Strengthening efforts are carried out from the standard project side, including conducting program audits.

Work orientation provides real benefits that can be felt by the community. The first step is to build institutions and establish cooperation with stakeholders. This can be started by preparing and developing an identification desk for deforestation conditions and biodiversity conditions; a community empowerment desk; and a financing desk to channel corporate social responsibility funds. The next challenge is to link the Vivo Certificate Plan with national banking commercial financing as a legal basis for establishing a green tax regime recognizing low business risks and cheap financing.

The carbon credit market for the forestry sector is open for development. There are at least three schemes that will be developed by the government. First, cap and trade, which allows cross-sector carbon trading. In this scheme, each ministry in its respective sector (energy, forestry and land use, agriculture, waste, and industry and production processes) sets emission limits (caps) for business actors. Businesses that produce higher emissions are obliged to buy credits from other companies with lower emission levels. Second, carbon offset is applied to sectors that do not have quotas, and those who provide emissions above the baseline can buy the excess emissions from those who provide carbon absorption efforts. The second scheme is called performance-based payment (result-based payment). For example, the Norwegian government paid USD54 million to prevent deforestation in Indonesia between 2016 and 2017, which is approximately USD5 per ton of CO_2 equivalent carbon (Megarani 2022). Then there are carbon levies on goods or activities that have the potential/content/emissions to hurt the environment. Carbon levies fall under the authority of the Minister of Finance and still need further regulation. All payments related to emissions trading must go through a respected agency such as the Environmental Fund Management Agency (BPDLH).

The Indonesian government has prepared a legal umbrella related to carbon trading which covers: Presidential Regulation (Perpres) no. 98

of 2021 concerning Implementation of Carbon Economic Value to Achieve Nationally Determined Contribution Targets and Control of Green House Gas (GHG) Emissions in National Development; Minister of Environment and Forestry (LHK) Regulation No 21 of 2022 concerning Procedures for Implementing the Economic Value of Carbon; Minister of Environment and Forestry Regulation No. 7 of 2023 concerning Procedures for Carbon Trading in the Forestry Sector; Ministerial Decree (KepMen) LHK No. 102/MENLHK/PHL/KUM.1/9/2023; Minister of Environment and Forestry Decree Number 1131/MenLHK/PPI.2/10/2023 concerning GHG Emission Reduction Certification Scheme; Decree of the Director General (KepDirjen) of Postal and Informatics Control (PPI) Number SK.30/PPI/IGAS/PPI.2/10/2023 concerning Methodological Framework for Calculating Emission Reductions/Increasing GHG Absorption in the FOLU sector; and Minister of Environment and Forestry Decree No. 1444/MENLHK/SETJEN/KUM.1/12/2023 concerning the Measurement, Reporting and Verification (MRV) Team and the Greenhouse Gas (GHG) Emission Calculation Methodology Panel. Meanwhile, several other regulations are in the preparation stage: the role of regional governments in implementing the economic value of carbon to achieve NDC targets; draft Minister of Finance regulations regarding tariffs and tax bases (DPP) for a carbon tax, and procedures and mechanisms for a carbon tax; and Minister of Environment and Forestry regulations regarding the implementation of NDCs and procedures for foreign carbon trading.

This regulatory umbrella is of course very important as a basis for ensuring that the carbon market in the forestry sector runs fairly and with integrity. There are at least four important aspects of the discussions taking place in the forestry sector carbon market. First, existing regulations should be able to encourage the development of good economies of scale, so that economically, the benefits of carbon can be felt optimally by the community. Second, in complex carbon market conditions, community groups need to know how this system works. Third, behind efforts to ensure the carbon market works well, a challenge that also needs to be addressed is how the system built can overcome coordination challenges and minimize transaction costs. Fourth, regulations must be able to overcome the hold-up problem, that is, the potential for market players to act opportunistically to pursue their profits. If these four things can be considered within the existing and future regulatory umbrella, then the integrity of the carbon market will be more inclusive, better developed, and deeper.

Figure 6.4 illustrates the business process for forestry sector carbon financing from site to market. Three major groups build business processes. First, conditions at the site level. The main agenda at the site level is controlling the economic, social, and environmental risks posed by the condition of degraded forest areas. This can be carried out by state and non-state actors. For example, the financing role carried out by the state has been led by the Forest and Land Rehabilitation (RHL) program. In 2022, 77,103 hectares of land were successfully rehabilitated, while in 2021 it could reach 203,386 hectares (Widi 2023b). The revenue sharing from reforestation funds (DBH DR) in Malinau Regency in 2023 was IDR 16 billion, and these funds were used to carry out six activities (three physical activities and three non-physical activities) (Widi 2023b). Physical activities include replanting trees, while non-physical activities include community empowerment. Other programs include Bang Pesona (Association for Social Forestry Development of the Archipelago) and the provision of productive economic tools for social forestry business groups, which were upgraded to gold or platinum class. This program is run under the Directorate General of Social Forestry and Environmental Partnerships (PSKL), with a budget allocation of IDR 37.08 billion in 2023 (Antara 2023). Likewise, village funds can be used to maintain the ecosystem in the village, including supporting the village forest program.

An important part of increasing community capacity is the collaboration between BPDLH and the Ford Foundation Program related to the Fund for the Welfare and Sustainable Economy of Indigenous Peoples and Local Communities (Terra Fund). The five major themes of

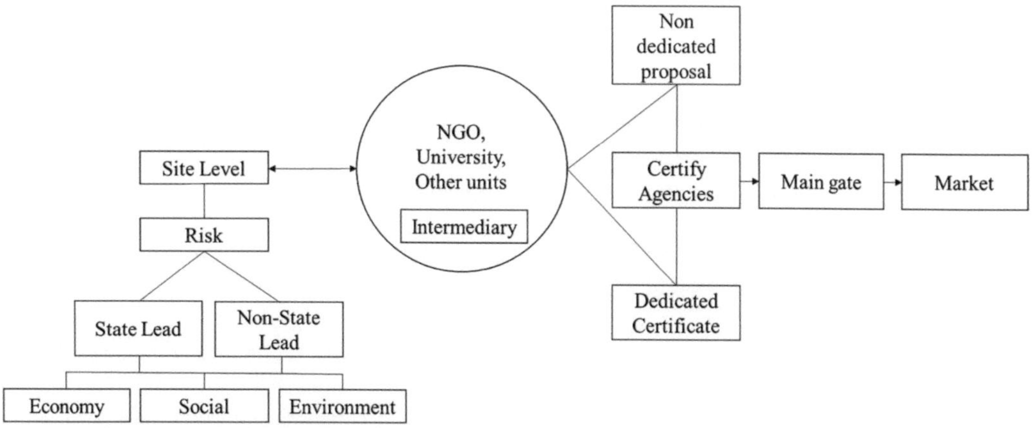

Fig. 6.4 Abstraction for inclusive carbon market

Dana Terra are: reducing the rate of deforestation and land degradation; sustainable agroecological practices; equitable access to the use of natural resources; increasing energy efficiency and utilization of renewable energy; and initiation of environmentally friendly small-scale community businesses and increased market access for farmers. Economic empowerment of communities (Indigenous communities and community groups living in areas around forests) develops the principles of sustainability. Among the target recipient groups are intermediary institutions and educational institutions that have experience in implementing programs in the field of environment and climate change. Intermediary institutions are aimed at community organizations and institutions working to improve the quality of the environment, especially in the land and energy sectors. The intermediary institution must be a legal entity and have administrative, institutional, financial, programming, and safeguarding capabilities. Fund allocation for intermediary institutions is 500 million–1.5 billion over 12 months. Anticipated results of this activity are increasing community income; increasing sustainability; and encouraging the use of renewable energy.

Meanwhile, the role of non-state actors is also important. Non-governmental organizations (NGOs) engage in advocacy for the community, while universities assist as part of community service programs. Another important aspect is the private sector, which is encouraged to produce sustainability reports and focus on the green economy (Dhaneswaro 2023). The current challenge is to build synergy from the resources and capabilities that reach the community living in forest areas to provide optimal benefits. The role of the village head, along with other figures, is important in overseeing incoming programs.

Furthermore, intermediaries, or intermediary institutions, play an important role as agencies that will help the community to make quality proposals to get funding. NGOs, community service institutions (LPM), or others can help become networks that prepare funding proposals (free proposals), including connecting with potential buyers (dedicated proposals). These proposals will be assessed by certification agencies, both from within and outside the country. The role of the National Professional Certification Agency (BNSP) is very important in preparing human resources capable of building a carbon market. In line with efforts to make carbon markets mandatory, it is necessary to think about providing incentives and affirmation programs to encourage more parties to have professional certification related to carbon market development. The skills certification scheme needs to include parties such as farmers' groups, cooperatives, and forest farmer groups to strengthen the carbon market.

The growing potential of national endowment funds for the environment needs to be supported by easy access without having to sacrifice project quality. National certification agencies become global subsystems in integrated carbon markets. A national certificate agency does not have to form a new institution but can optimize existing agencies and be recognized globally. For result-based payment, GHG emission reduction certification is required. Evidence of measurement, reporting, and verification, must be recorded in the National Registration System for Climate Change Control in the form of a registration number and/or code. The National Accreditation Committee will be an important institution to strengthen the roles of verifiers and validators. In this case, it is also important to have mutual recognition related to foreign carbon trading.

Once certification meets the clean and clear criteria, carbon trading traffic will enter the one-gate system. This is in line with the establishment of a designated national authority (DNA), which is the national focal point for the United Nations Framework Convention on Climate Change (UNFCCC). Likewise, in terms of performance-based payments, Regulation of the Minister of Environment and Forestry No. 21 of 2022, Article 32 paragraph 3 states: 'the implementation of performance-based payments after the payment agreement as intended in paragraph (1) and paragraph (2) is signed is deposited via BPDLH after obtaining approval from the Minister (MoEF). Furthermore, BPDLH distributes funding to beneficiaries. This benefit sharing is based on the proposal submitted by the beneficiary to BPDLH. Thus, DNA and BPDLH become goalkeepers related to carbon traffic both within and outside the country.

6.5 Conclusions and Recommendation

This chapter has discussed a 'new channel' for more sustainable forest management, and ways of simultaneously improving community welfare by being actively involved in the forest carbon market. Indonesia's large forest potential needs to be strategically directed toward optimizing the value of its environmental services, especially its carbon potential. This chapter has shown that the wealth of forest resources has the potential to improve the welfare of people living in and around forests. Economic difficulties and social pressures will lead to rising deforestation, and forest destruction will make it increasingly difficult for Indonesia to achieve its NDC target.

Connecting the two poles of forestry institutions and governance between sub-national, national, and global settings is certainly not easy. However, subtle differences in standards need to encourage stakeholders in the forestry sector to work together. While forest carbon markets often favor the interests of large investors, there is also a need to encourage carbon markets to work for small-scale forest farmers, and for Indigenous peoples who already feel protective of their forests. Carbon regulations need to be made simpler and easily understandable for various actors including small farmers, without compromising market integrity. Proper carbon governance is needed to ensure market discipline can be achieved optimally and avoid market failure. Carbon reductions are achieved by the participation of many groups of small farmers. On this basis, the influence of the distribution of benefits, recognition of the rights of local communities, and procedures that are simple and understandable to them are important. A more intensive process of public dialogue with farmer groups affected by the policy is needed, which will position them as the main agents for change in a better future.

References

Antara (2023) The Ministry of Environment and Forestry distributed IDR 1.67 trillion in funds for community activities. In: Antara News. https://www.antaranews.com/berita/3382170/kementerian-lhk-salurkan-dana-rp167-triliun-untuk-kegiatan-masyarakat. Accessed 11 Sep 2023

BPS (2023) Recapitulation of Area of Forest and Non-Forest Land Cover by Province 2014–2021 (Thousand Ha). In: bps.go.id. https://www.bps.go.id/statictable/2020/07/13/2110/rekapitulasi-luas-penutupan-lahan-hutan-dan-non-hutan-menurut-provinsi-tahun-2014-2020-ribu-ha-.html. Accessed 11 Sep 2023

BPS Papua (2022a) Important Indicators for Papua Province December 2022 Edition. Jayapura

BPS Papua (2022b) Papua Province Poverty Profile March 2022. Jayapura

Callon M, Law J, Rip A (1986) Putting Texts in Their Place. In: Mapp. Dyn. Sci. Technol. pp 221–230
Dhanesworo S (2023) Sustainability Report, a Form of Industrial Environmental Commitment. In: katadata.co.id. https://katadata.co.id/fitrianurhayati/infografik/63db849c9ce84/laporan-berkelanjutan-wujud-komitmen-lingkungan-industri. Accessed 11 Sep 2023
Hariyanto (2018) Lestari Capital Launches Innovative Financing Mechanism for Forest Protection and Restoration. In: www.industry.co.id. https://www.industry.co.id/read/43620/lestari-capital-luncurkan-mekanisme-pembiayaan-inovatif-untuk-perlindungan-dan-restorasi-hutan. Accessed 11 Sep 2023
Harmadi SHB, Suchaini U, Adji A (2020) Mismatch Of Village Development Indicators In Indonesia. Jakarta
Hidayat S (2016) Menimbang Ulang Konsep Good Governance: Diskursus Teoretis. Masy Indones 42:152–165
Hidayat S, Negara SD (2020) Special economic zones and the need for proper governance: empirical evidence from Indonesia. Contemp Southeast Asia. https://doi.org/10.1355/cs42-2e
Hidayat S, Indrayani I (2019) Reconsidering the concept of good governance: A theoretical discourse. Int. J. Innov. Creat. Chang. 9:
Hidayat S, Sambodo MT, Yuliana CI, Edi RY, Handoyo FW, Firdaus N, Novandra R, Farandy AR, Inayah I (2021) PANCUNG ALAS: Aktualisasi Community Forest Management Berbasis Proper Governance. In: Hidayat S, Sambodo MT (eds) PANCUNG ALAS Aktual. Community For. Manag. Berbas. Proper Gov., 1st ed. Airlangga University Press, Surabaya, pp 1–98
Indef, Green Peace (2022) Special Autonomy and the Curse of Papua's Natural Resources. Jakarta: Indonesia
Kompas (2020) Fighting Against Deforestation, This Woman Helps 5 Bujang Raba Forest Villages Earn IDR 1 Billion from Selling Carbon. In: kompas.com. https://regional.kompas.com/read/2020/11/11/07271111/berjuang-lawan-deforestasi-perempuan-ini-antar-5-desa-hutan-bujang-raba-raup?page=all . Accessed 11 Sep 2023
Kompas (2023) Carbon Trading Has the Potential to Trigger New Environmental Problems. Kompas
Lestari Capital (2023) Projects of Lestari Capital. In: lestaricapital.com. https://lestaricapital.com/projects/. Accessed 11 Sep 2023
LTS International Ltd (2019) Rimbak Pakai Pengidup Forest for Life: Sustainable Forest and Biodiversity Management in Nanga Lauk Village, Kapuas Hulu, West Kalimantan, Indonesia. Penicuik
MEF (2022) Indonesia REDD+ National Strategy 2021–2030. Jakarta
Megarani A (2022) Carbon Trading Procedures. In: www.forestdigest.com. https://www.forestdigest.com/detail/2063/perdagangan-karbon. Accessed 11 Sep 2023
Mustika PP (2022) Clean Carbon Absorption Needs to be Accompanied by the Establishment of Customary Forests. Kompas
Natarajan N, Newsham A, Rigg J, Suhardiman D (2022) A sustainable livelihoods framework for the 21st century. World Dev. https://doi.org/10.1016/j.worlddev.2022.105898
North DC (1994) Economic performance through time. Am Econ Rev 84:359–368
Plan Vivo Foundation (2019) Registration Certificate for Rimbak Pakai Pengidup Project. In: www.planvivo.org. https://www.planvivo.org/Handlers/Download.ashx?IDMF=7e9c3628-4f80-4e54-b5e9-ac6cc5ba4cba. Accessed 11 Sep 2023
Plan Vivo Foundation (2023a) Plan Vivo's Current Project in Bujang Raba Indonesia. In: www.planvivo.org. https://www.planvivo.org/bujang-raba. Accessed 11 Sep 2023
Plan Vivo Foundation (2023b) Plan Vivo's Current Project of Durian Rambun Indonesia. In: planvivo.org. https://www.planvivo.org/durian-rambun. Accessed 11 Sep 2023
Rodrik D (2003) In Search of Prosperity.
Susetyo PD (2022) How to Reduce Poor Groups Around the Forest. https://www.forestdigest.com/detail/1960/orang-miskin-sekitar-hutan. Accessed 4 Sep 2023
Susetyo PD (2023) Forest Carbon in Mitigating the Climate Crisis. https://www.forestdigest.com/detail/2131/karbon-hutan . Accessed 2 Sep 2023
Timulak L (2014) Qualitative Meta-analysis. In: SAGE Handb. Qual. Data Anal. pp 481–495
Widi S (2023a) Indonesia's Forest Area Reaches 125.76 Million Hectares. https://dataindonesia.id/sektor-riil/detail/luas-kawasan-hutan-indonesia-mencapai-12576-juta-hektare. Accessed 10 Sep 2023
Widi S (2023b) KLHK: Forest and Land Rehabilitation Covering 77,103 Ha by 2022. In: dataindonesia.id/. https://dataindonesia.id/ragam/detail/klhk-rehabilitasi-hutan-dan-lahan-seluas-77103-ha-pada-2022. Accessed 11 Sep 2023
Witoelar R (2023) Voluntary Forest Carbon Market: "Take It or Lose It."

7 A Cultural Ecology Perspective on Forest Resource Management in Indonesia

Asmanah Widarti, Sylviani, and Ari Nurlia

Abstract

As deforestation increases, so too does the ecological crisis. Ecological wisdom is among the noble values inherited by local communities through which the balance of natural resources is maintained, especially for communities around the forest. Local wisdom is the result of cultural ecological adaptation. Various local wisdoms in Indonesia describe harmony between humans and the environment with patterns of adaptive behavior towards the environment when it comes to utilizing and maintaining natural resources. The present study offers a basic idea of how academics can adopt local wisdom by adapting cultural ecology in customary forest management into a sustainable forest area management framework. The local wisdom of the Sasak tribe in Lombok, West Nusa Tenggara and the Serampas Traditional Forest Community in Jambi were chosen as research units. A mixed method with a sequential explanatory approach is used to analyze the cultural ecology of local communities in sustainable forest resource management. The research shows that local knowledge can be incorporated into forest management policies by applying cultural ecological adaptation as an effective strategy that is readily accepted by the community. Developing local wisdom is an important part of the sustainable development agenda, so it requires support and cooperation from various parties.

Keywords

Ecological crisis · Local wisdom · Cultural ecology · Sustainability

7.1 Introduction

Deforestation in various parts of the world is causing an increasing ecological crisis (Adeniyi 2021). Forest degradation affects the structure, composition and diversity of forests, causing climate change, increasing carbon emissions and reducing the function of providing ecosystem services and water sources (Gao et al. 2020; Pinheiro et al. 2016; Vásquez-Grandón et al. 2018). Degradation will reduce the capacity of forest landscapes to provide environmental goods

A. Widarti (✉) · Sylviani · A. Nurlia
Research Centre for Society and Culture - National Research and Innovation Agency, Jakarta, Indonesia
e-mail: asma006@brin.go.id

Sylviani
e-mail: sylv004@brin.go.id

A. Nurlia
e-mail: ari.nurlia@brin.go.id

H. Warsilah and A. H. Purnomo (eds.), *Approaching Natural Resources for Sustainable Development in Indonesia*, Sustainable Development Goals Series, https://doi.org/10.1007/978-981-96-6913-4_7

and services that support food security and other basic human needs (Amenu 2017).

Indonesia is a country that is experiencing significant forest degradation both in quantity and quality. Based on Forestry Law No. 5 of 1967, the total area of forest managed by the government is 140.4 hectares. However, the forest area is currently only 96 million hectares (MoEF 2023). For decades, forests have been exploited by prioritizing economic and political interests. The science and technology currently applied in forest resource management cannot guarantee sustainability in the future. Through their own culture, local communities in Indonesia and worldwide can wisely manage the forests that are their source of livelihood without endangering forest sustainability. Local knowledge carried by local communities develops through various experiences gained over the centuries under changing environmental, economic, political and social conditions. The process of human adaptation to the environment to meet needs based on cultural elements is called cultural ecological adaptation (Piccardo and Canepa 2021; Sutton and Anderson 2010).

The results of cultural ecological adaptation are passed down from generation to generation to become guidelines for utilizing natural resources in harmonious interaction with the environment, and this is considered local wisdom. Cultural ecological adaptation continues in a society as its environment changes. Local communities respond to various threats using local knowledge and skills that have grown over generations and will continue to adapt to changes in the future (Pokhrel et al. 2021). Local wisdom is increasingly recognized and considered influential in natural resource management. The results of studies by researchers both abroad and at home prove that nature conservation can be achieved through local community knowledge in the form of norms, culture or customs which effectively regulate community behavior in the management of natural resources (Abas et al. 2022; Mapedza et al. 2022; Schmidt et al. 2021). Local wisdom is a strategy for maintaining ecological balance while managing natural resources sustainably, and it has been tested for centuries by various disasters, obstacles and human errors (Fabiyi and Oloukoi 2013; Ncube 2022).

Government support for the implementation of local wisdom is contained in the 1945 Constitution, which states that "The state recognizes and respects customary law community units and their traditional rights as long as they are still alive and by the development and principles of the Regulated in Law of the Republic Indonesia". Local wisdom was formally introduced in Law Number 32 of 2009, which states that local wisdom is a noble value that applies in community life to, among other things, protecting and managing the environment sustainably. However, knowledge of local wisdom is increasingly being abandoned. At the same time, there is an urgent need for sustainable management of forest resources amidst the ongoing ecological crisis caused by climate change. Local wisdom, which is considered primitive knowledge, provides ample evidence that the adverse environmental impacts of disasters can be reduced through mitigation and adaptation of local wisdom practices (Ncube 2022).

Indonesia has considerable ethnic diversity, with its number of ethnic groups reaching 1340 (Indonesia.go.id 2017). The diversity of local wisdom, with its multicultural dimensions, originates in ethnic groups that differ in culture, language, values, customs and behavior when interacting with the environment where they live wisely. Even though plenty of community groups have shown evidence of their ability to manage forests sustainably, their capacity is also decreasing. The acceleration of globalization, accompanied by rapid technological developments, has changed lifestyles and habits that have lasted for thousands of years. Ancestral values, including those that are part of the heritage of the Indonesian nation, are slowly becoming forgotten and marginalized. This chapter will discuss how local wisdom, as a product of past cultural ecology adaptation, can adapt to the innovative interventions and dimensions of change that are occurring in the present. It is hoped that the research results can contribute to policy on maintaining and increasing the role of local wisdom in natural resource management.

7.2 Literature Framework

7.2.1 Principles and Role of Local Wisdom in Forest Resource Management

The norms or behavior of local communities in their wise interactions with the environment in which they live is known as local wisdom (Abas et al. 2022; Chikaire et al. 2012). Local knowledge about managing forest resources and the environment results from prolonged interactions and the ability of those communities to apply adaptation techniques to their environment. Even though they are local values, the values contained in each of these cultures have the same common thread: using and preserving natural resources is a universal value. Local wisdom emphasizes the importance of protecting natural resources. This is in line with sustainable development, which aims to preserve the environment by protecting forests, water, soil and biodiversity. Environmental sustainability is a crucial aspect of sustainable development that must be maintained so that ecosystems can continue to support the lives of humans and other creatures. While the language and approaches used may differ, the traditions and local knowledge conveyed have the same goal of protecting the natural environment (Grey 2014; Kartika et al. 2023; Murmu and Bhattacharya 2018).

Environmental values reflected in local wisdom practices include protection, sustainable use and maintenance. These values are directly related to the social system of a society. Natural ecosystems cannot be understood, preserved and managed without knowing the human culture that forms them, because biodiversity and culture are mutually reinforcing and interdependent (Pretty et al. 2010). All activities are intended to be carried out by community members and are aimed at the common interest and good. Society adapts to its environment by developing wisdom through knowledge, ideas and equipment, combined with traditional norms, cultural values and activities to manage natural resources to meet daily needs. Local wisdom has gone through the process of cultural ecological adaptation of Indigenous peoples to their environment. It continues to follow changes that occur in their current environment (Dhungana et al. 2023). This is because culture develops along with environmental changes. Sustainable development also requires innovation and adaptation to meet the challenges of climate change, population growth and environmental degradation. The ability of Indigenous peoples to adapt through local wisdom shows how traditional knowledge can contribute to sustainable solutions. In this way, the principle of cultural ecology is to make good use of natural resources, manage them appropriately and protect them for present and future generations (Grey 2014).

7.2.2 Traditional Practices: The Role and Practice of Local Wisdom in Natural Resource Management

Examples of local community practices that apply local wisdom in natural resource management activities can be found worldwide. Indigenous communities in the Northern Philippines provide timber and water management services to maintain agricultural productivity while helping to preserve biodiversity and ecosystem resilience from generation to generation at the landscape level (Camacho et al. 2016). In northern Ghana, traditional knowledge is studied through various forms, customs, rituals, rules and regulations. Traditional protected areas still play an important role in managing ecosystem services in these villages. However, this study also found that awareness of and adherence to traditional norms and practices is more associated with older villagers than the younger generation (Boafo et al. 2016).

Indigenous people in Ogun State, Nigeria, use their local wisdom to manage forest resources with a strong belief in gods, customs and traditions, myths and taboos. This traditional technique is effective in managing forest resources.

Studies also reveal that the further a community is from large cities, the more likely their Indigenous knowledge and practices will remain preserved. The study concludes that traditional practices have potential in forest management, and scientific intervention through an interactive framework with customary practices will be an effective way of innovating traditional practices and effective conservation management of forest resources (Adeniyi 2021).

In Indonesia, Indigenous communities have a forest management system with local wisdom that successfully maintains ecosystem balance and biodiversity. They adhere to traditional beliefs and rules prohibiting excessive exploitation and preventing environmental damage with the principle of sustainability. Indonesia has local wisdom that is spread across various regions, including the local wisdom of the Kajang tribe in South Sulawesi, which has the *Kamase mase* (modest) principle of life and firmly adheres to the *Pasang* (message) of the ancestors (Nurmala et al. 2022). The local wisdom of the Talang Mamak community in Riau manages natural resources based on beliefs/religious values. They believe that God is the owner of the universe, so they need to ask God's permission for every work related to forest use (Titisari et al. 2018). The sustainability of the Natural Tourism Park and water sources in the Sicike-Cike Natural Tourism Park (TWA), North Sumatra, is managed by cooperation between local communities and the government, supported by local wisdom that still survives today. The wider the community that is allowed to participate in managing their forests, the higher the sense of ownership of the forest (Situmorang and Simanjuntak 2015).

In Banten, West Java, the Baduy tribe maintains the preservation of natural resources with a traditional lifestyle steeped in cultural wisdom. Meanwhile, the Kotabaru Dayak Indigenous community in South Kalimantan successfully manage their forest based on local wisdom, maintaining environmental sustainability quite well because existing local traditions and customary laws are still inherent in the Indigenous community (Fahrianoor et al. 2020). Forest management using local wisdom is also practiced in Wae Rebo, Manggarai Regency, East Nusa Tenggara, where the community divides areas based on their function and use; thus *Beo* (village) is a place to carry out traditional ceremonies related to forest management activities, and *Lingko* (community garden) is an area to be managed so as not to disturb protected forest areas and *Cengit* (sacred area) (Uju et al. 2019). Tigawasa Village, Buleleng Regency, continues to maintain the local wisdom of its ancestors, including the community's conception of the existence of forest areas, the myths that develop in the community, ritual practices related to forests, and written and unwritten regulations regarding customary forests. The people of Tigawasa Village consider the traditional forest a sacred area that cannot be entered or used other than for traditional ceremonial purposes so that the forest ecosystem is maintained (Candraningsih et al. 2018).

Generally, the studies mentioned above confirm that each region essentially has local wisdom that is held in high regard as a value that guides and regulates the behavior of its people. There is a significant relationship between local wisdom and environmental conservation, and local wisdom practices have contributed to saving and preserving the natural environment. Local wisdom is spread across many Indigenous peoples and ethnic groups throughout Indonesia and the world. It is a valuable inheritance that must be protected and respected and constitutes social capital to preserve forest resources and the diversity of local wisdom worldwide.

7.3 Methodology

Research was conducted in Lombok, West Nusa Tenggara and Jambi (Fig. 7.1). The local wisdom of the Sasak tribe in Lombok, West Nusa Tenggara and the Serampas Traditional Forest Community in Jambi were chosen as research units.

The study used a mixed method with a sequential explanatory approach (Cresswell 2016), and descriptive qualitative analysis. The informants in this research were the community and traditional leaders of the Sasak Tribe in West Nusa Tenggara and the Serampas Indigenous Community in Jambi Province. All available data from

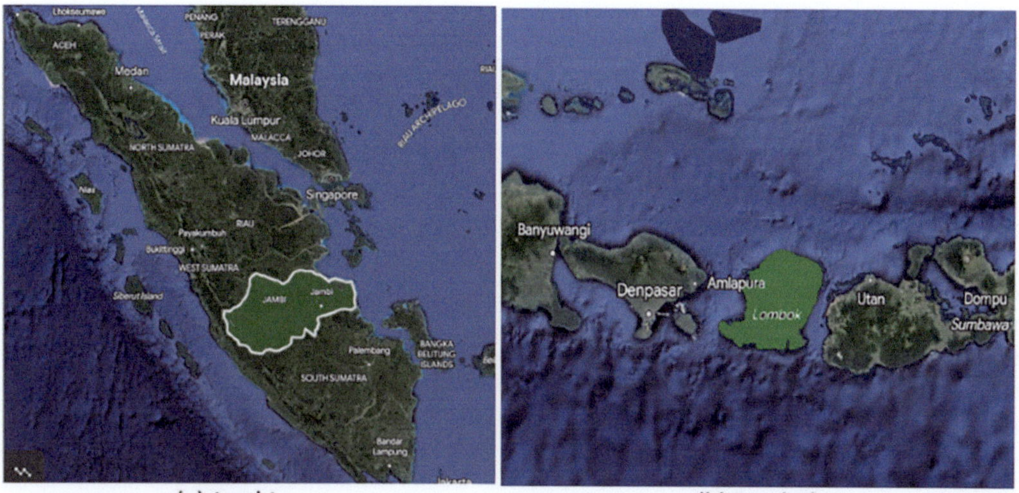

Fig. 7.1 Research locations: **a** Jambi and **b** Lombok

the sources was collected and analyzed, including interviews, observations, documents, reports, regulations and other supporting documents, which are elaborated with literature reviews. Next, the qualitative data was processed and analyzed using a qualitative descriptive approach (Yin 2011). Finally, the processed data are compared with existing phenomena to raise research findings to a higher conceptual level and to draw conclusions.

7.4 Results and Discussion

7.4.1 Using Local Wisdom to Implement Forest Resource Management in Indonesia

A participatory and inclusive approach to decision-making characterizes forest management with local wisdom. Governance systems incorporate local values, norms, and traditional knowledge, ensuring that decisions are guided by a deep understanding of the ecological, cultural, and social contexts in which forests exist and prioritize long-term well-being by considering ecological resilience and forest ecosystem sustainability. Through governance, local communities can effectively manage and regulate harvesting, hunting, and land use practices, ensuring that these activities are carried out sustainably and responsibly. Members generally actively contribute to the preparation of forest management plans and strategies. Decision-making processes are often based on building consensus and respecting traditional authority structures, with elders and community leaders playing an important role in facilitating discussions and reaching agreements.

Local communities have a deep sense of responsibility and concern for the forest, as they are closely connected to the forest and depend on it for their livelihoods and cultural practices. Customary rules and institutions are established within communities to regulate access to forest resources, resolve conflicts and ensure the equitable distribution of benefits derived from forests. The participatory nature of local community governance in local wisdom forest management fosters a sense of ownership, empowerment and self-determination among community members, strengthening their ability to protect and manage their forest resources sustainably (Niman 2022).

Customary forests are forests located within the territory of customary law communities with forest ecological functions, economic benefits and social functions that support the lives of customary forest communities. The local wisdom in forest resource management highlighted in this chapter is that of the Sasak Tribe in Lombok, West Nusa Tenggara and of the Serampas Community in the Serampas Traditional Forest, Jambi. These two cases will show that in the Mandala Customary Forest, the local wisdom of the Sasak Tribe is still applied purely to manage forest resources, while in the Serampas traditional forest, local wisdom remains the basis for forest resource management but cultural ecology adaptation linked to social innovation is occurring at the same time, to adapt to current community needs.

7.4.1.1 Local Wisdom of the Sasak Tribe in Lombok, West Nusa Tenggara

According to the local wisdom maintained by the Sasak tribal people in Lombok, West Nusa Tenggara, the forest is divided into two functions, *Pawang*, or guardian, and *Gawah*, which refers to production. *Pawang* is a protected (sacred) forest area for springs that must not be disturbed because it functions as a catchment area. *Gawah* is a forest area managed as a source of livelihood, and its resources are used according to agreed rules to ensure sustainability.

Pawang Mandala is one of the traditional forests that is protected by customary regulations (*Awig-awig*). All communities, whether Indigenous or otherwise, must obey customary rules for maintaining and preserving water springs. These include a prohibition on uprooting or cutting down trees, catching animals, polluting springs and poisoning rivers, which can result in the death of biotic animals that live in rivers, as well as grazing livestock in such a way as to ensure forest flora and fauna are not damaged. Water users, individuals and groups must pay fees to traditional forest water source managers. Anyone who violates this *awig-awig* will be subject to customary law sanctions in the form of fines. The most severe sanctions for violators are being ostracized or exiled and not being recognized as Indigenous people.

Indigenous communities have traditional institutions where *Pemangku Adat* leads the traditional discussion. *Penghulu Adat* and *Pembekel Adat* provide opinions, input and suggestions to traditional leaders regarding problems in the customary forest. Then there is *Lang-Lang Jagad*, a forest police officer. The Mandala Customary Forest exemplifies the harmonious relationship between the Sasak tribe and their environment, in which the environment is respected and maintained so that it remains a water source and sustainably provides various other benefits for the community. As well as its coolness and beautiful natural panorama, the Mandala Customary Forest also displays local traditions to attract the interest of local and international tourists.

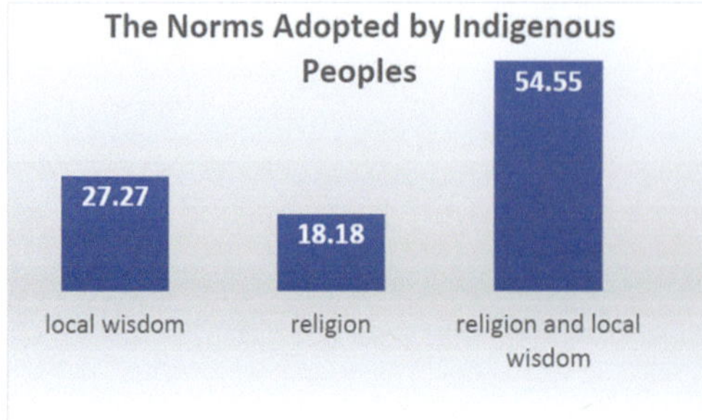

Fig. 7.2 Sources of the norms adopted by Indigenous peoples. (*Source* processed primary data)

Pawang Mandala is located in the Bayan Traditional Village in North Lombok Regency, one of the villages on Lombok Island that still adheres to the original customs of the Sasak-Bayan Tribe. The study area included 374 heads of families. The 45 sample respondents are representative of the family heads.

Respondents were asked about the source of the system of values or norms adopted in society, and stated that it was from religion and local customs/wisdom (54.55%), from local customs/wisdom (27.27%) and from religion (18.18%) (Fig. 7.2).

As many as 79% of the Indigenous people who were informants stated that they had made a big contribution to making the residential environment of the forest better and more peaceful. As many as 76.35% of the informants felt quite happy with their current situation. Even though only a few informants have detailed knowledge of norms, they still adhere to them because they fear the sanctions that may follow a violation.

Some of the traditional norms implemented by the Sasak tribe include rules for farming and felling trees in forest areas. Indigenous farming communities have a limited land processing system called OTM (minimum tillage) as a land conservation measure. Members who carry out intensive land processing or tree logging activities receive sanctions. The form of sanctions in each village is different; for example, in Bentek Village, the *Penyowok* sanction is applied, which takes the form of an obligation to slaughter a buffalo and plant trees in the forest in an amount decided by the group, while in Genggelang Village, the sanction applied can be a monetary fine, issued as a member of the forest management, and the obligation to plant ten times the number of trees cut. If these customary sanctions are not obeyed, they will be submitted through a legal process in accordance with government regulations.

Local wisdom in forest management is applied equally to followers of Islam, Hinduism and Buddhism to conserve forest resources and the environment. In customary forest management, the three main functions of forests are ecological and conservation, socio-cultural and economic. Ecological and conservation functions are aimed at maintaining biodiversity and protecting water springs. The socio-cultural function is for carrying out traditional ceremonies and sanctions. Meanwhile, the economic function uses non-timber forest products like bark, leaves and fruit.

7.4.1.2 Local Wisdom of the Serampas Traditional Forest Community in Jambi

The Serampas Customary Forest has a spatial layout pattern consisting of Tanah Ajum, land that may be opened for residential areas; directional land, which can be used for gardening/farming; wetland which is intended for rice fields; and dry land intended for farming. Ecologically, the Serampas Customary Forest meets forest health parameters, such as protected forest conditions with *lubuk larangan* and irrigation water sources built to meet the community's drinking water needs. The animal habitat ecosystem is maintained, and the vegetation structured. The population generally make their living as farmers. The economy of Rantau Kermas Village is very dependent on plantation and agricultural products such as cinnamon, coffee, chillis, rice, potatoes and sweet potatoes.

The Serampas Customary Forest Community (MHA) in Rantau Kermas Village does not encroach on customary forest areas because the community still adheres to its customs and complies with what is laid down by *Depati Payung*, community leaders, or village officials, so minimal violations occur in forest management. Every Serampas Indigenous Community managing its forest must first ask permission from the *Depati*, and a "customary *kenduri*" is held.

Technical, social and economic factors all affect the success of forest management. Technical factors include farming activities, types of commodities, pest and disease attacks, soil processing, the availability of irrigation water, and planting patterns. Indicators of social factors include the level of education, government policy, and the role of companions and institutions. Economic indicators include the availability of production facilities, markets, human resources, and forest sustainability factors, including soil fertility, natural disasters, tree populations, group collaboration, community conflicts and community participation in forest protection.

These variables meet the criteria for external environmental factors for Serampas Customary Forest management. Land management is carried out in a mutually cooperative and participatory manner involving gender. Internal land ownership conflicts can be resolved using customary law and deliberation. Production and market facilities are already available, while social and public facilities are available for kindergarten and elementary school education. Community empowerment is carried out through training for permanent farmers.

Community empowerment in forest conservation efforts was considerably enhanced when a Village-Owned Enterprise (*Bumdesd*) named

Bumdes Alam Depati Payung was formed, and a customary forest was then established. Previously, customary forest management had been carried out by existing customary rules without support from various parties. Once the *Bumdesd* had been formed, the regional government, various ministries and institutions began to play many roles in training and joint patrols of forest areas.

@@There are four business units in Rantau Kermas Village:

Serampas coffee business unit Coffee is one of the commodities favored by the Serampas Indigenous Community, and it has contributed to changing the image of Rantau Kermas Village from that of a mystical village into one that produces premium robusta coffee. The idea of empowering community land through coffee planting is also being developed, the results of which can later be sold to the *Bumdes.*

Customary Forest Business Unit Various sustainability programs are aimed at improving the community's economy in the hope that the community will be prosperous and the forest will be sustainable. The community carries out craft activities, sewing and moving coffee processing machines at the Serampas coffee house. In Adopted Trees, a voluntary donation scheme which receives assistance from KKI WARSI, the proceeds from the donations are allocated in a ratio of 60% for village development and 40% for customary forest management.

Micro Hydro Power Plant business unit The micro hydro power plant (MHP) is very beneficial for the Serampas Indigenous Community of Rantau Kermas Village because it can meet the electricity needs of the entire village 24 h a day and be a lever for the village economy. It also increases empowerment.

Ecotourism Business Unit Serampas Indigenous Community has formed a tourism awareness group (*Pokdarwis*) to empower the community in the field of tourism. Tourism potential in Rantau Kermas Village includes building tourist roads in customary forests, *Lubuk Larangan* tourism, *Depati* Alam IV lake tourism, and other cultural tourism.

7.4.2 Cultural Ecological Adaptation and Efforts to Increase the Role of Local Wisdom

Cultural ecology explains how human culture, including values, beliefs, traditions and institutions, interacts with the surrounding physical and social environment. Culture is adaptive and develops as the environment changes (Sutton and Anderson 2010). Developing cultural ecology adaptation in the application of local wisdom is very important because it is in line with environmental and cultural changes that are currently occurring. Climate change, urbanization, globalization and technological innovation have significantly impacted ecosystems and culture in various regions. Cultural ecological adaptation can help formulate strategies that suit current conditions (Huong and Huong 2018). It is important to realize that cultural ecological adaptation does not mean a complete return to past practices. Rather, it is a dynamic process in which traditional knowledge is integrated with new understandings of the environment and culture. Identifying relevant elements and adapting to change is key to ensuring that local wisdom remains effective in facing challenges. By combining traditional knowledge and technological innovation, society can develop new ways to overcome the problems of sustainable natural resource management (Huang and Cao 2022).

Management of forest resources with local wisdom uses various methods or norms in different places including: (i) involving local communities in all decision-making processes related to forest management; (ii) recognizing, respecting and preserving customary laws, local community traditions that can help preserve resources and ecosystems well; (iii) applying the principle of sustainable use of forest resources, where the resources taken must be renewable so that they can continue to be utilized in the future; (iv) identifying and mapping forest areas for various uses, including conservation, economic development and other community needs; (v) providing education and training to local communities regarding sustainable practices in forest management and building potential and capacity to manage resources wisely, including access rights and control over natural resources in the community's jurisdiction; and (vi) open dialogue with parties involved in forest management, including the government, local communities and non-governmental organizations, to develop sustainable local economies such as agroforestry or ecotourism which can help meet community needs without destroying forests (Huang and Cao 2022; Iswandono et al. 2015; Ngo et al. 2021).

To increase the role of local wisdom in managing forest resources, a holistic approach is needed that combines local wisdom values with modern knowledge and technology. Education, awareness campaigns, appropriate regulations and support from various parties can encourage and facilitate the application of local wisdom in resource management. The state's recognition of Indigenous peoples and their local wisdom should be complemented by making the values underlying that local wisdom the basis for national forest management arrangements.

7.5 Conclusions and Recommendations

7.5.1 Conclusion

This study concludes that the noble values of local wisdom within the community are important social capital for maintaining balance with the environment and have been successfully applied in customary forest management. This is in line with the principle of sustainable development, which emphasizes the necessity of satisfying the requirements of the present without compromising the ability of future generations to meet their own needs. Local wisdom that develops from various factors and influences human behavior towards the environment shows the importance of cultural ecological adaptation, which integrates traditional knowledge with dynamic environmental changes. The development of local wisdom has many positive aspects and can contribute to enriching and empowering local communities in a more equitable manner. The rapid environmental changes, including social, economic, technological and political, that are occurring in many regions and the threat to the survival of local wisdom mean that it is time to reposition Indigenous communities as subjects or actors and refocus on utilizing their local wisdom to increase community participation in effective sustainable forest management.

7.5.2 Recommendations

Sustainable forest management involving local communities as active subjects is the key to achieving sustainable development. The government must have the courage to implement sustainable forest management policy by involving local communities as subjects who actively participate in managing forests, and by promoting the wealth of local wisdom knowledge and social capital so that forest resource management at the local level is effective. Local wisdom-based forest management can create a balance between human needs and environmental sustainability, and ensure sustainability for future generations.

References

Abas A, Aziz A, Awang A (2022) A systematic review on the local wisdom of indigenous people in nature conservation. Sustainability 14(6):3415

Adeniyi TD (2021) The use of indigenous knowledge in forest management by the aboriginal people in Ogun State, Nigeria. Wageningen University and Research, Wageningen, Netherlands

Amenu BT (2017) Review: forest management and conservation practices in Ethiopia: Opportunities and constraints

Boafo YA, Saito O, Kato S, Kamiyama C, Takeuchi K, Nakahara M (2016) The role of traditional ecological knowledge in ecosystem services management: the case of four rural communities in Northern Ghana. Int J Biodiversity Sci Ecosyst Serv Manag 12(1–2):24–38. https://doi.org/10.1080/21513732.2015.1124454

Camacho LD, Gevaña DT, Carandang AP, Camacho SC (2016) Indigenous knowledge and practices for the sustainable management of Ifugao forests in Cordillera, Philippines. Int J Biodiversity Sci Ecosyst

Serv Manag 12(1–2):5–13. https://doi.org/10.1080/21513732.2015.1124453
Candraningsih I, Pujaastawa G, Sudiarna I (2018) Konservasi Hutan Berbasis Kearifan Lokal di Desa Tigawasa, Kecamatan Banjar, Kabupaten Buleleng. Humanis. https://doi.org/10.24843/JH.2018.v22.i02.p06
Chikaire J, Osuagwu O, Oguegbuchulam M, Ngozi EE-O (2012) Indigenous knowledge system: the need for reform and the way forward
Cresswell JW (2016) Research Design Pendekatan Metode Kualitatif, Kuantitatif dan Campuran (Edisi Keem, ed.). Pustaka Pelajar
Dhungana G, Pal I, Ghimire R, Dhungana RK, Tuladhar N (2023) Foundation of indigenous knowledge theory for disaster risk reduction. In: Pal I, Shaw R (eds) Multi-hazard vulnerability and resilience building. Elsevier, pp 347–361
Fabiyi OO, Oloukoi J (2013) Indigenous knowledge system and local adaptation strategies to flooding in coastal rural communities of Nigeria. J Indigen Social Dev 2:1–19
Fahrianoor, Windari T, Taharuddin T, Ruslimar'I, Maryono (2020) The practice of local wisdom of dayak people in forest conservation in South Kalimantan. J Wetlands Environ Manag 1:33–42. https://doi.org/10.20527/jwem.01.01.01
Gao Y, Skutsch M, Paneque-Gálvez J, Ghilardi A (2020) Remote sensing of forest degradation: a review. Environ Res Lett 15(10):103001. https://doi.org/10.1088/1748-9326/abaad7
Grey S (2014) Indigenous knowledge. In: Michalos AC (ed) Encyclopedia of quality of life and well-being research. Springer Netherlands, pp 3229–3233
Huang L, Cao Y (2022) Review of the integrated development of ecological and cultural forestry. Sustainability 14(11):6818
Huong D, Huong N (2018) Ecological culture and educational issue of ecological culture—motivation for the human development. Am J Educ Res 6:694–702. https://doi.org/10.12691/education-6-6-17
Indonesia.go.id (2017) Suku Bangsa. Indonesia.go.id. Retrieved January 31, 2023 from
Iswandono E, Zuhud E, Hikmat A, Kosmaryandi N (2015) Integrating local culture into forest conservation: a case study of the Manggarai Tribe in Ruteng Mountains, Indonesia. Jurnal Manajemen Hutan Tropika (j Tropical for Manag) 21:55–64. https://doi.org/10.7226/jtfm.21.2.55
Kartika A, Silvi E, Sepriadi I, Hermaya I, Valva Y, Razak A, Anhar A (2023) Literature review: forest conservation by local communities integrated local wisdom. Jurnal Program Studi Magister Pendidikan Biologi 1(1):9–16
Mapedza E, Rashirayi T, Xueliang C, Haile AT, van Koppen B, Ndiyoi M, Sellamuttu SS (2022) Indigenous knowledge systems for the management of the Barotse Flood Plain in Zambia and their implications for policy and practice in the developing world. In: Sioui M (ed) Current directions in water scarcity research, vol 4. Elsevier, pp 209–225
MoEF (2023) Laju Deforestasi Indonesia Tahun 2021–2022 Turun 8,4%. In: Ministry of Environment and Forestry. Retrieved August 18, 2023
Murmu SC, Bhattacharya S (2018) Indigenous knowledge on forest management: an approach towards sustainable development
Ncube B (2022) Indigenous knowledge perspectives on water management and its challenges in South Africa. In: Sioui M (ed) Current directions in water scarcity research, vol 4. Elsevier, pp 227–242
Ngo T, Nguyen T, Duong T, Ly T (2021) Forest-related culture and contribution to sustainable development in the northern mountain region in Vietnam. For Soc 5:32–47. https://doi.org/10.24259/fs.v5i1.9834
Niman EM (2022) Symbolic interaction of the indigenous communities of Manggarai, Indonesia (Study of Penti Cultural Local Wisdom in The Context of Environmental Conservation). Jurnal Seni Budaya 20(2):144–154
Nurmala, Dassir M, Supratman (2022) "Pasang", knowledge and implementation of local wisdom in the Kajang Traditional Forest Area, South Sulawesi. J Tourism, Hosp Travel Bus Events 4(1):40–47
Piccardo C, Canepa M (2021) Cultural ecology: paradigm for a sustainable man-nature relationship. In: Leal Filho W, Marisa Azul A, Brandli L, Lange Salvia A, Wall T (eds) Partnerships for the goals. Springer International Publishing, Cham, pp 248–258
Pinheiro TF, Escada MIS, Valeriano DM, Hostert P, Gollnow F, Müller H (2016) Forest degradation associated with logging frontier expansion in the amazon: the BR-163 region in Southwestern Pará, Brazil. Earth Interact 20(17):1–26. https://doi.org/10.1175/EI-D-15-0016.1
Pokhrel K, Khatiwada S, Paudel N, Dhakal K, Chidi C, Timilsena N, Mahat D (2021) Innovative practices for the promotion of local/indigenous knowledge for disaster risk reduction management in Sudur Paschim Province, Nepal. J Geogr Res. https://doi.org/10.30564/jgr.v4i3.3223
Pretty J, Adams B, Berkes F, Athayde S, Dudley N, Eugène H, Maffi L, Milton K, Rapport DJ, Robbins PF, Samson CM, Eleanor, Sterling, Stolton S, Takeuchi K, Tsing AL, Vintinner EC, Pilgrim S (2010) How do biodiversity and culture intersect?
Schmidt MVC, Ikpeng YU, Kayabi T, Sanches RA, Ono KY, Adams C (2021) Indigenous knowledge and forest succession management in the Brazilian Amazon: contributions to reforestation of degraded areas [original research]. Front for Global Change. https://doi.org/10.3389/ffgc.2021.605925
Situmorang R, Simanjuntak E (2015) Forest management through local wisdom of the community around the Sicike-Cike Nature Park, North Sumatra. Widyariset 18:145–153. https://doi.org/10.14203/widyariset.18.1.2015.145-154
Sutton MQ, Anderson EN (2010) Introduction to cultural ecology. AltaMira Press
Titisari PW, Elfis E, Zen IS, Khairani K, Janna N, Suharni N, Sari TP (2018) Local wisdom of Talang Mamak

Tribe, Riau, Indonesia in supporting sustainable bioresource utilization. Biodiversitas Journal of Biological Diversity.
Uju GF, Bhuja P, Boro TL (2019) Kearifan Lokal Masyarakat dalam Pengelolaan Hutan Lindung di Kampung Wae Rebo, Desa Satar Lenda, Kabupaten ManggaraI. Jurnal Biotropikal Sains 16(1):1–11
Vásquez-Grandón A, Donoso PJ, Gerding V (2018) Forest degradation: when is a forest degraded? Forests 9(11):726. https://www.mdpi.com/1999-4907/9/11/726
Yin RK (2011) Qualitative research from start to finish. The Guilford Press, New York/London

8 Tradition-Based Rural Socio-Ecological Crisis Mitigation: The Experience of the Baduy Community (*Urang Kanekes*), Banten-Indonesia

Rusman Nurjaman and Enkin Asrawijaya

Abstract

The Baduy Indigenous community faces environmental degradation, threats to biodiversity, and socio-cultural disruption. They have implemented sustainable living principles through tradition-based mitigation initiatives, such as the *Kawalu* ceremony, to overcome these challenges. This chapter explores the sustainability practices of the tradition-based Baduy community and their relevance for overcoming socio-ecological crises. The research was conducted using qualitative methodology with an eco-populist approach. By studying *Kawalu* traditions, we can understand how traditional practices can effectively mitigate the crisis. Eco-populism in the *Kawalu* tradition is demonstrated by protecting nature from various threats. These measures include limiting tourist visits, preventing over-exploitation, and preserving sustainable agricultural techniques. During the *Kawalu* period, the entire Baduy area is closed to outside visitors and the Baduy community is busy cleaning the entire village of objects prohibited by custom. Likewise, community members are prohibited from traveling long distances and observe certain taboos. This tradition is an effort to mitigate the socio-ecological crisis from the threat of environmental degradation and ensure the availability of natural resources for future generations. The novelty explored in this chapter is the role of the *Kawalu* tradition in mitigating contemporary "disasters" due to the socio-ecological crisis along with the increasing phenomenon of mass tourism, the climate crisis, population pressure, and socio-cultural changes which do not always have a positive impact on the future of society, especially the Baduy community.

Keywords

Sustainability · The Baduy community · Socio-ecological crisis mitigation · Eco-populism

8.1 Introduction

The Baduy Indigenous community, or *Urang Kenekes*, in Lebak Regency, Banten Province, Indonesia, are known for their determination to continue traditional practices, and have strong

R. Nurjaman (✉) · E. Asrawijaya
Research Centre for Society and Culture - National Research and Innovation Agency, Jakarta, Indonesia
e-mail: rusm006@brin.go.id

E. Asrawijaya
e-mail: enki001@brin.go.id

H. Warsilah and A. H. Purnomo (eds.), *Approaching Natural Resources for Sustainable Development in Indonesia*, Sustainable Development Goals Series, https://doi.org/10.1007/978-981-96-6913-4_8

connections with nature. They rely on agriculture as a means of subsistence and cultural traditions that reflect closeness to the natural world. However, now they face the threat of a socio-ecological crisis characterized by environmental degradation, threats to biodiversity, and socio-cultural disruption.

According to Shohibuddin (2016), there are currently at least two kinds of crises faced by rural communities, including Indigenous communities. First, the agrarian crisis is characterized by inequality and limited access to land and other agrarian resources. There have been overlapping agrarian relations since the feudal and colonial eras, leading to increasingly sharp social differentiation. Data show that 68% of land is controlled by corporations (Walhi 2022). In contrast, most farming households (56%) own less than 0.5 ha of land (BPS 2013). The rise of land conflicts today reflects this inequality, which is becoming a social crisis. Second, the ecological crisis is characterized by a decline in carrying capacity and even destruction of the natural environment. As shown by many studies, the green revolution that has taken place since the 1970s has radically changed the condition of the land and environment in Indonesia, making it increasingly vulnerable and causing biodiversity loss (Shiva 1991; Taylor 2021).

Indigenous communities also face a socio-ecological crisis due to climate, population pressure, and mass tourism (Amico et al. 2019). Climate change has caused unpredictable weather patterns, affecting crop yields and food production. The population boom has increased resource demand while putting pressure on limited land availability. Additionally, mass tourism has disrupted local ecosystems, leading to environmental degradation. The latter integrates villages into the modern economy, potentially threatening the traditional practices and way of life of those who greatly value nature and the environment.

In response to these challenges, the Baduy community has implemented tradition-based mitigation initiatives that can be traced to a series of traditional practices, such as the *Kawalu* ceremony. These customary provisions are central to their way of life and efforts to mitigate their socio-ecological crisis. *Kawalu* traditions are important for expressing gratitude for agricultural success and encouraging sustainable practices. In addition to sticking to these traditions, the community has taken steps to protect nature and limit tourist visits to preserve their way of life. Tradition-based mitigation is an approach that combines cultural traditions with ecological knowledge to address environmental challenges while preserving cultural practices. Unlike conventional crisis management methods that prioritize technological interventions or impose solutions from external actors, tradition-based mitigation recognizes Indigenous knowledge systems and local wisdom as valuable resources for sustainable development.

This chapter explores tradition-based sustainability practices and their relevance in overcoming the socio-ecological crisis in rural areas of the Baduy community in Kanekes Village. By studying the *Kawalu* traditions, we can understand how traditional practices can effectively mitigate such crises. Discussions of *Kawalu*, as one of the important traditions in the Baduy community culture, often appear in books or other literature discussing the Baduy community's lives. To name a few examples, *Kehidupan Masyarakat Kanekes* (*The Life of the Kanekes Community*), by Danasasmita and Djatisunda (1986), *Agroekosistem Orang Sunda* (*Sundanese Agroecosystem*) (2011) and *Ekologi Perladangan Orang Baduy* (*The Baduy Cultivation Ecology*) (2012), both by Iskandar, *Tata Ruang Masyarakat Baduy* (*The Baduy Community Spatial Planning*) (2006) and *Kearifan Lokal Masyarakat Baduy dalam Mitigasi Bencana* (*The Baduy Community Local Wisdom in Disaster Mitigation*) (2010), both works by Permana, as well as several recent works such as *From Respects Hermits to Ordinary Citizens: The Conversion of The Baduy, Ethnicity, and Politics of Religion in Indonesia (1977–2019)* by Suryani (2021), which captures more than four decades of the dynamics of the development of Baduy society, and *Reforming Group Legal Personhood in Indonesian Land Law: Towards Equitable Land Rights for Traditional Customary Communities* (Mulyani 2020). However, so far, the role of tradition in preventing the threat of a socio-ecological crisis in contemporary villages, which is the focus of the present discussion, has never been the main focus of discussion in the literature mentioned above. Therefore, the novelty raised in this paper is the role of the *Kawalu* tradition in mitigating contemporary "disasters" caused by the socio-ecological crisis along with the increasing phenomenon of mass tourism, climate crisis, population pressure, and socio-cultural changes which do not always have a positive impact on the future of Indigenous communities, especially the Baduy community.

Furthermore, to discuss the important role played by tradition in mitigating the rural socio-ecological crisis and sustainability, especially in the context of the Baduy community customary area, this chapter considers eco-populism as a conceptual framework that is deemed relevant to explain the interrelationships between various elements which are the main focus of the discussion.

8.2 Methodology

This research used a qualitative descriptive approach to explain the environmental management of the Baduy community from the eco-populism perspective. Secondary data sources were collected from the literature review. Primary data collection was carried out by interviewing several informants as supporting data. Data processing included reduction, presentation, and conclusions (Huberman and Miles 2002).

8.3 Theoretical Background: Eco-Populism and Indigenous Peoples: A Conceptual Exploration

Indigenous people have always played an important role in preserving and conserving their land. Over the past few decades, there has been increasing recognition of their environmental activism and the emergence of what is often called eco-populism in their communities (United Nations 2021). Eco-populism is a socio-political movement driven by the concern of marginalized groups, especially Indigenous peoples, for environmental protection and sustainable living (Middeldorp and Billon 2019). This is characterized by a grassroots approach that seeks to overcome ecological injustice and protect traditional knowledge while demanding social and economic justice (Asrawijaya 2020).

To understand the rise of eco-populism in Indigenous communities, it is important to consider the historical context. Indigenous peoples have long faced colonial oppression, resource exploitation, and cultural marginalization, leading to deep-rooted connections between their identity, land, and environment. This historical background has laid the foundation for environmental activism and environmental populist movements among Indigenous groups (Aguilar et al. 2021; Arce 2017; Robles 2001).

Several factors have contributed to the emergence of eco-populism in Indigenous communities. First, increasing recognition of the changing crisis and its devastating impact has forced Indigenous groups to become vocal advocates for environmental protection (Maldonado et al. 2013). Second, the threat of extractive industries and large-scale development projects encroaching on customary territories has become a rallying point for Indigenous communities to demand rights and protect their land (Asrawijaya and Hudayana 2021). Third, the erosion of traditional knowledge and practices and the loss of biodiversity have pushed Indigenous communities to mobilize to defend their environments (Congretel and Pinton 2020).

The concept of eco-populism among Indigenous peoples has profound implications for sustainable development and social justice. It challenges mainstream narratives by amplifying the voices and perspectives of Indigenous peoples, advocating alternative models of sustainable development and highlighting the importance of preserving nature-based cultures. However, eco-populism also faces challenges, such as a lack of recognition from the political system, the need to navigate power imbalances, and the risk of co-optation by external actors (Asrawijaya and Hudayana 2021).

8.4 Result and Discussion

8.4.1 Baduy Communities

The Baduy community currently occupies around 5101.85 ha of customary land, of which 3000 ha is "forbidden forest". Around 2136.85 hectares are fields and residential areas of Kanekes Village, Banten with a total population of around 11,669 people or 3413 families (BPS 2018). The Baduy tribe coexists with nature. They are recognized for abiding by their customs and living apart from the growth of the surrounding village. Figure 8.1 shows the Baduy settlement.

Baduy communities are grappling with several interrelated crises threatening their region's social well-being and ecological integrity. The Baduy community has always lived by cultivating their land through farming in line with agro-ecological principles. This means that their way of farming prioritizes the principles of ecological sustainability (subsistence-oriented, low external input, and the use of appropriate technology) and the principles of social benefit (McCune and Rosset 2021). However, for almost two decades now, population growth has had a major impact on the availability of land for farming. This has implications for the fallow period becoming shorter (from an average of seven to three years) and the area of land cultivated by each family becoming smaller (Suryani 2021).

Short fallow periods and climate change reduce soil fertility, causing crop yields to decrease. This shift has led to the Baduy community relying on market supplies for food, making them vulnerable to fluctuating food supply chains. Despite this, no famine has occurred due to food shortages, and the Baduy community is accustomed to consuming tubers and hunting animals. The increasing number of tourist visits to Baduy and the influx of other cultivated plants with high commercial value have given new hope. For some of the Baduy community, this is an opportunity to earn cash by becoming a tour guide, selling souvenirs, honey, or other local specialty products, and cultivating commercial crop commodities. However, mass tourism, which is their main source of income, creates other problems, which we will discuss in the next section.

8.4.2 Kawalu Month Traditions

Kawalu comes from the word *walu* which means balik, kabali or return. The *Kawalu* ceremony, which takes place after the rice harvest, aims to welcome Nyi Pohaci back to the rice barn (*leuit*) after a long time in "her husband's house" at *huma* (Danasasmita and Djatisunda 1986). The *Kawalu* ceremony is usually held three times, taking place in Inner Baduy (Cibeo, Cikeusik, Cikertawana) in the months of Kasa, Karo, and Katilu or between mid-January and April in the Gregorian calendar. The *kawalu kahiji* (the first *kawalu*) ceremony is carried out after the rice harvest in *huma serang* (commonly owned customary fields in the Inner Baduy). The ceremony is held in Cikeusik and Cikartawana on the 17th and in Cibeo on the 18th of Kasa. On those two days, the Baduy community, especially those who have reached puberty, will fast. Second, the *Kawalu* ceremony is carried out after the rice harvest at *huma puun* (the fields of the *puun*/leader). The ceremony is held in Cikeusik and Cikartawana on the 18th, while in Cibeo, it is on the 19th of Karo. Third, the *kawalu tutug* (third *kawalu*) ceremony occurs on the 17th in Cikeusik and Cikartawana and on the 18th in Cibeo.

On each of the specified dates, all members of the Baduy communities (both Inner Baduy and Outer Baduy) who are adults or have reached the age of puberty (15 years) fast. Some people from the Outer Baduy villages come to Inner Baduy to participate in the *Kawalu* ceremony. In practice, they follow each of their *puun*, depending on their ancestors' origins, whether in Cibeo, Cikeusik, or Cikartawarna. Each Inner Baduy and Outer Baduy family brings harvested rice to give to the *puun*, as an offering to the ancestors, and to try rice from the new harvest. The latter is also known as *nganyaran,* meaning renewal.

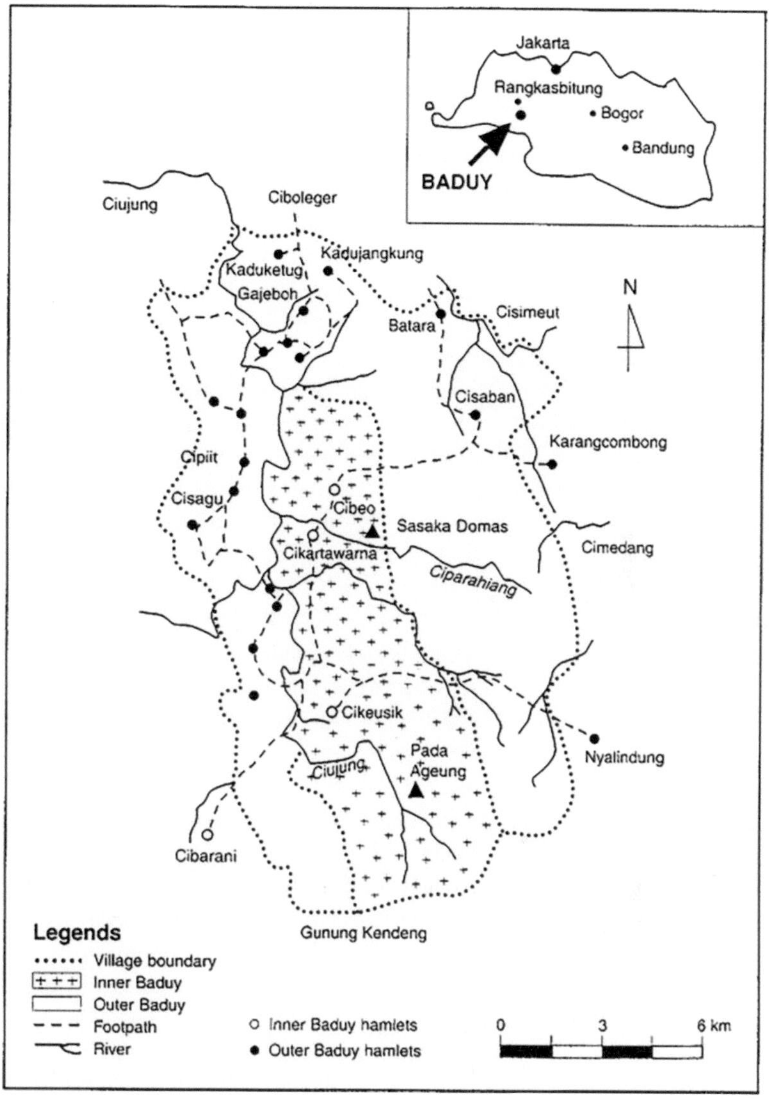

Fig. 8.1 Inner and Outer Baduy region maps. (*Source* Iskandar & Ellen (2000))

At the last *Kawalu* (*kawalu tutug*) ceremony, each Inner Baduy village holds a *Ngalaksa* ceremony on a predetermined date. This is an important and most sacred ceremony in the *Kawalu* tradition series, which is held on the fifth day of *kawalu tutug,* or the 21st of the month of Katiga, for up to seven consecutive days. *Ngalaksa* is the activity or ceremony of making *laksa*, a kind of *lontong* (rice cake) made from rice flour. Starting at *Kapuunan,* the *Ngalaksa* ceremony is carried out in a chain throughout the Outer Baduy village. The sacred element of this ceremony is centered on the laksa flour, made of rice from the seven best rice fields grown in *pupuhunan huma serang* in *kapuunan* and *huma tuladan* (fields intended for Outer Baduy people) in *panamping* (Outer Baduy). In the Baduy community belief, the best substances contained in the earth are collected there. In the *pupuhunan*, there is "earth magic" power that will seep into the rice planted there. Apart from that, the sacredness of the *Ngalaksa* ceremony refers to the laksa-making officers, selected as people known for their good behavior and experience who fast while carrying out their duties (Permana 2010).

Kawalu ceremonies, which last for three months, are a major event in the life of the Baduy community. During the *Kawalu* period, the entire Baduy area in Kanekes Village is declared closed to visitors. They also clean the entire village of items or objects classified as *buyut* (taboo, prohibited) by custom. Within one month, all villages in Baduy must be declared clean of all objects or goods made in factories or another modern production chain (Permana 2010). Closing access for outsiders to Baduy territory and clearing Baduy territory of "modern goods" aims to cleanse Baduy in the month of *Kawalu*, which is considered sacred.

Table 8.1 shows that the *Kawalu* month has a special meaning in the calendar system and in the cultural heritage of the Baduy community. The planting calendar indicates that the Baduy community, apart from having a philosophy of spatial planning through the division of land use or utilization—e.g., *leuweung larangan* (forbidden forests), *leuweung tutupan* (a kind of secondary forest area that can be cultivated into fields), and *leuweung garapan* (forest areas that are cleared become land that is cultivated by the community for farming)—also have a philosophy of understanding time or *tata wayah* (seasonal rules) (Sobirin 2007). Table 8.1 shows the stages of *huma* in the fields or *leuweung garapan.* This confirms that the Baduy community has a high level of civilization because they are familiar with astronomy as a guide in determining the calendar system for farming.

At the time of *Kawalu,* not only is the Baduy area closed to outsiders, the Baduy community themselves (especially those from Inner Baduy) are prohibited from traveling far. At this time, the Kanekes area looks deserted, with residents preferring to stay in their houses. They live the *Kawalu* with great solemnity. The furthest they can travel is to the Outer Baduy area and around the village. This is because there are many activities during the *Kawalu* ceremony, and the Baduy community must participate. At the beginning of the *Kawalu* month, Outer Baduy women will attend Inner Baduy as family representatives. Later, they pound rice together for the *Kawalu* ceremony and conduct other activities, including deer hunting, rice offerings, and prayers. Meanwhile, there are also taboos: the Baduy community must adhere to not eating cassava, bamboo shoots, and chicken's eggs. If the Baduy community does not carry out the *Kawalu* tradition, it will result in *kabendon* or customary sanctions that will bring disaster to people who violate it.

These various taboos are a reflection of the wisdom of the Baduy community and a form of conservation that aligns with knowledge developed among the Baduy community. Put into context with various environmental problems in the contemporary world, it represents Indigenous peoples' answer to ecological issues related to resource limits (Magdoff and Foster 2011). Renewable resources are increasingly being depleted due to massive and continuous exploitation by industrial players, causing scarcity and even extinction for several certain species. Scarcity occurs due to overshooting, a condition where the exploitation of renewable resources exceeds their capacity to renew themselves.

The *Kawalu* month is also an opportunity to purify oneself from wrongdoing and express gratitude from the Baduy community to God Almighty by fasting. Fasting takes place from the time of waking until 6 p.m. The *Kawalu* ceremony also involves praying for the blessings of life, universal safety, and increased gratitude. The tradition is mostly carried out by the Indigenous Inner Baduy community in Cibeo, Cikawartana, and Cikeusik Villages. When breaking the fast, a special ritual is the eating of betel leaves.

8.4.3 Ngawalu and the Urgency of Mitigating the Rural Crisis

The *Kawalu* tradition is deeply rooted in physical-spiritual behavior aimed at expressing gratitude for agricultural success. It involves rituals such as planting ceremonies and offerings made by farmers during the harvest season. Through this practice, members of the Baduy community seek harmony between human activities and nature. This activity has been practiced for generations. It originates from ancient beliefs and practices that emphasize harmony with nature. The *Kawalu* tradition combines efforts to protect nature with agricultural practices. People realize the importance of protecting their natural environment to

Table 8.1 *Kawalu* in the Baduy Community Ceremonial and Cultivation Calendar

No	Month	Economic activity (Farming stages)	Traditional activities
1	*Kapat*	*Narawas huma serang* (Cutting the bushes in the future fields)	*Seba* and *Ngalaksa*
2	*Kalima*	*Nyacar in huma serang* (Clearing the fields)	*Muja* on the 16th, 17th/, & 19th. *Geseran* dan *sunatan* (Venerating the Sasaka in Ageung, marriage ceremonies)
3	*Kanem*	*Nukuh* in *huma serang* (Cutting the branches of trees, burning the bushes that were already cut)	*Hajatan* dan *selamatan* (marriage ceremonies)
4	*Katujuh*	*Ngaduruk* and *ngaseuk* in *huma serang*; Scattered in *huma puun* (Burning the bushes that were already cut, planting rice in the fields of *Serang*)	Playing the music of *Angklung*
5	*Kadalapan*	*Ngored* (clearing weeds) in *huma serang*; *Nukuh*, *ngaduruk*, and *ngaseuk* (cutting, burning, and planting) in *huma puun*; *Nukuh* and *ngaduruk* in huma *tangtu* (fields located in Inner Baduy); Planting rice in the *puun* 's fields	–
6	*Kasalapan*	*Ngored* (clearing weeds) in *huma serang* and *huma puun*; *Ngaseuk* at huma tangtu (Planting rice in the Outer Baduy fields)	–
7	*Kasapuluh*	*Ngored* and *meuting* (Clearing weeds and to stay all night)	–
8	*Hapit Kayu*	*Ngored* and *meuting* (Clearing weeds and to stay all night)	–
9	*Hapit Weak*	*Ngored* and *meuting* (Clearing weeds and to stay all night)	–
10	*Kasa*	Harvest in *huma serang*	*Kawalu tembeuy (Kawalu mitembey)*; Fasting on the 18th in Cikeusik and Cikertawana, the 19th in Cibeo
11	*Karo*	Harvest in *huma puun*	*Kawalu tengah* (second *kawalu*); Fasting on the 18th in Cikeusik and Cikertawana, the 19th in Cibeo
12	*Katiga*	Harvest at *huma tangtu*	*Kawalu tutug* and *Ngalaksa*; Fasting on the 17th in Cikeusik and Cikertawana, the 18th in Cibeo

Source (Iskandar 2012; Permana 2006; Suryani 2021)

maintain their way of life. As part of their conservation efforts, they have taken special measures such as limiting tourist visits, preventing overexploitation, and promoting sustainable farming techniques.

Practicing *Kawalu* traditions has had a positive impact in mitigating certain aspects of the socio-ecological crisis in Kanekes Village. Community conservation efforts have resulted in the preservation of local ecosystems, ensuring the availability of vital resources for future generations. By limiting tourist visits, they have been able to reduce the impact of ecological damage caused by mass tourism and protect their cultural heritage. Implementing and maintaining *Kawalu* traditions to overcome the socio-ecological crisis is not without challenges. Several aspects of the Baduy community's lives are starting to be integrated with the market economic system. For example, apart from rice, they also grow other commodity crops that the market needs, such as aromatic ginger, at high prices. This determination of a market economic system can conflict with traditional practices, as development pressures often prioritize short-term profits over long-term sustainability. Additionally, limited resources and external influences can hinder the effectiveness of conservation efforts.

Integrating traditional practices such as *Kawalu* into contemporary discourse on socio-ecological issues is important for several reasons. First, as a form of local wisdom, *Kawalu* contains valuable insights about sustainable living and resource management that can underpin mainstream approaches to environmental conservation. Second, paying attention to traditional practices means recognizing the wisdom embedded in local culture and fostering a sense of community ownership regarding the socio-ecological crisis.

The *Kawalu* tradition plays a crucial role in addressing the rural socio-ecological crisis faced by the Baduy community. It promotes an appreciation for nature and encourages responsible actions towards natural resources. The Baduy community ensures ecological balance and resource conservation by following traditional techniques and respecting natural cycles. However, traditional practices alone cannot address complex socio-ecological problems. The Baduy community has a long track record of rejecting the implementation of the green revolution in their region. Farming is their main activity, and they view their area as sacred and forbidden, with customary rules prohibiting chemical fertilizers, pesticides, and actions like digging or plowing the land. Their rejection of the green revolution is based on farming rationality and environmental ethics that have been applied for centuries (Iskandar 2012).

The various taboos in farming adhered to by the Baduy community are similar to the spirit and practices that have developed in the modern agricultural system. At the end of the 1980s, the green revolution prompted Western agricultural experts to abandon the agricultural development paradigm of the HEIA (high external input agriculture) system (Reijntjes et al. 1992). Instead, they implemented the LEISA (low external input and sustainable agriculture) system. The LEISA system uses local resources with no external inputs, such as organic farming, to achieve sustainable agriculture, or ecological and social benefits that align with agroecological principles (McCune and Rosset 2021). While it appears to be an innovation in farming, the LEISA system is nothing new, because almost all rice farmers in Indonesia practiced it before the green revolution. In this case, the Baduy community has been implementing the system from the past until the present.

At this point, it is important to know that if the Baduy community modernizes their way of farming and managing land according to the recommendations of the green revolution, namely intensification, mechanization, and technological intervention, various traditions or customs that have long been their identity will disappear. The reason is that various existing traditional practices, such as the *Kawalu* ceremony, follow the calendar, cycles, and stages of the rice planting process for one year or 12 months (see Table 8.1).

8.4.4 Discouraging Mass Tourism for Environmental Sustainability

The geographical location of the Baduy community villages is relatively far from the center of the capital city of Jakarta. This makes it easier for outsiders, especially from big cities like Jakarta or the surrounding Bodetabek (Bogor, Tangerang, Bekasi) suburb area, to visit Baduy. It is not surprising that every week, Baduy villages are visited by thousands of tourists. Baduy is far removed from modern life and rejects modernization in almost all aspects of life, especially in farming, lighting (electricity), and road infrastructure. Even though some Kanekes people have accommodated cultivated plants or commodities such as aromatic ginger, they traditionally plant and manage their fields with low or no external input. All villages in the Baduy region, both in Outer Baduy and especially Inner Baduy, rely on simple lighting without electricity or other electronic devices (TV, refrigerator, radios, etc.). Likewise, the alley road infrastructure that connects villages within the Baduy area and surrounding villages outside the area still consists of hardened rock/gravel roads that are difficult for vehicles to negotiate.

However, the Baduy community is not an exclusive community that completely closes itself off from interaction with outside society. The Baduy community themselves, like Sundanese people in other parts of the world in general, know the concept of *saba* (or *nyaba*), which means going, traveling, and visiting (Erwinantu 2012; Rigg 2009) to get to know other cultures. So, from the perspective of the Baduy community, they perceive the arrival of guests from outside Baduy in the context of the *saba* – as they also travel to places outside Baduy on foot to learn about the culture of people in other places (*silaturahmi budaya*, cultural friendship) – rather than taking a vacation or traveling in the currently popular sense. On the other hand, the condition of the natural environment, which is still beautiful and well maintained, far from noise, and protected from air pollution, actually attracts outsiders to visit Baduy. Even though it helps encourage the economic development of the surrounding community, for the Baduy community, this trend creates risks and challenges. The heavy flow of tourists to Baduy, which continues to increase yearly, is causing serious problems. Annually, the Baduy region is visited by tens of thousands of tourists. The peak was in 2019 when visitors reached 42,228, almost double the previous year. Even in 2020, when the pandemic began to spread throughout the country, the number of visits to Kanekes Village, 20,327, was still relatively high, exceeding the number of visits in 2017, which was still normal. In the year following the pandemic, the number of visits decreased again due to the effects of mobility restriction policies. In other words, only a pandemic could control people's interest in visiting Kanekes Village, with mass tourism subsiding momentarily.

As shown by many studies, mass tourism raises a dilemma. Without ignoring its contribution to economic growth, establishing this area as one of the tourist destinations in Banten Province has given rise to new problems in its development. The phenomenon of mass tourism causes environmental damage and other socio-cultural problems. Environmental problems can be seen in the increasing volume of waste. Meanwhile, socio-cultural problems occur due, among other things, to increasing urbanization around tourist locations and the growth of uncontrolled informal activities. In other words, economic growth driven by mass tourism, especially through increasing income and expanding employment opportunities, only gives rise to false prosperity. More than that, mass tourism worsens the socio-ecological crisis in rural areas.

Table 8.2 shows that mass tourism has occurred in the last few years. According to tourism reviewers, the negative impacts of mass tourism can be anticipated by encouraging sustainable tourism, well-behaved tourism, and targeting tolerable tourism planning growth rates, as well as prioritizing a culture of conservation. Some argue that ecotourism can be an alternative solution to disturbing mass tourism. However, as more critical studies also show, ecotourism is not always effective in overcoming the negative impacts of mass tourism.

Table 8.2 Number of tourists visiting Kanekes Village, Baduy

No	Year	Number of visitors
1	2017	19,374
2	2018	21,390
3	2019	42,228
4	2020	20,327*
5	2021	6247*
6	2022	770**

Source Katadata.co.id
*Pandemic year
**January and February

To a certain extent, ecotourism, which means tourism in protected areas that does not damage the environment and provides (material) benefits for local communities, is considered too utopian. In reality, the costs and benefits of economic and social change are rarely shared equally, and improvements in the welfare of communities living near protected areas have yet to be achieved. Brennan and Allen's (2001) study of the Zulu community in South Africa, where ecotourism has been implemented, shows that opportunities for local communities to participate in decision-making procedures in ecotourism initiatives are uneven and limited. One indicator is that there is little evidence that the lives of Zulu women have improved thanks to the state's intention to build an institutional framework that supports ecotourism there.

However, the Baduy community do not need to look abroad to overcome the dilemma of mass tourism. Moreover, various concepts, schemes, or alternatives offered from outside are not always suitable and effective in anticipating problems in the local area. On the other hand, the Baduy community has its own way of stemming the negative impact of tourism, which is rooted in existing traditions. In this context, the *Kawalu* period can be seen as an important effort to counteract the impacts of mass tourism. During *Kawalu*, which lasts for three months (Kasa, Karo, Katiga – fourth week of January to third week of April), the entire Baduy region closes access to outside visitors. During the *Kawalu* period, local and foreign tourists, regional officials, and state officials are prohibited from entering the Inner Baduy area so that the Baduy community can carry out the *Kawalu* tradition smoothly and without being disturbed.

Limiting tourist visits is important for mitigating the Baduy community's socio-ecological crisis. Mass tourism can disrupt the balance between human activities and nature, leading to deforestation, pollution, and loss of cultural identity (Amico et al. 2019). By implementing regulations that limit access to certain areas of their villages, Baduy communities are protecting their environment from degradation caused by excessive tourism (Khozhaniyazova 2022). This approach allows them to maintain sustainable practices while preserving their cultural heritage. Apart from the spatial aspects, they are still loyal to Baduy's *pikukuh* (guidelines) (Permana 2006). The practice of "self-discipline" through the *Kawalu* tradition is also part of Baduy-style mitigation and conservation practices in dealing with the impact of the socio-ecological crisis in their environment. However, it is important to understand both the benefits and challenges associated with limiting tourist visits. On the one hand, this strategy helps protect the environment and maintain socio-ecological resilience. On the other hand, it may result in economic hardship for residents who rely on tourism as a source of income (Barkin 2022). Balancing these factors is critical for long-term sustainability.

The *Kawalu* tradition, as a form of local wisdom, plays a crucial role in reducing the socio-ecological crisis of the environment and local natural resources. This traditional practice, deeply rooted in the community's culture and beliefs, promotes sustainable living and conservation. The tradition emphasizes the importance of harmony between humans and nature and instills a sense of responsibility towards nature among community members. It recognizes that human activities have a direct impact on the environment and local resources. The connection with nature fosters an understanding of the need to protect and preserve it for future generations. Furthermore, the *Kawalu* tradition promotes sustainable resource management practices. The community has developed knowledge and techniques over generations that ensure the responsible use of natural resources. Therefore, this tradition contributes to reducing environmental degradation and ensuring long-term sustainability. We must recognize and support such traditions as they hold immense potential for mitigating the environmental challenges we face today.

8.5 Conclusions

The *Kawalu* tradition has an important meaning for the Baduy community in their efforts to mitigate their social and ecological crises. This tradition protects nature and maintains a harmonious relationship with the environment. The community's additional efforts to preserve its traditions amidst modern influences further strengthen its socio-cultural and ecological resilience. Limiting tourist visits has proven effective in maintaining sustainable practices while preserving cultural heritage.

The Baduy community's adherence to traditional practices such as *Kawalu* shows their commitment to mitigating socio-ecological crises. This ancient ritual is a powerful reminder of humanity's connection to nature and emphasizes gratitude for the natural resources that sustain life. Through these traditions, we can learn important lessons about how tradition-based approaches can effectively address various aspects of the socio-ecological crises faced by Indigenous peoples. *Kawalu* addresses environmental challenges and contributes to the preservation of cultural practices, ensuring the continuity of traditional knowledge. We must learn from these examples and support such communities in their efforts to create a more sustainable future for all.

References

Aguilar-st M, Bull B, Sveinsd AG (2021) Resistance, repression and elite dynamics: unpacking violence in the Guatemalan mining sector. Geoforum 118:117–129. https://doi.org/10.1016/j.geoforum.2020.12.011

Amico LA, Ituarte-Lima C, Elmqvist T (2019) Learning from the social-ecological crisis for legal resilience building: multi-scale dynamics in the coffee rust epidemic. Sustain Sci 15(2):485–501. https://doi.org/10.1007/s11625-019-00703-x

Arce BC (2017) From Zapata to Zapatismo: Mesoamerican epistemology in the Zapatista imaginary. Hisp J 38(2):125–144

Asrawijaya E (2020) Gerakan Ekopopulisme komunitas Samin melawan perusahaan semen di pegunungan Kendeng. Jurnal Sosiologi Pendidikan Humanis 5(1):35–47

Asrawijaya E, Hudayana B (2021) The power of a leader in the Samin people's opposition movement to the development of a cement factory in the North Kendeng Mountains. Jurnal Humaniora 33(1):26. https://doi.org/10.22146/jh.56224

Barkin D (2022) Shaping a communitarian ethos in an era of ecological crisis. Front Sustain. https://doi.org/10.3389/frsus.2022.944252

BPS (2013) Laporan Hasil Sensus Pertanian Tahun 2013

BPS (2018) Leuwidamar dalam Angka 2018

Brennan F, Allen G (2001) Community-based ecotourism, social exclusion and the changing political economy of KwaZulu-Natal, South Africa. In Harrison D (ed) Tourism and the less developed world: issues and case studies. CABI Publishing

Congretel M, Pinton F (2020) Local knowledge know-how and knowledge mobilized in a globalized world: a new approach of indigenous local ecological knowledge. People Nature 2:527–543. https://doi.org/10.1002/pan3.10142

Danasasmita S, Djatisunda A (1986) Kehidupan Masyarakat Kanekes

Erwinantu (2012) Saba Baduy: Sebuah Perjalanan Wisata Budaya Inspiratif. Gramedia

Huberman M, Miles MB (2002) The qualitative researcher's companion. Sage

Iskandar, J. (2012). *Ekologi Perladangan Orang* Baduy*:* *Pengelolaan Hutan Berbasis Adat*. Alumni.

Iskandar J, Ellen RF (2000) The contribution of Paraserianthes (Albizia) falcataria to Sustainable Swidden Management Practices among the Baduy of West Java. Hum Ecol 28(1):1–17

Iskandar J, Iskandar B (2011) Agroekosistem Orang Sunda. Kiblat

Khozhaniyazova GM (2022) The tradition of mutual assistance kömek in the agrarian culture of the Karakalpaks in an environmental crisis. Histor Ethnol 7(3):388–397. https://doi.org/10.22378/he.2022-7-3.388-397

Magdoff F, Foster JB (2011) What every environmentalist needs to know about capitalism: A citizen's guide to capitalism and environment. Monthly Review & NYU Press

Maldonado JK, Shearer C, Bronen R, Peterson K, Lazrus H (2013) The impact of climate change on tribal communities in the US: displacement, relocation, and human rights. Clim Change 120:601–614. https://doi.org/10.1007/978-3-319-05266-3

McCune N, Rosset P (2021) Agroecology. In: Handbook of Critical Agrarian Studies. Edward Elgar Publishing

Middeldorp N, Billon PL (2019) Deadly environmental governance: authoritarianism, eco-populism, and the repression of environmental and land defenders. Ann Am Assoc Geogr 109(2):324–337

Mulyani L (2020) Reforming group legal personhood in Indonesian land law: towards equitable land rights for traditional customary communities. The University of Melbourne

Permana RCE (2006) Tata Ruang Masyarakat Baduy. Wedatama Widya Sastra

Permana RCE (2010) Kearifan Lokal Masyarakat Baduy dalam Mitigasi Bencana. Wedatama Widya Sastra

Reijntjes C, Haverkort B, Waters-Bayer A (1992) Farming for the future. An introduction to low-external-input and sustainable agriculture. Macmillan

Rigg J (2009) A dictionary of the Sunda language of Java. Pustaka Jaya.

Robles W (2001) The landless rural workers movement (MST) in Brazil. J Peasant Stud 28(2):146–161

Shiva V (1991) Biodiversity: a third world perspective. Third World Network

Shohibuddin M (2016) Peluang dan Tantangan Undang-undang Desa dalam Upaya Demokratisasi Tata Kelola Sumber Daya Alam Desa: Perspektif Agraria Kritis. MASYARAKAT: Jurnal Sosiologi 21(80):1–33

Sobirin (2007) Tragedi Kawasan Lindung dan Hilangnya Hak Asasi Alam. In: Seri Sundalana 6 Menyelamatkan Alam Sunda dan Kajian lainnya mengenai Kebudayaan Sunda. Pusat Studi Sunda

Suryani A (2021) From respected hermits to ordinary citizens: The conversion of the Baduy, ethnicity, and politics of religion in Indonesia (1977–2019) [Leiden University]. https://hdl.handle.net/1887/3134744

Taylor M (2021) Ecological crisis in the rural world. In Handbook of Critical Agrarian Studies. Edward Elgar Publishing

United Nations (2021) Challenges and opportunities for indigenous peoples' sustainability. Department of Economic and Social Affairs Social Inclusion. https://www.un.org/development/desa/dspd/2021/04/indigenous-peoples-sustainability/

Walhi (2022) Membangkang Konstitusi, Mewariskan Krisis Antar Generasi. Walhi

The Socio-Ecological Crisis of Natural Resources Due to the Food Estate Program in Central Kalimantan

9

Henny Warsilah

Abstract

Indonesia is currently being hit by a food crisis. Through the State Revenue and Expenditure Budget, which was realized from the 2022 defense budget, the Ministry of Agriculture has allocated funds amounting to IDR92.2 trillion for national food security. Since the COVID-19 epidemic, the average consumption of rice per capita per week in Indonesia has increased, reaching 1404 kg since 2018. In response to this problem, the government has taken the initiative to create a Food Estate Program. However, this initiative has not produced the expected results. On the contrary, it is causing a social and ecological disaster that is detrimental to natural resources and residents in the Gunung Mas forest, Central Kalimantan. A growing problem is that the Food Estate Program has implications for weak sustainable development. Moreover, although cassava is not a staple food for residents, the community's customary forests have already been cut down to make way for food estate areas for cassava farming. The traditional wisdom of planting vegetables in intercropping forests has been disrupted, so that the Dayak community is becoming food insecure and may become poor. This chapter analyzes the socio-ecological crisis caused by the Food Estate Program and shows how it links to the food crisis in the context of sustainable development. This research uses qualitative methodology with analytical descriptive analysis.

Keywords

Socio-ecological crisis · Food estate · Dayak food habits · Local wisdom · Dayak Central Kalimantan

9.1 I. Introduction: Indonesia's Food Problem

The primary challenge in achieving food security in Indonesia is the fact that the increase in food demand is outpacing the growth and supply of food. This surge in demand can be attributed to population growth, economic expansion, rising purchasing power, and shifts in dietary preferences. Additionally, rice-focused food policies tend to be implemented broadly, disregarding the existing diversity of staple foods.

Despite the heightened demand, however, national food production seems to be stagnating. Growth in the national food production capacity has been sluggish and stagnant, primarily due to:

H. Warsilah (✉)
Research Centre for Society and Culture - National Research and Innovation Agency, Jakarta, Indonesia
e-mail: hennywarsilah@gmail.com

H. Warsilah and A. H. Purnomo (eds.), *Approaching Natural Resources for Sustainable Development in Indonesia*, Sustainable Development Goals Series, https://doi.org/10.1007/978-981-96-6913-4_9

(a) competition for land and water resources, (b) a lack of growth in land productivity and agricultural labor, (c) the looming threats of global warming, global farming, and natural disasters. This imbalance between demand growth and national production capacity growth has led to a reliance on imported rice, making the nation's food supply increasingly dependent on external sources (Warsilah 2013: 98–99).

Indonesia's rice dependence is not unique; globally, rice is a key food source for approximately 3 billion people, with around 600 million people in Asia relying on it as a staple. More than 60% of the world's population, roughly one billion people residing in Asia, experience poverty and malnutrition while relying on rice (Warsilah 2013: 99). Since 2004, the Indonesian government, led by President Soesilo Bambang Yudhoyono, has been implementing a policy to diversify staple foods. This initiative has led to the establishment of the Food Security Agency and the Regional Food Security Council at various administrative levels.

The 2022 UN *Food Security Report* highlights climate change as a significant driver of global food insecurity, with extreme weather events cited as the primary catalyst for global food insecurity (FAO 2022). Climate change-induced droughts, floods, and heatwaves are becoming more frequent and severe, significantly impacting food production systems. Recent research reveals that human-induced climate change has curtailed global agricultural productivity growth by approximately 21% since 1961. The adverse effects are particularly pronounced in warmer regions, including Indonesia, where productivity reductions are estimated at 30–33% (Ortiz-Bobea et al. 2021).

The UN Food Agency reports that global food prices surged by over 14% throughout 2022 compared to the previous year. December 2022 marked the ninth consecutive month of price increases, with the international price index for the most traded food commodities averaging 132.4 points, down from a revised 135 points in November. For the whole of 2022, the benchmark index averaged 143.7 points, reflecting an 18-point increase (14.3%) from 2021, the most significant annual rise in food prices since records commenced in 1990 (Economy ecozone.com 2023).

In 2020, the World Food and FAO issued a warning to global leaders regarding the potential impact of the COVID-19 pandemic on food and income supply chains. Responding to this, President Joko Widodo initiated the Food Barn program on a national scale, aiming to repurpose millions of hectares of forests and peat areas for national food storage through rice fields and cassava cultivation, primarily in customary forests.

With Indonesia ranking as the world's third-most populous country, housing 277.43 million people, approximately 42% of whom reside in rural areas (BPS 2023), ensuring food security becomes imperative. To address these concerns, the Ministry of Agriculture has allocated IDR92.2 trillion for national food security within the 2022 State Revenue and Expenditure Budget (Databoks 2022). Food sector subsidies, including rice subsidies for impoverished individuals, rose from IDR6.4 trillion in 2005 to IDR13.0 trillion in 2009. As both population and agricultural demands continue to escalate, the requirement for rice continues to surge, along with the need for additional land (see Table 9.1).

Table 9.1 Prediction of Indonesian Rice and Agricultural Land Needs in 2030

Year	Total population (Million)	Rice needs (Million Tons)	Land requirements (Million Hectares)
2007	220	30.58	11.6
2011	237. 641	33.32	13.566
2022	277.43	54.5	13.17
2030	425	59	23.4
	Lack of rice	26.66	9.9

Source James (2023); Directorate General of Land and Water Management

The Indonesian state has often faced food crises: in the 1966 crisis the price of rice surged by 600 times; the 1998 crisis was precipitated by a financial downturn that triggered a food crisis;

Table 9.2 Average rice consumption in Indonesia, 2018–2021

No	Year	Value/Kg/Capita/Week
1	2018	1.404
2	2019	1.374
3	2020	1.379
4	2021	1.451

Source BPS (2021)

the 2003 crisis prompted the liberalization of the Logistics Agency's role; and the 2007–2008 crisis was marked by a sharp increase in rice prices. Beyond the challenge of food crises, the demand for national food supplies continues to rise due to growth in food consumption and population expansion. Efforts to address this demand through agricultural sector development have been undertaken; however, these endeavors have yet to significantly impact community welfare or alleviate the persisting issues of poverty and food insecurity in rural areas.

According to a report from the Central Statistics Agency (BPS 2021), the average rice consumption among the Indonesian population has increased since the onset of the pandemic. In 2018, the average per-capita weekly rice consumption stood at 1.404 kg. This figure slightly decreased to 1.374 kg per capita per week in 2019. Subsequently, rice consumption continued to climb into the second year of the pandemic, reaching 1.451 kg per capita per week in 2021 (see Table 9.2).

Table 9.3 shows Indonesian rice production for the last five years. Table 9.4 shows both production and consumption from 2018 to 2022. The production level is stable and does not increase, but the consumption level continues to increase significantly. This shows that the need for staple food rice continues to rise. The government plans to import 2 million tons of rice by the end of 2023 to maintain government rice reserves and maintain price stability.

Dependence on rice as a primary food source stems from the rice self-sufficiency policies of the 1970s in Indonesia, which gradually shaped the dietary habits of rural Indonesians from non-rice foods to rice-based meals. In response to these circumstances, the government is now actively promoting food diversification to alleviate the reliance on carbohydrate-heavy diets by introducing more cost-effective and affordable food options. Implicitly, the Indonesian government's policy seeks to focus on fortifying food security through self-reliance, diversification of domestic production, fostering a conducive agricultural business environment, and ensuring sustainable agricultural development. These policies collectively contribute to establishing and enhancing food security within the nation, a crucial and strategic effort aimed at nourishing the population and bolstering food sustainability through the development of a production system founded on local resources, food institutions, and cultural practices.

The government initiated the Food Estate Program, an integrated initiative encompassing agriculture, plantations, and even animal husbandry, and aimed at cultivating food production within a designated area. This program is a core component of the 2020–2024 National Strategic Programs. Regrettably, the program's outcomes fell short of expectations. Particularly in Central Kalimantan, the program led to an adverse socio-ecological crisis of local natural resources. Indigenous customary forests were felled and repurposed as food estate land for cassava cultivation. Moreover, the failure of the cassava crops exacerbated social challenges. This outcome arose, firstly, from misalignment with the food habits of the Dayak people of Central Kalimantan, whose staple food is not cassava. Second, the disruption of the local wisdom associated with intercropping, a traditional method involving simultaneous cultivation of multiple crops in the same forest field, compounded the issue. Third, the Dayak community's well-being suffered, with the program leaving them food insecure and economically disadvantaged. Ecologically, the transformation of customary forests into food estate land has contributed to the environmental crisis and disrupted the government's sustainable development programs. This chapter examines in detail the socio-ecological crisis resulting from the Food Estate Program in Central Kalimantan.

Table 9.3 Indonesian rice production in tons, 2014–2019

Island	2014[1]	2015[1]	2016[2]	2017[2]	2018[3]	2019[3]
Sumatera	16,628,893	18,358,385	20,217,575	21,564,080	11,920,816	10,931,056
Jawa	36,663,049	38,970,026	40,893,658	40,055,249	32,557,778	30,328,885
Bali & Nusa Tenggara	3,800,309	4,219,190	3,865,079	4,250,619	3,027,344	2,793,227
Kalimantan	4,847,679	4,830,069	4,839,755	5,098,145	2,949,814	2,921,474
Sulawesi	8,508,020	8,615,127	9,095,960	9,704,344	8,331,420	7,227,908
Maluku Arch	174,835	193,056	181,301	188,753	165,276	136,200
Papua	223,680	211,988	261,439	287,404	248,087	265,283
Indonesia	70,846,465	75,397,841	79,354,767	81,148,594	59,200,534	54,604,033

Source Ministry of Agriculture. Production of Paddy in Indonesia, 2014–2019 (Ton)

Table 9.4 Indonesian rice production and consumption, 2018–2022

Year	Rice Production (million tons)	Rice Consumption (million tons)
2018	33.94	29.57
2019	31.31	28.93
2020	31.50	29.37
2021	31.36	30.04
2022	31.54	30.20

Source CNN Indonesia, 2022a, b. Infographic

9.2 Methodology

This chapter analyzes the extensive practice of procuring land for large-scale food production within the Food Estate Program in Central Kalimantan, focusing specifically on the Gunung Mas Forest. The chosen methodology involves descriptive qualitative analysis employing a content analysis approach. The study of land acquisition practices and the subsequent marginalization of local communities is conducted using two principal paradigms: the ethnoecological paradigm and the ethno-development paradigm. Descriptive analysis, as outlined by Clarke (2001), forms the basis for assessing both these paradigms. The decision to employ these paradigms as analytical frameworks is rooted in their shared perspective on the intricate relationship between humans and nature.

9.3 Sustainable Development Approach for Food Estate Program

Indonesia is a country rich in various types of natural resources, and natural resources and human resources are the most important components of national development. As the basic capital for national development, natural resources must be utilized as effectively as possible to benefit the entire population of Indonesia. The potential of these natural resources is also used as a driver of development. However, it seems that the natural resources currently available in Indonesia cannot be utilized optimally for the welfare of the Indonesian people. Indonesia still exports a lot of natural resources in raw form so prices tend to be lower.

To maintain the availability of natural resource raw materials, the use of natural resources should be supported by the principles of sustainable development. Sustainable development is development carried out to meet the needs of the present without reducing the ability of future generations to meet their needs. A development will not occur if it is not driven by the availability of sufficient development capital. Indonesia is very fortunate to have abundant natural resources. Apart from abundant natural resources, human resources also play an important role in the implementation of national development.

The exploitation of natural resource wealth in Indonesia provides enormous benefits for the country's economy. However, excessive exploitation has caused environmental and social damage, especially in areas rich in natural resources, for example, Kalimantan. Companies often do not care about the conditions of local communities and the environment. This happens because the granting of rights to manage natural resources is not based on the principles of sustainable development or on benefits for society. Unilateral control of natural resources without the involvement of local communities in various regions has created conflict, both open conflict and latent conflict that can explode at any time.

The government has established a vision for sustainable natural resource management in national policy, namely "the realization of natural resources that are managed in a fair, comprehensive, integrated and environmentally sound manner for the welfare of society". Natural resource management is intended so that these natural resources can provide benefits in the long term. Natural resource governance is carried out by utilizing natural resources wisely and sustainably so that their availability can be exploited by current and future generations. Natural resource management includes all stages of management: planning, implementation, and monitoring and evaluation. Natural resource management must also have an environmental perspective, which means paying attention to the balance of the ecosystem and the carrying capacity of the environment. Sustainable means aimed not only at the interests of the current generation but also at those of future generations. Comprehensive means covering all areas of management including conservation and utilization of natural resources (Warsilah 2020).

The implementation of the agricultural development paradigm through the Food Estate Program signifies a shift in the agricultural paradigm, emphasizing holistic agricultural development as its principal focus. The agricultural paradigm's objective is to ensure independent and sustainable food security, both within households and across the nation. Agricultural development aims to elevate the strategic role of agriculture in the national economy, encompassing its contributions to food provisioning, industrial raw materials, feed supply, bioenergy, capital formation, employment generation, foreign exchange resources, income sources, and environmental preservation.

The 1996 World Food Summit defined food security as the condition where all individuals (at household, national, and global levels) possess physical and economic access to sufficient, safe, and nutritious food that meets their dietary needs and preferences. The fundamental definition of food security was incorporated into the Indonesian Food Law in 1996 through Law No. 7 of 1998 Concerning Food, and a more comprehensive definition was introduced in the 2012 version (Law No. 18 of 2012 Concerning Food). In essence, the ability to afford food is crucial, extending beyond mere availability for purchase; food must contain the necessary nutritional elements for a healthy and active life. Additionally, the concept of 'food choice' encompasses religious, ethical, and cultural suitability, with this aspect particularly significant for minority groups in Indonesia, including Indigenous communities.

The Food Estate Program, a long-term initiative by the Indonesian government, serves the purpose of maintaining domestic food security. It embraces the concept of comprehensive food development, spanning agriculture, plantations, and even animal husbandry within specific regions. This collaborative program involves the Ministry of Agriculture and local governments across various districts in Indonesia. The notion of a food estate emerged from the perspective or paradigm of developing expansive agricultural land, initially introduced during the administration of President Susilo Bambang Yudhoyono (SBY). The developmentalism paradigm in agriculture, originally proposed during World War II, continued during the Suharto era under President SBY, and remains foundational to the creation of food estates in Indonesia, persisting through the administrations of subsequent presidents, including President Joko Widodo.

While the negative impacts of a food estate include environmental degradation, the most prominent challenge is the budgetary one, due to the substantial financial requirements of such endeavors. On a positive note, a food estate can contribute to stabilizing food security and enhancing human resource quality. One solution involves the development and responsible utilization of dormant natural resources to prevent environmental harm (Sabar and Dewangga 2021). President Joko Widodo initiated the Food Estate Program in multiple regions, including Central Kalimantan. After two years, however, the program is regarded as unsuccessful in this region. A 600-hectare cassava plantation yielded no harvest, and 17,000 hectares of newly established paddy fields remained unharvested. Apart from cassava, the government has also established food estates or rice storage areas in peatlands, specifically in the Kapuas and Pulang Pisau Regencies.

The food storage approach mandated by the central government replaces forests and intricate food landscapes with monoculture crops. This transition diminishes the potential for sustaining the natural food systems of Indigenous peoples and obstructs autonomous local production by small-scale farming communities or Indigenous groups. This shift, coupled with the emphasis on starchy monoculture crop commodities in food storage, poses constraints on households' access to diverse and nutritious food sources.

The NGO Greenpeace Indonesia has vehemently criticized the launch of the Food Estate Program, alleging that it has triggered a socio-ecological crisis of natural resources, notably within forest areas and peatlands. Greenpeace's recent report, titled "Food Estate: Planting Destruction, Reaping Climate Crisis", underscores how one of the government's national strategic projects has exploited vast forests and peatlands, imperiling Indonesia's crucial Indigenous territories and biodiversity. The overall area designated for the food estate project is estimated to endanger around 3 million hectares of forests if the project is continued. The program also serves as a platform for oligarchies and capital interests to legitimize the covert maintenance of state control by the ruling elite.

9.4 Discussion on the Food Estate Program in Central Kalimantan

9.4.1 Population and Ethnic Groups in Central Kalimantan

Based on the population census data for 2010, this province has a population of 2,207,367 people, comprising 1,147,878 males and 1,054,721 females. Data from Central Kalimantan's BPS for 2023 indicate that the province's population was projected to increase to 2,741,100 in 2023. Central Kalimantan encompasses 13 districts and one city. According to the 2010 Indonesian Population Census data, three major ethnic groups in Central Kalimantan are the Javanese, accounting for 21.68%, the Banjar at 21.03%, and the Dayak at 20.42%. Other ethnic groups from Kalimantan, excluding the Dayak, constitute approximately 26.67%. The Dayak tribe predominantly inhabits inland areas, the Banjar tribe is primarily concentrated in urban regions, while the Javanese population is found in transmigration zones. Other transmigrant ethnic groups in Central Kalimantan include the Madurese, Sundanese, Balinese, and various ethnic groups from East Nusa Tenggara. There are also Malay tribes residing along the coasts of Sukamara and West Kotawaringin, Batak tribes, and other Indigenous groups from diverse regions across Indonesia.

As stated by Lontaan and Sanusi (1976: 2–3 in the Color of Indonesia 2015), there are ten Indigenous tribes inhabiting the West Kotawaringin region, based on the Ngaju Dayak Tribe. These are the Mendawai Tribe, Ruku Mapaan Tribe, Darat Tribe, Lamandau Tribe, Bulik Tribe, Mentobi Tribe, Belantikan Tribe, Batang Kana/Kawak Tribe, Delang Ulu and Ilir Tribe, and Banjar Tribe. The Banjar Tribe's presence (also the tribe of the Sultan) in the Kotawaringin area, which forms a part of Central Kalimantan, is linked to the establishment of the Kotawaringin Duchy approximately five centuries ago (Table 9.5).

Table 9.5 Ethnic composition in Central Kalimantan based on the 2000 Census

No	Tribes	Amount	Concentration (%)
1	Dayak	742.729	41.24
2	Banjar	435.758	24.20
3	Jawa	325.160	18.06
4	Madura	62.228	3.46
5	Sunda	24.479	1.36
6	Other Tribes (Melayu, Tionghoa, Arab, Batak, Bugis, Bali, Flores (NTT))	210.359	11.69
	Total	1800.713	100.00

Source Population Census, 2000

In the 2000 census, the ethnic composition in Central Kalimantan included Banjarese (24.20%), Javanese (18.06%), Dayak Ngaju (18.02%), Sampit Dayak (9.57%), Dayak Bakumpai (7.51%), Madura (3.46%), Dayak Katingan (3.34%), and Dayak Maanyan (2.80%). The substantial proportion of Banjar and Javanese ethnic groups in Central Kalimantan can be attributed to the migration of Banjar people from South Kalimantan and the significant transmigration from Java to Central Kalimantan. The Dayak ethnic group represents the largest demographic in Central Kalimantan, with the combined total of Dayak tribes accounting for 41.24%. Following an inter-ethnic conflict involving the Dayak, Malay, and Madurese ethnic groups in Central Kalimantan, the Madurese population no longer reside in the region and have been relocated to their area of origin.

9.4.2 The Food Estate Program in Indonesia

According to data released by the FAO, 6% of Indonesia's population of all ages experienced "moderate or severe food insecurity" in 2021, that is, 16.8 million Indonesians lack sufficient physical, social, and economic access to adequate, safe, and nutritious food (FAO report 2012, *Coming to Terms with Terminology*). In response, the Government of Indonesia initiated the National Food Program (Fig. 9.1).

On March 29, 2021, the Minister of Environment and Forestry delivered a presentation before Commission IV of the House of Representatives of the Republic of Indonesia. The documentation submitted included specific details of the proposed food storage area outlined by the Inspector General of the Ministry of Defense (MoD), encompassing a total area of 775.757 hectares (Table 9.6). The presentation emphasized the fact that several requests from the MoD lacked a location map or had an "unclear" location map. It was also said that virtually all of the specified regions and maps aligned with the Forest Moratorium Map, conserved peat areas, or existing forestry concession territory.

Table 9.6 Food storage areas proposed by the Ministry of Defense

Provinces	Proposed food storage area (hectares)
Bangka Belitung	32.603
West Sumatra	43.806
Jambi	152.545
Riau	42.340
North Sumatra	11.968
Aceh	28.966
West Kalimantan	36.000
East Kalimantan	21.000
Central Kalimantan	113.827
Maluku	47.395
West Sulawesi	37.222
South Sumatra	36.069
Papua	172.016
TOTAL	775.757

Source The Ministry of Defense and KLHK, 2021

9.4.3 Local Food Habits of the People of Kalimantan

Local food habits represent socio-cultural values that are transmitted across generations, imparting significance to the development of individual or group behavior, attitudes, and beliefs rooted in local cultural principles. Consequently, dietary practices within distinct community groups vary

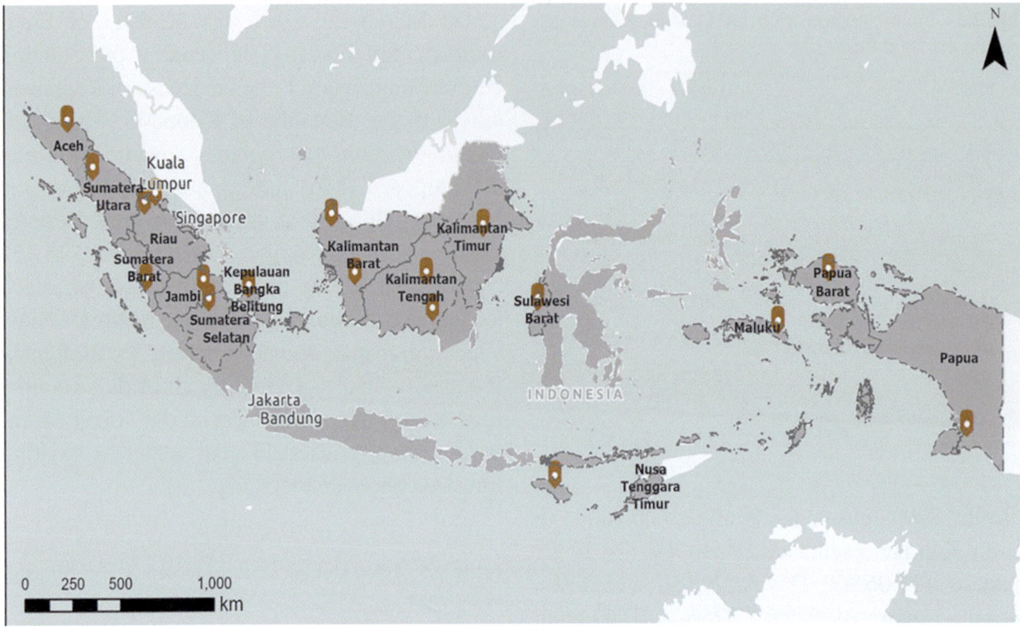

Fig. 9.1 Map showing provinces with Areas of Interest for food estates in Indonesia. (*Source* Ministry of Environment, Indonesian Forestry, 2021.)

between different regions, encompassing diverse food preferences and consumption patterns (Warsilah 2013: 111). The inhabitants of Kalimantan rely on producing rice through a shifting cultivation method as their staple food, while supplementary foods like corn, cassava, and other tubers complement their diets.

By the end of the fourth week of December 2022, the rice stock in Central Kalimantan province stood at 16,803 tons, which includes reserves at the Central Kalimantan government's logistics warehouse totaling 3369 tons. According to preliminary data from KSA BPS, rice production in 2022, based on grain yield, amounted to 353,864.63 tons of paddy, equivalent to 233,338.34 tons of rice (Sunarti 2022).

The government is pursuing food security through the promotion of new dietary habits rather than endorsing the concept of "local food sovereignty." This means that regions with cultural diversity in terms of non-rice staple food habits could gradually become reliant on rice and other staple foods. Consequently, certain areas are at risk of encountering rice shortages, contributing to food insecurity. Since food habits are integral components of community culture, achieving local food sovereignty necessitates the collective engagement of all stakeholders (Warsilah 2012).

9.4.4 Natural Resources in Central Kalimantan

Natural resources in Central Kalimantan predominantly comprise forested areas, which account for 80% of the total land mass. According to Regional Regulation No. 8/2003, the forested expanse in Central Kalimantan spans 10,294,853.52 hectares, or 64% of the entire area (BPS Central Kalimantan 2021). The remaining portion consists of primary forests, occupying approximately 25% of the territory. Land is being progressively taken over by oil palm plantations extending to 700,000 hectares (2007 data). Rubber and rattan plantations cultivated by local communities are dispersed across nearly all regions, particularly within the districts of Kapuas, Katingan, Pulang Pisau, Gunung Mas, and Kotawaringin. Abundant untapped natural resources include coal, gold, zircon, and iron mines, alongside copper, kaolin, and gemstones.

Central Kalimantan boasts substantial fisheries potential – especially freshwater – owing to the extensive network of rivers, lakes, and swamps that characterizes the region. The mining potential within Central Kalimantan primarily revolves around gold mines and zircon sand. However, mining activities within the community remain traditional, resulting in suboptimal outcomes (The Color of Indonesia 2015).

The Dayak community, traditional custodians of vast peat forest landscapes, preside over approximately 1.5 million hectares of peatlands situated amidst several major rivers, many of which house their original settlements. Employing a sustainable and small-scale approach known as 'swidden' or shifting cultivation, the Dayak community utilizes shallow peat found within a few hundred meters of riverbanks. Conversely, a 'handle' hydrological management system is implemented in coastal peatland regions (Limin Suwido et al. 2007: 291–301).

For countless millennia, the Dayak people have harnessed the Gunung Mas Forest landscape to procure and harvest food through sustainable methods. This forest is also a vital repository of biodiversity, and includes the habitat of the Bornean orangutan (Greenpeace Indonesia 2022a, b).

9.4.5 The Food Estate Program in Central Kalimantan and the Socio-Ecological Crisis of Natural Resources

The Ministry of Agriculture and Forestry, in conjunction with the Ministry of Defense, has cleared forests for the monoculture Food Estate Program, a move considered disastrous by local communities. As revealed by a Dayak tribe member in an interview with BBC News Indonesia (2022a, b), the forest once served as a source of wood for building houses, hunting small animals such as mouse-deer and pigs, and gathering traditional herbs and vegetables. All these resources have now disappeared, including four hectares of land that had been utilized for generations to cultivate rubber trees to support family economies and grow vegetables used to fulfill food needs, such as eggplants, long beans, and kundur. "The forest has never been stepped on; it was a place where Dayak people went. Now it looks like a field. Who wouldn't be upset? Rubber trees have been planted for decades and were ready to be tapped, so why were they uprooted?", the same tribe member exclaimed in anger when interviewed by BBC News Indonesia at his home (BBC News Indonesia 2022a, b).

The situation in Central Kalimantan exemplifies how the food estate project, deemed to disregard the right to a healthy environment, is inconsistent with efforts to ensure the right to food, causing an ecological crisis regarding natural resources and a social crisis marked by the erosion of local wisdom, culminating in a new food crisis. Additionally, the project has contributed to an expansion of flood-prone areas in Central Kalimantan in recent years, which has been attributed to forest clearance for the food estate initiatives. The government must halt the Food Barn project in Central Kalimantan and undertake measures to rectify the environmental degradation resulting from forest clearing for cultivation.

BBC News Indonesia, in collaboration with the NGO Peatland Monitoring, discovered that the National Food Barn project in this region has only exacerbated existing issues, intensifying the scope and duration of flood disasters and compelling the Dayak people to adapt their planting practices. Aryo Nugroho, Director of LBH Palangkaraya, emphasized that the Food Estate project overlooks the right to a healthy environment and runs counter to efforts aimed at ensuring the right to food (BBC News 2023).

In his research published in *Kompas* (2022), Joesep Budianto elucidated that following the implementation of the Food Estate Program in Central Kalimantan, Indigenous peoples are encountering mounting challenges to satisfying their nutritional requirements by sourcing directly from nature. This predicament arises from government restrictions on traditional farming practices thought to jeopardize the food security of these

Indigenous communities. The findings are rooted in a survey conducted by the Kompas Research and Development Institute in July 2022, with support from the Rainforest Journalism Fund Pulitzer Center. Survey results indicated that 75% of respondents were finding increasing difficulty in locating wild animals for hunting, as well as in finding fruits (68%), vegetables (47%), medicinal plants (40%), and fish in rivers (35%).

NGOs contend that the National Strategic Project poses a heightened risk of exacerbating the situation. Sacrificing additional forests and peatlands for large-scale monoculture agriculture is likely to exacerbate, rather than enhance, Indonesia's food security, while increasing the incidence of peat fires and emissions of carbon dioxide that are detrimental to the climate. It is noteworthy that nearly a third of global greenhouse gas emissions attributed to agriculture result from the conversion of new agricultural land. Furthermore, these new Food Barns are expected to promote the production of less nutritious staples like rice, while curtailing access to a varied and healthy array of forest foods for local communities.

Currently, two new food storage initiatives are underway in Central Kalimantan. First, a 31,719-hectare cassava plantation has been proposed by the Ministry of Defense in Gunung Mas Forest. Figures 9.2 and 9.3 show the Gunung Mas Forest area in Central Kalimantan where trees have been cut down in preparation for the Food Estate Program and cassava has been planted.

Second, a Ministry of Agriculture Food Barn project will especially target the cultivation of rice on swampland covering an area of 165,000 hectares in Kapuas and Pulang Pisau Regencies, with the majority of this land previously designated for the Peatland Development Project. Less than 1% of this area (473 hectares) is allocated for horticultural production, predominantly for oranges, durian, and chilies, while a mere 23 hectares are dedicated to cultivating green leafy vegetables (Ministry of Agriculture of the Republic of Indonesia 2020).

Multiple reports suggest that the National Food Barn Program will cover an expansive area of 2.3 million hectares (data from the Ministry of Environment and Forestry 2021). The establishment of this cassava plantation within Gunung Mas Forest is just one instance of various regions being transformed into extensive agricultural zones under the government's Food Estate Program. Regrettably, the monoculture approach adopted within this program has not only fallen short of delivering the expected cassava yield but has also contributed to the marginalization of local wisdom and knowledge.

Traditional Dayak wisdom was either ignored or misunderstood when the central government attempted to expand hydrological management and cultivation away from the riverbanks and into the peat domes on a large scale, utilizing a network based on mega-canals. A main canal 187 km in length was excavated to connect the Kahayan, Kapuas, and Barito rivers, accompanied by 958 km of primary canals across four 'blocks,' and nearly 2000 km of secondary and tertiary canals (Houterman and Ritzema. 2009).

Bayu Herinata, Director of Walhi Central Kalimantan, argues that the Food Estate project must be halted, particularly given the history of similar projects that have faltered. For instance, the million-hectare peat land was transformed into the Peat Land Development project during the New Order era. "Almost all food estate projects in Indonesia that rely on large-scale development and substantial capital have consistently encountered failure. The destruction of forests and peatlands will trigger socio-economic losses, impoverishing not only the people but also draining the state's finances" (Walhi Central Kalimantan 2021).

President Susilo Bambang Yudhoyono initiated a food storage program in Delta Kahayan, Bulungan Regency, North Kalimantan in 2011, and in Ketapang, West Kalimantan in 2013. However, both these plantations are regarded as failures, in terms of providing both employment and rice production. Only 0.1% of the 100,000-hectare project in Ketapang was successfully cultivated, and in Bulungan, less than 100 ha of the 50,000-hectare project area reportedly yielded

Fig. 9.2 The Gunung Mas Forest in Central Kalimantan planted with cassava for the Food Estate Program. (*Source* https://www.ekuatorial.com/2022/01/ugal-ugalan-lumbung-pangan/)

rice. These projects have exacerbated food insecurity due to environmental harm stemming from forest clearance and peat drainage. The Ketapang Food Barn also faced allegations of corruption within the government elite, leading to fines and imprisonment for companies and government officials (as reported in various local media outlets, such as Koran Kaltara 2019; Republika Online 2020).

This chapter aligns with various alternatives proposed by previous researchers that can enhance household food security in Indonesia without compromising forests, ameliorate local food habits and consumption patterns, and mitigate greenhouse gas emissions. These alternatives not only bolster food security but also, more importantly, enhance food sovereignty and resilience to price and climate fluctuations. They ensure that the food chain remains under the control of producers and consumers within communities, rather than being dominated by companies. Sustainable traditional Indonesian agroforestry approaches can be coupled with ecological agriculture that enhances soil fertility without reliance on chemicals. This is achieved by augmenting soil organic matter, improving water retention, preventing land degradation, shielding soil from erosion, and safeguarding waterways from pollution.

9.5 Recommendations and Conclusions

9.5.1 Recommendations

First, to make the sustainable development program a success, the central government should encourage ecological agriculture and traditional agroforestry rather than rushing to implement the Food Estate Program. The sustainable development approach offers a dual solution to the food crisis, the challenge of climate change, and the implementation of sustainable development programs.

Second, the government should halt the Food Barn project in Central Kalimantan which does not adopt local food, and take steps to restore damage to the forest environment so that local food can be fulfilled again.

Fig. 9.3 The Gunung Mas Forest being cleared for the Cassava Food Estate Program. (*Source* Domain Public- Image: Greenpeace Indonesia, 2022a, b)

9.5.2 Conclusion

This research indicates that the implementation of the Food Estate program reflects a disregard for the Dayak community's perception of forests as a source of life, and makes no reference to sustainable development programs. Food Estates are an example of a development strategy that is less sensitive to the needs of local communities and fails to implement sustainable development programs that highlight environmental, social, and economic variables. This failure highlights the need to re-evaluate the continuation or implementation of Food Estate programs, and wide-scale land acquisition policies in various regions in Indonesia that ignore sustainable development programs.

References

Journals and Books

Anonymous (2021) Central Kalimantan food estate instant policy full of controversy. www.pantaugambut.id. Accessed 30 Jan 2021

Budianto Y (2022) Asa Masyarakat Adat Kalteng Memenuhi Ketahanan Pangan. https://www.kompas.id/baca/riset/2022/08/31/asa-masyarakat-adat-kalteng-memenuhi-ketahanan-pangan

Clarke G (2001) From Ethnocide to Ethnodevelopment? Ethnic minorities and indigenous peoples in Southeast Asia. Third World Q 22(3):413–436

Houterman J, Ritzema P (2009) Land and water management in the ex-mega rice project area in Central Kalimantan

James (2023) Directorate General of land and water management

Limin Suwido J, Yunsiska E (2007) History of the development of tropical Peatland in Central Kalimantan Indonesia. Tropics 16(3):291–301

McCarthy JF, Krystof O (2015) Responding to food security and land questions: policy principles and policy choices in Kalimantan Indonesia. BRICS Initiatives for Critical Agrarian Studies (BICAS), May 2015

Muh Ilmi Ikhsan S, Ahmad D (2021) The paradigm of agricultural development through food estate, MISEKTA. https://misekta.id/news/paradigama-pembangun-pertanian-melalui-food-estate

Nasution M, Bangun OV (2020) Challenges of the food estate program in maintaining food security. APBN Bull 5(16):7–10

Ortiz-Bobea A, Toby RA, Carlos MC, Robert GC, David BL (2021) Anthropogenic climate change has slowed global agricultural productivity growth. Nature Clim Change 11(4):306–312. https://doi.org/10.1038/s41558-021-01000-1.

Sunarti (2022) Stok Pangan Kalteng Aman. Multi Media, https://mmc.kalteng.go.id/berita/read/40105/kepala-dinas-tphp-prov-kalteng-pastikan-nataru-tahun-2022-stok-pangan-kalteng-aman. Kaltara Newspaper. "The Food Estate Program in Bulungan is considered a failure", 6 Dec 2019

Warsilah H (2012) Quo Vadis Kedaulatan Pangan Lokal : Mengembalikan Kedaulatan Pangan Lokal, Berita Manado. http://pangan.banglikab.go.id/

Warsilah H (2013) The rural-coastal communities food-habits in supporting food security. Jurnal Masyarakat Budaya, 15(1) Tahun. file:///Users/henny/Downloads/143-Article%20Text-277-1-10-20150921%20(1).pdf

Warsilah H (2020) Maleh Dadi Segoro. Krisis Sosial Ekologis Kawasan Pesisir Semarang-Demak.PT.Lintas Nalar, Yogyakarta

Warsilah H (2023) Reconstruction of ethnic identity in the development of a New Capital City in East Kalimantan: participation and access to city resources. In Assembling Nusantara. Springer

Documents

BBC News Indonesia (2023) Food Estate in Gunung Mas Regency, Central Kalimantan, is not the responsibility of the Ministry of Agriculture. https://www.bbc.com/indonesia/indonesia-64975831

CNN Indonesia INFOGRAPHICS: RI Rice Production and Consumption 2018–2022 in full here: https://www.cnnindonesia.com/ekonomi/20230330013441-95-931061/infographics-production-dan-konsumsi-beras-ri-2018 -2022.

Dw. Com (2020) Food estate development plan in Central Kalimantan threatened to fail?. 26 June 2020

Greenpeace Indonesia (2022a) The Indonesian Government Only Feeds the Climate Crisis Through Food Estates. https://www.greenpeace.org/indonesia/siaran-pers/55727/government-indonesia-only-memi-makan-krisis-iklim-dinding-food-estate/?utm_term=&utm_campaign=G.

Greenpeace Indonesia (2022b) Food Estate Feeding The Climate Crisis. https://www.greenpeace.org/indonesia/food-estate-feeding-the-climate-crisis/

Ministry of Agriculture of the Republic of Indonesia (2020) Central Kalimantan Food Estate, Turning Swamps into Fruit and Vegetable Areas

Republika Online (2020) The previous food estate failed, how about the Central Kalimantan Food Estate?. October 22, 2022

The Color of Indonesia (2015) https://www.thecolourofindonesia.com/2015/10/demografi-kalimantan-tengah.html.

The Gecko Project (2021) Rainforests Fall for Indonesia's Food Estate Programme. October 14, 2021

10 The Transformation of the Samin Community's Resistance to the Ecological Crisis in the Kendeng Mountains

Sulistya Ekawati, Dewi Retna Indrawati, Sigit Andy Cahyono, and Dewi Ratna Kurniasari

Abstract

The Kendeng Mountains are an important element of the karst area, which is a fragile, vulnerable water resource for agriculture and households. The construction of a cement factory in the Kendeng Mountains is an example of the application of modernization theory to improve the economy and people's welfare through the extraction industry. However, it is thought that, locally, this development will cause economic collapse and an ecological crisis. The government regards cement companies as an opportunity to improve the economy. The Samin people, however, believe the cement company will destroy the nature and water sources on which their livelihood depends. The Samin people adhere to the teachings of Saminism and their local wisdom consists in managing nature with spiritual values to meet their daily needs by cultivating crops or farming. The Samin community's resistance has changed from a passive resistance movement to colonialism to an active form of resistance through taking to the streets, hearings, and legal channels. The present research uses a qualitative approach with literature studies and content analysis, and finds evidence of conflicts between the government, cement companies, and the Samin community.

Keywords

Resistance · Conflict · Samin · Karst · Cement mine · Kendeng Mountains

10.1 Introduction

The karst area in Indonesia covers 15.4 million hectares, or equivalent to 0.08% of the country's land area (Aprilia et al. 2021). Kendeng is a karst mountain in the northern part of Java Island. The Kendeng Mountains are located in seven districts in two provinces: Pati, Grobogan, Blora, and Rembang Districts in Central Java Province; and Lamongan, Bojonegoro, and Tuban Districts in East Java Province (Hadi et al. 2019; Pravitasari et al. 2020). Karst areas are environments with great potential but low carrying capacity (Budiyanto 2015). Ecologically, karst areas have

S. Ekawati (✉)
Research Centre for Society and Culture - National Research and Innovation Agency, Jakarta, Indonesia
e-mail: suli018@brin.go.id

D. R. Indrawati · D. R. Kurniasari
Research Center for Population - National Research and Innovation Agency, Jakarta, Indonesia

S. A. Cahyono
Research Center for Ecology and Ethnobiology - National Research and Innovation Agency, Jakarta, Indonesia

H. Warsilah and A. H. Purnomo (eds.), *Approaching Natural Resources for Sustainable Development in Indonesia*, Sustainable Development Goals Series, https://doi.org/10.1007/978-981-96-6913-4_10

an important function as large groundwater reservoirs and habitats for various flora and fauna (Aprilia et al. 2021; Hartono 2016; Nugroho and Kristanto 2020).

Limestone is a natural resource that can be mined in karst areas. However, mining activities can damage cave ecosystems, underground rivers (Nugroho and Kristanto 2020), and community water sources (Asrawijaya 2021; Hadi et al. 2019). Water is an important element for the Samin people (Hadi et al. 2018) because Samin teachings only allow them to become farmers and live independently (Mojo et al. 2017). The existence of a cement company is a threat to the livelihood of the Samin community because it will damage the local ecosystem, causing conflict (Asrawijaya 2020; Hartono 2016).

The social movement pioneered by the Samin community is unique because it uses symbols of local cultural traditions (Bräuchler 2020; Filipus 2023). To date, the Samin people have been synonymous with backwardness, social deviation, and lagging behind the progress of civilization, but their refusal to protect the environment can send shock waves across the country and the world (Supratikno and Adi 2021).. This chapter aims to analyze the gaps in karst area management policies in Indonesia, the transformation of the conflict in the Samin community, and the actors involved.

10.2 Theoretical Background

Modernization theory measures the progress of society through stages of economic and technological development leading to increased industrialization and urbanization (Pogosyan 2021). Three paradigms influence development: conservation, eco-populism, and developmentalism (Witter and Bitmer 2005). Conservationists argue that areas need to be legally protected and undisturbed by human activities to achieve ecological balance. It is assumed that residents pose a threat to natural resource conservation efforts. Eco-populist groups argue that Indigenous and local communities are the groups most at risk and in need of protection. Local communities can conserve natural resources better than the government. This ideology rejects the presence of private parties and conservation activists who ignore local communities. Meanwhile, developmentalism argues that damage to natural resources is caused by poverty, so solutions and policies are more 'developmentalist' in nature. They think that eco-populists are too romantic and exploit local communities, while conservationists are thought to not pay attention to the problems of poverty in communities around conservation areas (Kartodihardjo and Jhamtani 2006).

If conservationists think that natural resources should not be changed, then for developmentalism, natural resources are economic assets that must be exploited for development. For eco-populists, natural resources are part of a unified ecosystem between humans, animals, and plants. The tendency of the economic growth development pattern, which always emphasizes economic growth and is more characterized by a develop- mentalistic paradigm, is to regard natural resources only as economic assets.

The conflict between the Samin community and the cement industry in the Kendeng karst mountains (Pati and Rembang) is an example of this paradigm contestation (Gare 2021). For the government and investors, the Kendeng karst mountains have economic potential and are an asset that must be exploited for economic growth. However, for the farming community and *Sedulur Sikep*, Mount Kendeng is a protected area for water, a source of living space, and an ecosystem.

This theoretical concept shows that the current ecological crisis is rooted in the logic of environmental modernization, homogeneity, and control. This approach not only reveals a more nuanced and comprehensive analysis of the role of mining in the current global ecological crisis (Kotzé 2019) but also sheds light on the ongoing struggles of the Samin people and the possibility of future ecological justice (McGregor et al. 2020). The fundamental solution is to resolve the profound contradiction between conservation and economic development (Tian et al. 2019) and overcome the socio-economic crisis (Fanning et al. 2022) that is triggering the ecological crisis. Sustainable development seeks equitable economic development, social welfare, and environmental preservation.

The SDGs, global and national commitments to improve society's welfare, include 17 global goals and targets for 2030 that were declared by both developed and developing countries at the UN General Assembly in September 2015.

10.3 Methodology

This study uses a qualitative research approach to understand the meaning symbolized in people's behavior (Busetto et al. 2020) and its explanation (Hamilton and Finley 2019). The data used is naturalistic with an inductive method using descriptive and narrative reporting models. Data were collected through a literature study, and selected and analyzed according to the theoretical framework (Vindrola-Padros and Johnson 2020). Data validity was carried out using triangulation techniques. Descriptive qualitative data analysis was carried out using content analysis (Roller 2019), gap analysis, stakeholder analysis, conflict analysis, and studies on transforming the Samin community's resistance. This process is done in two ways: first, mapping to look for similarities and differences between the available data; and second, connecting the results with theory. In the analysis, data reduction, data presentation, and conclusion drawing or verification are carried out (Fig. 10.1).

10.4 Results and Findings

10.4.1 Karst, Cement Mining, and the Ecological Crisis

10.4.1.1 Karst Potency

Karst is an area with unique topographic characteristics formed by dissolved limestone such as quartzite, dolomite, limestone, salt, gypsum, and anhydrite (Parise et al. 2018). The Kendeng Mountains karst area has geological, ecological, cultural, and tourism potential, including:

Limestone and geological landscape The North Kendeng Mountains have abundant limestone resources for cement, cosmetics, paint, and other commercial products (Hadi et al. 2020a, b). Kendeng karst is a unique karst landscape formation consisting of caves (Darmastuti et al. 2016), stalactites, stalagmites, and limestone landscapes. This uniqueness is attractive to geologists, researchers, and tourists (Nugroho and Kristanto 2020) in ecotourism and agrotourism (Qomariah 2020).

Clean water source The Sukolilo and Grobogan karst areas function as water storage for the entire Pati and Grobogan regions (Wacana et al. 2011), especially for clean water (Hadi et al. 2018), drinking water, and irrigation (Hadi et al. 2020a, b; Wacana et al. 2011). In the Tegaldowo and Timbrangan areas of Rembang Regency, there are 29 springs and 44 ponors (Nugroho and Kristanto 2020).

Biodiversity The Kendeng Mountains karst ecosystem is a habitat for endemic species (Geekiyanage et al. 2019; Goldscheider 2019; Nitzu et al. 2018), for example, 33 land snail species (Nurinsiyah 2015), 33 bird species (Setiani 2012), 16 orders of arthropods (Narulita 2012), six species of bats (Tamasuki et al. 2016), and the habitats of mammals (Rahayu 2012).

10.4.1.2 Socio-Economic and Environmental Impacts of Karst Mining

The karst ecosystem is damaged by natural and anthropogenic factors, for example, cement mining (Aprilia et al. 2021). The following are some of the mining activity impacts on the karst ecosystem in the Kendeng Mountains:

Social impact Construction of the cement factory led to the eviction of local communities (Harsasto 2020) and triggered changes in social structure and community life. Local communities are disturbed by the cement factory's presence because it can damage the Watu Putih Groundwater Basin (Mustofa et al. 2022), springs, caves, ponor, and subterranean river, and cause drought on 58,368 hectares of land in Rembang (Mustofa et al. 2022).

Economic impact Mining can damage the environment and sources of income for farmers (Khusnia 2018), but the cessation of mining also impacts the economy of the people who agree

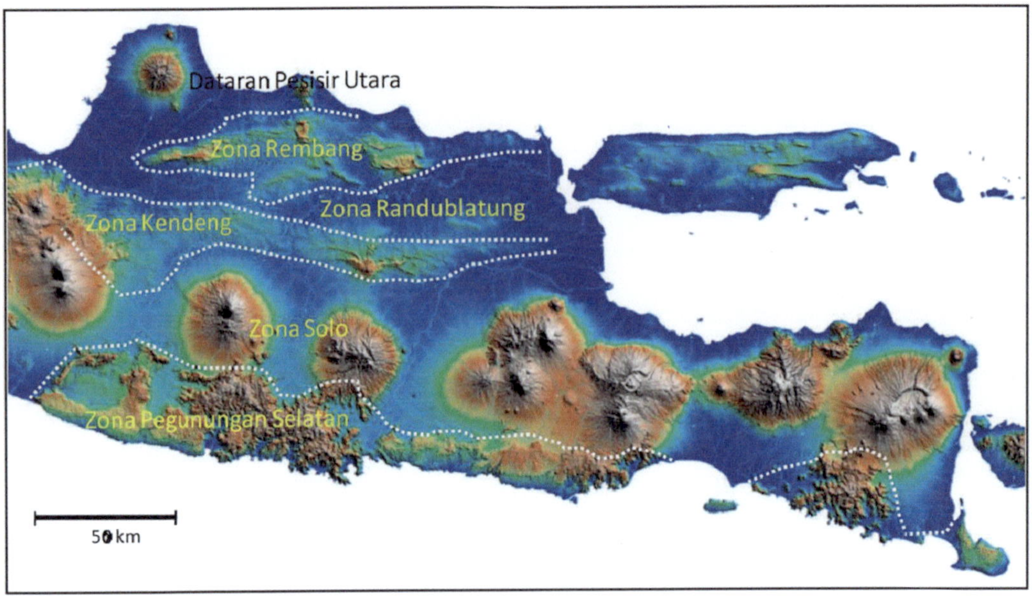

Fig. 10.1 Regional physiographic zoning of central and Eastern Java Island. (*Source* Hussen 2016; Wibowo and Nurani 2021 (division based on Pannekoek 1949 and van Bemmelen 1949)

(Aprilia et al. 2021) with the cement factory construction.

Environmental impact The characteristics of karst – unique, fragile, and hard to recover – make cement mining activities disrupt the hydrological function of water sources, land productivity, habitat existence (Rokhmad 2020), water supply, water distribution, absorption function, and water storage (Nugroho and Kristanto 2020), and have the potential for flash floods and landslides, especially in Kayen (Qomariah 2020). Besides that, during limestone processing into cement, dust and toxic gases are usually released into the air (Rokhmad 2020).

The participants at the Earth Summit in Rio de Janeiro in 1992 agreed to define sustainable development as development oriented towards meeting human needs through using natural resources wisely and efficiently and considering the sustainability of their use for current and future generations. These goals emphasize the relationship between sustainable development's environmental, social, and economic aspects. Constructing a cement factory in a karst area has several negative impacts on the surrounding community, both environmentally and socio-economically. These activities contradict the principles of sustainable development goals: SDG1 (no poverty), SDG2 (zero hunger), SDG3 (good health and well-being), SDG6 (clean water and sanitation), SDG13 (climate action), and SDG16 (peace, justice, and strong institutions).

10.4.2 Karst Ecosystem Management Policy in Indonesia

Law No.26/2007 on Spatial Planning and Government Regulation No.26/2008 on National Regional Spatial Planning (RTRW) declare karst areas geologically protected areas (articles 52 and 53) in the form of Karst Natural Landscapes Areas (KBAK), regulated by the Ministry of Energy and Mineral Resources (ESDM). Article 19 of Law No.32/2009 on Environmental Protection and Management states that every regional spatial plan must be based on strategic environmental studies (KLHS) and environmental permits (Rellua 2013). Based on the

Ministry of Environment Regulation No.17/2012, the community has the right to participate in the AMDAL and environmental permit process. However, with the enactment of Government Regulation in lieu of Law No.2/2022 on Job Creation, the Environmental Permit becomes an Environmental Approval, namely a decision on environmental suitability or a statement of environment management capability that has received approval from the central government. Besides that, the only people included in decision-making are those who are affected.

The Job Creation Law makes the environmental approval process simple and efficient. Environmentalists fear that the speed of the environmental agreement process will encourage environmental damage and trigger conflict with the community. The Job Creation Law has also transferred the issuance of mining permits from regional authorities to central authorities with Presidential Decree 55/2022. However, the Ministry of Energy and Mineral Resources circular No.1.E/HK.03/MEM.B/2022 states that non-metal mining permits are transferred from the central government to the regions, which makes karst management regulations from the central government ineffective.

One example occurred in the Sukolilo KBAK stipulated through Minister of Energy and Mineral Resources Regulation No.2641.K/40/MEM/2014, which covers an area of 20,079 hectares in three regencies (Pati, Grobogan, and Blora). In 2012, Central Java Province granted mining permits in parts of the Kendeng Mountains to PT Semen Gresik through Central Java Governor Decree No.660.1/30/2016. In 2016, the Supreme Court issued a decision directing the Governor of Central Java to revoke the permits. The Central Java Governor's Decree issued following the Supreme Court's decision did not revoke the license but postponed its implementation until the Governor's Decree concerning the Environmental Permit for Mining Activities and Construction of a Cement Factory by PT Semen Gresik (Persero) Tbk had been issued. This is despite the fact that Article 40, Point (2), of Law No.32/2009 on the Environment states that if the environmental permit is revoked, the business/activity permit is canceled.

10.4.3 Conflict and Resistance of the Samin Community

10.4.3.1 The Samin People: History and Culture

The Samin community originates from the Javanese peasant movement against Dutch colonial taxes led by Samin Surosentiko, who created life values for his followers (Safitri 2018). The Samin people are also known as Sedulur Sikep. In Javanese, *sedulur* means brothers, sisters, and kinship, and *sikep* means embracing everyone (Mojo et al. 2017), where the attitude or actions must follow the words spoken (Pinasti and Lestari 2018). In ancient Samin beliefs, Mount Kendeng was believed to be a source of strength for the Samin people and the place of origin of their ancestors, so it must be maintained (Supratikno and Adi 2021). Samin Surosentiko's teachings spread to villages and regencies in the Kendeng Mountains area (Endrayadi 2020; Wibowo and Nurani 2021). In 1890, Samin began spreading his teachings in Klopodhuwur Village, Blora Regency (Endrayadi 2020; Pinasti and Lestari 2018).

Ki Samin Surosentiko's teachings are full of life values, including *gotong-royong* (mutual cooperation), helping each other, honesty, simplicity, patience, and truth (Endrayadi 2020; Pinasti and Lestari 2018; Wibowo and Nurani 2021). His core teachings contain three laws: the Law of Action (*Angger-angger Pratikel*), the Law of Speaking (*Angger-angger Pangucap*), and the Law of Things that Must Be Carried Out (*Angger-angger Lakonana*) (Endrayadi 2020). Samin's teachings encourage his followers to become farmers and protect nature (Endrayadi 2020; Pinasti and Lestari 2018). For them, the land is the mother that provides food, so it must be protected and respected (Wibowo and Nurani 2021).

10.4.3.2 Resistance of the Samin Community: Chronology and Transformation

The Samin movement was a covert resistance movement against the Dutch with moral teachings in the form of, among other things, withdrawing from society, not paying taxes, refusing to give to the village barn, refusing to corral cows in public pens, not trading, only farming (Mustofa et al. 2022), using the *ngoko* language (speaking informally), freely managing forests (Purwanto 2016), and not schooling (Kirom 2020). After Indonesia's independence, Samin children began going to school and swiftly accepted modernization (Hadi 2017; Lestari 2015) while still adhering to Samin teachings (Anisin 2019).

The Samin movement is a community reaction to facing the most difficult economic situation and its leaders act as role models (Korver 1976; Supratikno and Adi 2021). In the colonial era, tax policies and colonial regulations that were detrimental to farmers were the cause of resistance from the Samin people. Currently, the ecological crisis, cultural existence, and environment are the basis for the Samin community's resistance to the establishment of a cement factory because it is contrary to their environmental teachings and practices (Supratikno and Adi 2021), causes environmental damage (Asrawijaya and Hudayana 2021; Purnaweni et al. 2019) and new problems (Wibowo and Nurani 2021), harms livelihoods, social norms, and local wisdom, and causes natural disasters (Asrawijaya and Hudayana 2021; Purnaweni et al. 2019).

The threat of environmental damage due to the cement factory establishment in the Kendeng mountains covers the Pati, Rembang, Blora, Bojonegoro, Kudus, Grobogan, and Tuban areas. In Tuban, the PT Holcim (Dynamix) and PT Semen Indonesia cement factories have already been established. In Pati and Rembang, efforts are ongoing to fight against the proposed building of a cement factory. Initiated by the Sedulur Sikep community (Putri 2017b; Setiadi et al. 2017), this resistance is regarded as a 'hidden transcript resistance' within the framework of the millenarianism and messianism movements (Korver 1976) and

Table 10.1 Forms of resistance in the Samin community

Year	Forms of resistance
1905	Resistance to the Dutch with self-isolation, not paying taxes, refusing to give to the village barn, refusing to put cows in public pens, not trading, *ngoko*, free forest management, not going to school
2014	The Kendeng Mountains Concerned Community Network (JMPPK), accompanied by WALHI and the Legal Aid Institute (LBH) Semarang, sues the State Business Court (PTUN)
	The community blocks the entrance to the cement factory
	100 women sit on the roadside leading to the location where the first stone has been laid for PT Semen Indonesia
	Lamporan ritual is performed by farmers using a torch to drive away 'pests', i.e., the cement factory
2015	Praying to God and praising the Prophet
	Appeal submitted to the Supreme Court
	Residents walk from Pati to Semarang
2016	Kendeng Long March
	Lawsuit granted by the Supreme Court
2017	Governor of Central Java issues a new permit for PT Semen Indonesia
	The Indonesian Legal Aid Foundation (YLBHI) reports allegations of maladministration by Central Java Governor Ganjar Pranowo to the Indonesian Ombudsman
	Access road to the factory blocked with sharpened bamboo and mortar
	Kendeng women demonstrate in front of the Central Java Governor's office building
	Feet cast with cement in front of the state palace
2018	Nine women wearing *kebayas* with caped batik cloth and carrying mortars take action at the Tani Monument
2020	Ten Kendeng women question the cement factory, which continues to operate during the Coronavirus pandemic
2022	JMPPK holds an *istigasah* (joint prayer) commemorating the sixth anniversary of winning a mining environmental permit lawsuit against PT Semen Indonesia

a love of harmony (Darmastuti et al. 2016), or according to Filipus (2023) as resistance from a traditional spirituality perspective based on the teachings of Saminism (Purnaweni et al. 2019). This resistance was similar to Gandhi's rebellion in India (1908). The difference is that Samin resistance is accompanied by the rhetoric of *nggendeng* (pretending to be crazy) in every action (Mukodi and Burhanuddin 2015).

The hidden transcript resistance of the Samin community is a form of defiance (Scott 1977) that rejects government regulations, especially in paying tribute or land tax. Putri (2017a) describes the movement as a grassroots movement against the global market and modernization. Covert resistance is no longer a trend and has been replaced by open resistance (Buechler 1995; Diani and McAdam 2003) with demonstrations, organizational formation, legal resistance, advocacy resistance, cultural action, social media, *macapat* songs (Iriyanto 2020), and films (Yusuf 2021). However, some Samin people state that demonstrations and open resistance are not authentic resistance (Novianto 2020). Table 10.1 presents the forms of resistance of the Samin community.

The Samin resistance movement has transformed from socioeconomic-based resistance in the context of colonial exploitation to an environmental-based resistance movement (Aziz 2011). While the old social movement was passive, limited/closed, and excluded the masses, the new social movement is active, inclusive, open to all social classes, and has mass support (Darmastuti et al. 2016). Alongside the new issues being confronted, another original quality of the current Saminist resistance movement can be seen in the appearance of Saminist women who are active at the forefront of resistance actions (Rosyid 2014). The movement has also built alliances with other groups concerned about environmental issues (Purnomo 2020; Zarkasi and Rahardian 2022), for example, JMPPK (Asrawijaya 2021) (Fig. 10.2).

A number of attitudes and principles typify the Samin community resistance movement in the Kendeng Mountains. First, it represents resistance from members of civil society who feel oppressed and threatened with with an ecological crisis and the loss of their rights (Adam and Bagir 2022). Second, the movement is confronting global capitalism, modernization (Pinasti and Lestari 2018), and state power, as seen in the transnational company PT Sahabat Mulia Sakti, which is a subsidiary of a German company. Third, it is resisting the tyranny of power in the form of a state-owned company (PT Semen Indonesia) and the local government (Suharko 2017). A weak state ruling order gave the Samin people the political opportunity to resist (McAdam et al. 2004). They were also driven by environmental-based local economic interests in maintaining food, water, and environmental sovereignty.

Fourth, the many women involved in the action were mothers who had no organizational experience (Ernanda 2023; Freudenberg and Steinsapir 1991). In Samin culture, husbands are prohibited from divorcing their wives except because of death, which is implied in the phrase *siji kanggo sak lawase* (one woman forever). The Samin tribe sees women (mothers) as the manifestation of the Earth, which protects, cares for, and loves humans (Putri 2017b), as implied in *tembung gunem* (Sugihardjo et al. 2012). Human relations with nature (Filipus 2023), land, and water (Setiadi et al. 2017) are encapsulated in the phrase '*Ibu Bumi kang maringi, ibu bumi dilarani, ibu kang ngadili*' (Mother Earth, which gives, Mother Earth, which is hurt, Mother Earth, which judges). The Samin people see the Kendeng Mountains as human stomachs (Mojo 2017) which, if damaged, will destroy life, and they realize that the destruction of nature will have a more significant impact on women (Pratama 2018). Fifth is the belief that local residents have the right to participate in policy-making regarding environmental issues (Putri 2017b). Sixth is a rejection of the hegemony of science and academics, especially those employed by industry or government. Seventh, the belief that economic growth is not always good because local residents are often sacrificed for the benefit of the general public (Freudenberg and Steinsapir 1991).

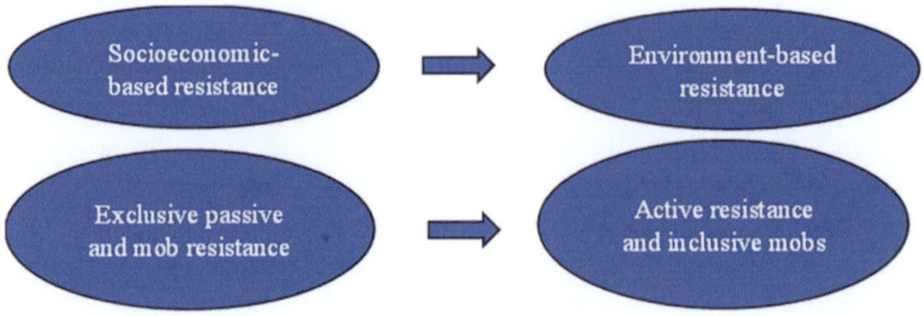

Fig. 10.2 Conflict transformation in the Samin community (*Source* Aziz, 2011; Darmastuti et al., 2016; Rosyid, 2014; Purnomo, 2020; Zarkasi & Rahardian, 2020; Asrawijaya. 2021, *illustrated by the Authors*, 2024)

10.4.3.3 Actors, Interests, and Influence

The Samin community movement has received support from many parties. There are at least 39 actors involved in the conflict between the Samin community and the mining company (Fig. 10.3).

Currently, the Samin community is at a critical juncture in its struggle to protect its environment and cultural heritage. The cultural resilience and spirit of resistance of the Samin people amidst the challenges of modernization, the ecological crisis, and internal and external pressures show that Samin teachings are still relevant, even though conflicts still occur. The conflicts have also caused changes to the structure of Samin society, especially the economic structure and internal divisions of the Samin community (Subekti 2016), changes that will be further embedded by modernization, illegal mining, and the ecological crisis. Samin people's experience shows that eco-populism thinking remains resilient and resistant to the developmentalism thinking that is supported by modernization and capitalism. Developmentalism, which promises prosperity but is followed by an ecological crisis, tries to hold back changes in the eco-populism struggle, which campaigns more for environmental concern, sustainability, human rights, and network development.. The Samin people believe in preserving traditional values and respecting human relationships with nature.

10.5 Conclusions and Recommendations

10.5.1 Conclusions

The Samin community faces the pressure of modernization, development, and exploitation of natural resources, which threatens their residential environment, livelihood, and cultural existence. The community has experienced a strategic transformation from passive and exclusive resistance to inclusive active resistance, from local practices and rituals to becoming more organized and collaborative. This shift in strategy demonstrates adaptability when facing increasingly complex challenges. The struggle of the Samin people is inseparable from conflict and the contestation of interested parties, both economically, socially, and culturally. While prolonged conflicts can damage Samin teachings, modernity will also accelerate their abandonment by the younger generation, who are more open and educated and are not willing to bear the burden of Samin teaching.

10.5.2 Recommendations

The government and relevant stakeholders should recognize and respect community rights and resources. The use of Kendeng mountain resources requires the community's involvement in decision-making if inclusive and sustainable solutions are to be found. Samin thought can be

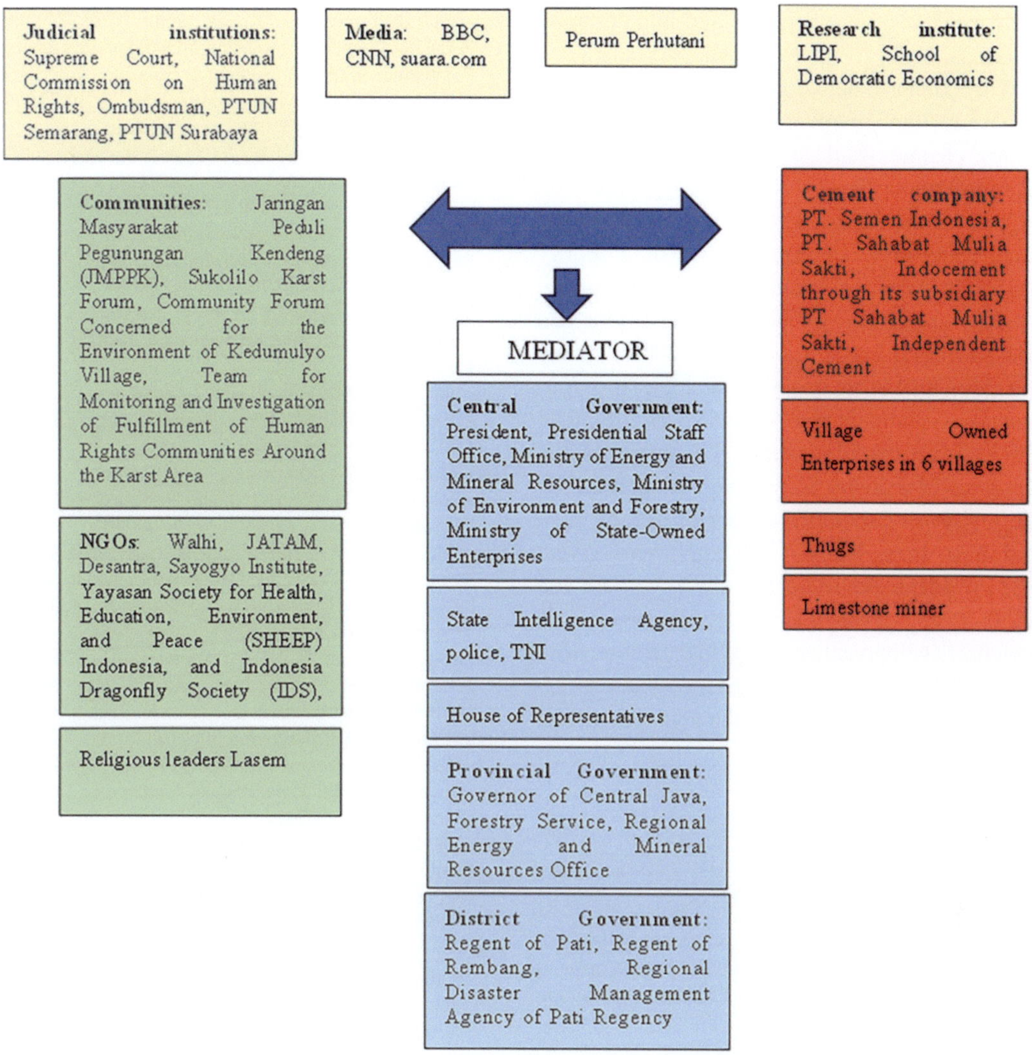

Fig. 10.3 Actors involved in the conflict between the Samin community and the cement company

re-invented if they can adapt to changing times and remain an example in the future. The eco-populism approach should be a priority when decisions on natural resource management are being made, to mitigate social crises, ecological crises, conflicts, and unsustainable development.

To support the resilience of the Samin community and ecological preservation in the Kendeng Mountains, it is necessary to develop sustainable alternative livelihoods such as ecotourism, organic farming, and handicrafts, accompanied by financial support and training. Advocacy for land and environmental rights should be strengthened through cooperation with NGOs and legal institutions, and the government should be encouraged to implement sustainable development policies and supervise the cement industry. The spreading of Samin cultural values through school curricula, cultural festivals, and public awareness campaigns is important to preserve cultural heritage and increase community participation in environmental conservation. Leadership training and the facilitation of dialogue between communities, governments, and industry are needed to reduce conflicts and find common solutions.

The extensive network of the Samin community has resulted in forceful pressure to encourage policy change (Zarkasi and Rahardian 2022). Even though the Samin people's struggle received support from many parties, it needed to be more successful. Decentralization of authority has made karst management regulations issued by the central government ineffective (Beynen and Townsend 2005). Law No. 32/2009 requires the government to adopt a Strategic Environmental Assessment (SEA/KLHS) to formulate and evaluate spatial planning. The results of the Kendeng Mountains SEA were not included in the revision of spatial planning in Central Java and East Java, but no sanctions were given. The President should instruct the governors and relevant districts to internalize SEA studies in regional development planning. In the future, SEA studies should involve local governments which can apply the study results in the field (Hadi et al. 2019).

References

Adam R, Bagir ZA (2022) The indigenous politics of justice: the case of the Sedulur Sikep movement in Central Java. Jurnal Kawistara 12(2):181. https://doi.org/10.22146/kawistara.67991

Anisin A (2019) Violence, resistance, and social transformation in anarchist thought and practice. JRSSW 38(4):383–400. https://doi.org/10.1080/15426432.2019.1672608

Aprilia D, Arifiani KN, Sani MF, Jumari J, Wijayanti F, Setyawan AD (2021) Review: a descriptive study of karst conditions and problems in Indonesia and the role of karst for flora, fauna, and humans. Intl J Trop Drylands 5(2):4. https://doi.org/10.13057/tropdrylands/t050203

Asrawijaya E (2020) Gerakan ekopopulisme komunitas Samin melawan perusahaan semen di Pegunungan Kendeng. JSPH 5(1):35–47

Asrawijaya E (2021) Gerakan Petani Kendeng dalam Mempertahankan Kedaulatan Pangannya. Citizen 1(3):137–147. https://doi.org/10.53866/jimi.v1i3.16

Asrawijaya E, Hudayana B (2021) The power of a leader in the Samin people's opposition movement to the development of a cement factory in the North Kendeng mountains. Humaniora 33(1):26. https://doi.org/10.22146/jh.56224

Aziz M (2011) Identitas Kaum Samin Pasca Kolonia Pergulatan Negara, Agama dan Adat dalam Pro-Kontra Pembangunan Pabrik Semen di Sukolilo, Pati Jawa Tengah. Jurnal Kawistara. https://doi.org/10.22146/kawistara.3937

van Beynen, P, Townsend K (2005) A disturbance index for karst environments. Environ Manag 36(1):101–116. https://doi.org/10.1007/s00267-004-0265-9

Bräuchler B (2020) Bali Tolak Reklamasi: The local adoption of global protest. Convergence 26(3):620–638. https://doi.org/10.1177/1354856518806695

Budiyanto E (2015) Peran teknologi penginderaan jauh untuk kajian desertifikasi batuan karst. Jurnal Geografi 13(2):104–115

Buechler SM (1995) New social movement theories. Sociol Q 36(3):441–464. https://doi.org/10.1111/j.1533-8525.1995.tb00447.x

Busetto L, Wick W, Gumbinger C (2020) How to use and assess qualitative research methods. Neurol Res Pract. https://doi.org/10.1186/s42466-020-00059-z

Darmastuti R, Bajari A, Martodirdjo HS, Maryani E (2016) Samin women's resistance movement against cement factory construction: a case of Sukolilo in Indonesia. Man India 96(12):4903–4918

Diani M, McAdam D (2003) Social movements and networks: relational approaches to collective action. Oxford University Press

Endrayadi EC (2020) Ki Samin Surosentiko dan ajaranya dalam tradisi lisan masyarakat di kawasan pegunungan Kendeng Provinsi Jawa Tengah. Humaniora Dan Era Disrupsi 1(1):556–568

Ernanda E (2023) Challenging the patriarchal culture; Feminist critical discourse analysis of the Indonesian environmental heroines. Wacana. https://doi.org/10.17510/wacana.v24i1.1173

Fanning AL, O'Neill DW, Hickel J, Roux N (2022) The social shortfall and ecological overshoot of nations. Nature Sustain 5(1):26–36. https://doi.org/10.1038/s41893-021-00799-z

Filipus JC (2023) Spiritualitas Politik sebagai Dasar Perjuangan Masyarakat Adat Samin Melawan Ekonomi Kapitalisme Pabrik Semen di Pegunungan Kendeng, Sukolilo Pati. Sagacity 3(2):16–30

Freudenberg N, Steinsapir C (1991) Not in our backyards: the grassroots environmental movement. Soc Nat Resour 4(3):235–245. https://doi.org/10.1080/08941929109380757

Gare A (2021) The eco-socialist roots of ecological civilisation. Capital Nat Social 32(1):37–55. https://doi.org/10.1080/10455752.2020.1751223

Geekiyanage N, Goodale UM, Cao K, Kitajima K (2019) Plant ecology of tropical and subtropical karst ecosystems. Biotropica 51(5):626–640. https://doi.org/10.1111/btp.12696

Goldscheider N (2019) A holistic approach to groundwater protection and ecosystem services in karst terrains. Carbonates Evaporites 34(4):1241–1249. https://doi.org/10.1007/s13146-019-00492-5

Hadi N (2017) Local wisdom of Samin community in tradition and modernization frame. In: Proceedings of the 2nd ICSE2017. Vol 2, pp 41–45

Hadi SP, Purnaweni H, Prabawani B (2019) The Powerless of Strategic Environmental Assessment (SEA): a case studies of North Kendeng Mountain Area, Central Java Indonesia. E3s Web Conf. https://doi.org/10.1051/e3sconf/201912502014

Hadi SP, Purnaweni H, Hadiyanto, Prabawani B (2018) Sustainable use of water resources at North Kendeng Mountain, Central Java, Indonesia. IOP Conf Ser: Earth Environ Sci 191:012005. https://doi.org/10.1088/1755-1315/191/1/012005

Hadi SP, Purnaweni H, Prabawani B, Hamdani RS (2020a) Community movement for sustainable use of natural resources: a case study of North Kendeng Mountain Area, Central Java, Indonesia. IOP Conf Ser: Earth Environ Sci 448(1):012069. https://doi.org/10.1088/1755-1315/448/1/012069

Hadi SP, Buchori I, Purnaweni H, Prabawani B, Adiwibowo S (2020b) Strategic environmental assessment (SEA) in a conflict situation: a case study on North Kendeng mountain area, Central Java, Indonesia. Asian Jr Microbiol Biotech Environ Sci 22(1)

Hamilton AB, Finley EP (2019) Qualitative methods in implementation research: an introduction. Psychiatry Res. https://doi.org/10.1016/j.psychres.2019.112516

Harsasto P (2020) Membedah Diskursus Modal Sosial dan Gerakan Sosial: Kasus Penolakan Pabrik Semen di Desa Maitan, Kabupaten Pati. Politika 11(1):18–30. https://doi.org/10.14710/politika.11.1.2020.18-30

Hartono MD (2016) Perlindungan Hak atas Air dalam Pembangunan Pabrik Semen di Provinsi Jawa Tengah. Jurnal HAM 13(13):111–146

Hussen (2016) Buku Panduan Fieldtrip Geologi Cekungan Jawa Timur Utara 16–17 Desember 2016. Department of Geological Engineering UGM

Iriyanto E (2020) Tembang Macapat: Kritik Sosial Sedulur Sikep terhadap Ekspansi Industri Semen di Pegunungan Kendeng. Sutasoma 8(2):70–79. https://doi.org/10.15294/sutasoma.v8i2.43185

Kartodihardjo H, Jhamtani H (2006) Politik lingkungan dan kekuasaan di Indonesia. Equinox Pub

Khusnia K (2018) Resolusi Konflik Pembangunan Pabrik Semen Di Kecamatan Gunem Kabupaten Rembang. JPGS 7(2):111–120

Kirom S (2020) Sejarah Masyarakat Samin 1890–1919: Perspektif Wilhelm Dilthey. Tamaddun 8(1):159–172

Korver APE (1976) The Samin movement and millenarism. J Hum Social Sci Southeast Asia 132(2):249–266. https://doi.org/10.1163/22134379-90002642

Kotzé L (2019) A global environmental constitution for the Anthropocene? Transnatl Environ Law 8(1):11–33. https://doi.org/10.1017/S2047102518000274

Lestari P (2015) Analisis Perubahan Sosial pada Masyarakat Samin. Dimensia. https://doi.org/10.21831/dimensia.v2i2.3403

McAdam D, McCarthy JD, Zald MN (2004) Comparative perspectives on social movements: political opportunities, mobilizing structure, and cultural framings. Cambridge University Press

McGregor D, Whitaker S, Sritharan M (2020) Indigenous environmental justice and sustainability. Curr Opin Environ Sustain 43:35–40. https://doi.org/10.1016/j.cosust.2020.01.007

Mojo E (2017) Kearifan Lingkungan Sedulur Sikep dalam Merespon Bencana Pembangunan Pabrik Semen di Kawasan Pegunungan Kendeng Utara Kabupaten Pati. Universitas Diponegoro

Mojo E, Hadi SP, Purnaweni H (2017) Sedulur Sikep's environmental wisdom in conservation of North Kendeng Mountains Sukolilo. Adv Sci Lett 23(3):2504–2506. https://doi.org/10.1166/asl.2017.8657

Mukodi M, Burhanuddin A (2015) Domestifikasi Perempuan Samin dalam Khasanah Masyarakat Islam Modern. Al-Tahrir 15(2):411. https://doi.org/10.21154/al-tahrir.v15i2.271

Mustofa MU, Raudya MDK, Nurdini FM, Sulaeman KM (2022) Radikalisasi Grassroots Movements Dalam Politik Ekologi di Indonesia Pasca Reformasi. IJPP, pp 83–100

Narulita R (2012) Keanekaragaman arthropoda di kawasan eksokarst Gunung Kendeng, Kabupaten Pati, Jawa Tengah. FST UIN Syarif Hidayatullah Jakarta

Nitzu E, Vlaicu M, Giurginca A, Meleg I, Popa I, Nae A, Baba Ş (2018) Assessing preservation priorities of caves and karst areas using the frequency of endemic cave-dwelling species. Int J Speleol 47(1):43–52. https://doi.org/10.5038/1827-806X.47.1.2147

Novianto A (2020) Berebut Saminisme: Artikulasi Politik Masyarakat Adat dalam Konflik Pembangunan Pabrik Semen di Pegunungan Kendeng. Gadjah Mada University Press

Nugroho NE, Kristanto WAD (2020) Karakter dan Potensi Risiko Kerusakan Ekosistem Karst Cekungan Air Tanah Watuputih Kabupaten Rembang, Provinsi Jawa Tengah. JILK 2(1):34. https://doi.org/10.31315/jilk.v2i1.3288

Nurinsiyah AS (2015) Land Snail Fauna of the Sukolilo karst in Java (Indonesia). Am Conchol 43(3)

Parise M, Gabrovsek F, Kaufmann G, Ravbar N (2018) Recent advances in karst research: from theory to fieldwork and applications. Geol Soc Lond Special Publ 466(1):1–24. https://doi.org/10.1144/SP466.26

Pinasti SVI, Lestari P (2018) Masyarakat samin ditinjau dari sejarah dan nilai nilai pendidikan karakter. Istora. https://doi.org/10.21831/istoria.v13i2.17737

Pogosyan V (2021) Updating social theory: redefinition of modernization. Wisdom 19(3):182–193. https://doi.org/10.24234/wisdom.v19i3.486

Pratama AY (2018) Perlawanan politik petani Kendeng Utara dalam menolak pembangunan pabrik semen di Kabupaten Rembang, Provinsi Jawa Tengah. Universitas Muhamadiyah Surakarta

Pravitasari AE, Rustiadi E, Adiwibowo S, Wardani IK, Kurniawan I, Murtadho A (2020) Dinamika dan

Proyeksi Perubahan Tutupan Lahan serta Inkonsistensi Tata Ruang di Wilayah Pegunungan Kendeng. J Region Rural Dev Plann 4(2):99–112. https://doi.org/10.29244/jp2wd.2020.4.2.99-112

Purnaweni H, Kismartini, Budy Pratama A, Wulandari C (2019) Conflict among stakeholders in karst area management of Pati Regency. IOP Conf Ser: Earth Environ Sci 314(1). https://doi.org/10.1088/1755-1315/314/1/012082

Purnomo AB (2020) A model of interreligious ecotheological leadership to care for the earth in the Indonesian context. EJST 16(4):15–25

Purwanto A (2016) Samin dan kehutanan Jawa abad XIX. Jurnal Sosiologi Reflektif 10(1):139. https://doi.org/10.14421/jsr.v10i1.1140

Putri PS (2017a) Re-claiming lost possessions: a study of the Javanese Samin (Sedulur Sikep) Movement to Maintain their Peasant Identity and Access to Resources. University of Oslo

Putri PS (2017b) The meaning making of an environmental movement: a perspective on Sedulur Sikep's narrative of anti-cement movement. PCD J 5(2):297. https://doi.org/10.22146/pcd.30471

Qomariah LN (2020) Agrowisata dI Pegunungan Kendeng Utara, Kabupaten Pati sebagai Penanganan Tanggap Bencana (Banjir Bandang & Tanah Longsor). Universitas Muhammadiyah Surakarta

Rahayu SAR (2012) Pemanfaatan habitat karst gunung kendeng kabupaten pati, jawa tengah oleh biodiversitas vertebrata kelas mamalia. FST UIN Syarif Hidayatullah Jakarta

Rellua O (2013) Proses Perizinan dan Dampak Lingkungan terhadap Kegiatan Reklamasi Pantai. Lex Administratum 1(2):158–167

Rokhmad A (2020) Configuration and the role of community leaders in the conflict of natural resources of limestone mining for the cement industry in Rembang Indonesia. Int J Energy Econ Policy

Roller MR (2019) A quality approach to qualitative content analysis: similarities and differences compared to other qualitative methods. Forum Qualit Sozialforschung. https://doi.org/10.17169/fqs-20.3.3385

Rosyid M (2014) Perempuan Samin dalam Tantangan Politik Lokal di Kudus Jawa Tengah. Palastren 7(2):394–418

Safitri MA (2018) Social movement in Indonesian mining law enforcement: the case of peasants-scholars nexus in karst mining dispute in Java. ICEML 2018:227–231

Scott JC (1977) Protest and profanation: agrarian revolt and the little tradition. Part II. Theory Soc 4(2):211–246

Setiadi S, Saraswati AR, Rosyid N (2017) Geger Sikep: environmental (re)interpretation among the contemporary anti-cement movement in Kendeng, Central Java. Komunitas 9(1):13–28. https://doi.org/10.15294/komunitas.v9i1.8673

Setiani W (2012) Keanekaragaman jenis burung dan pemanfaatan habitat pada ekosistem karst Gunung Kendeng Kabupaten Pati, Jawa Tengah. FST UIN Syarif Hidayatullah Jakarta

Subekti T (2016) Konflik Samin vs PT Semen Indonesia. Jurnal Transform 2(2):189–202

Sugihardjo S, Lestari ENY, Wibowo A (2012) Strategi Bertahan dan Strategi Adaptasi Petani Samin terhadap Dunia Luar. SEPA 8(2)

Suharko S (2017) Masyarakat Adat versus Korporasi: Konflik Sosial Rencana Pembangunan Pabrik Semen di Kabupaten Pati Jawa Tengah Periode 2013–2016. Jurnal Ilmu Sosial Dan Ilmu Politik 20(2):97. https://doi.org/10.22146/jsp.24776

Supratikno A, Adi S (2021) The Samin community in the globalization era: role of the religion of Adam in the post-cement industry advocacy. IJASS 4(1):47–55

Tamasuki K, Wijayanti F, Fitriana N (2016) Komunitas Kelelawar (Ordo Chiroptera) di Beberapa Gua Karst Gunung Kendeng Kabupaten Pati Jawa Tengah. AL-Kauniyah: Jurnal Biologi. https://doi.org/10.15408/kauniyah.v8i2.2694

Tian D, Xie Y, Barnosky AD, Wei F (2019) Defining the balance point between conservation and development. Conserv Biol 33(2):231–238. https://doi.org/10.1111/cobi.13221

Vindrola-Padros C, Johnson GA (2020) Rapid techniques in qualitative research: a critical review of the literature. Qual Health Res 30(10):1596–1604. https://doi.org/10.1177/1049732320921835

Wacana P, Rodialfallah A, Chandra F, Mesah D, Raimon R (2011) Study of the potential of Karst Region in North Kendeng, the Districts of Grobogan and Pati. Proc Asian Trans-Discipl Karst Conf 1:169–180

Wibowo H, Nurani IA (2021) Overview of karst in Rembang and Blora: their prehistoric cave settlement potential. Berita Sedimentologi 47(3):22–28

Witter H, Bitmer R (2005) Between conservation, ecopopulism and developmentalism discourse in biodiversity policy in Thailand and Indonesia. (37; CAPRI Working Paper)

Yusuf M (2021) Saminisme dalam Sengketa Pembangunan Pabrik Semen: Analisis Semiotik Film Sikep Samin Semen. Jurnal Pemikiran Sosiologi 8(2):152–184

Zarkasi IF, Rahardian R (2022) Actor-network theory in policy change of environmental permit policy for the construction of a cement factory in Pati Indonesia. Otoritas 12(1):41–51. https://doi.org/10.26618/ojip.v12i1.5998

Peat Ecosystems in Indonesia: Assessing the Multi-Crisis and a Sustainable Path

11

Surati, Kushartati Budiningsih, and Agung Wibowo

Abstract

Peat ecosystems are crucial for maintaining natural balance, carbon sequestration, and water systems, and providing habitats for diverse flora and fauna. However, inadequate policies in the past have resulted in a multi-crisis on peat land. This chapter examines the multifaceted crisis peat ecosystems face in Indonesia, focusing on social, economic, and cultural perspectives. The research employs a qualitative methodology with analytical descriptive and discourse analysis. Drawing on cases from Sumatera and Kalimantan, this study highlights the profound impact of the peat crisis on the livelihoods of communities relying on these ecosystems. The destructive consequences of forest and land fires, coupled with declining water quality, have resulted in habitat destruction, jeopardizing the sustainability of local communities and eroding social cohesion. Additionally, the cultural and spiritual significance of peat ecosystems for specific ethnic groups magnifies the potential loss of cultural heritage and ancestral knowledge related to resource management. Addressing these challenges necessitates holistic and integrated approaches involving local communities, government, and the private sector. Prioritizing social and agricultural improvements, effective fire management, and long-term economic sustainability are essential. The insights from this study provide a basis for policymakers to develop sustainable policies and practices, mitigating the crisis while safeguarding both peat ecosystems and the communities that depend on them.

Keywords

Peat ecosystems · Multi-crisis · Livelihoods · Community distrust · Sustainable policies · Indonesia

11.1 Introduction

Peat ecosystems are pivotal in preserving ecological equilibrium, serving as carbon sinks, regulating water systems, and creating habitats for

Surati (✉)
Research Centre for Society and Culture - National Research and Innovation Agency, Jakarta, Indonesia
e-mail: surati@brin.go.id

K. Budiningsih
Research Center for Behavioral and Circular Economics - National Research and Innovation Agency, Jakarta, Indonesia
e-mail: ra.kushartati.budiningsih@brin.go.id

A. Wibowo
Department of Forestry, Faculty of Agriculture, University Palangka Raya, Palangka Raya, Indonesia
e-mail: agungwi@for.upr.ac.id

H. Warsilah and A. H. Purnomo (eds.), *Approaching Natural Resources for Sustainable Development in Indonesia*, Sustainable Development Goals Series, https://doi.org/10.1007/978-981-96-6913-4_11

diverse flora and fauna. Within Indonesia, these vital ecosystems encompass a substantial landmass, spanning from 13.4 to 14.9 million hectares (Abdurrahim et al. 2023; Anda et al. 2021; Yuwati et al. 2021). Their significance extends beyond ecological considerations, as they hold profound socioeconomic, environmental, and cultural value, influencing the overall well-being of local communities.

Regrettably, the inadequacies of previous policies have given rise to persistent crises in peatlands that demand urgent and long-term interventions to rectify the situation comprehensively. It was the Peat Land Development Project (Proyek Lahan Gambut or PLG) established by the Orde Baru regime under Presidential Decree No. 82 of 1995 that aimed to convert peatlands in Central Kalimantan province into agricultural fields for food crops, particularly paddy cultivation (Surahman et al. 2018). This initiative is called the "1 million hectares paddy fields project." In its execution, the construction of 1 million hectares of rice fields on peatlands ignored soil conditions, as the primary canals cut through dense peat areas, connecting the entire project zone into a single water system, assuming uniform underground, topographical, and hydrological conditions throughout the project area. Later, this approach posed environmental risks such as fires in the dry season and floods during the rainy season, leading to rice production failures over the years (Goldstein 2016).

Since the early 2000s, the government has taken several measures to restore the function of the damaged peat. Among the most prominent is Presidential Directive 2 of 2007 to accelerate the rehabilitation and revitalization of the PLG area in Central Kalimantan, which covers an area of 1.4 million hectares used as a settlement for transmigrants and consisting primarily of peat. After several years, the programs of the Presidential Directive have not yet been implemented. The Peat Restoration Agency (Badan Restorasi Gambut or BRG), established in 2016, became the Peat Restoration Agency and Mangrove (Badan Restorasi Gambut dan Mangrove or BRGM) in 2020. The BRG(M)'s mission includes accelerating the implementation of peat restoration and improving the welfare of people in the peat restoration area. BRG(M) has a 3R program: rewetting peatlands, revegetating or planting burned areas, and economic revitalization, i.e., improving people's livelihood near peatlands. These programs are still being implemented today, although their effectiveness is questionable.

In 2020, the government launched a program known as Food Estate to reuse degraded ex-PLG 1995 peatlands that had been abandoned for a quarter of a century. These are generally dry peatlands in large hydrological areas. The soil undergoes rapid biogeochemical oxidation and releases large amounts of carbon dioxide into the atmosphere, which is highly susceptible to fire (Goldstein 2016). Some doubt the success of this program and fear even more severe environmental damage. The destructive effects of forest and wildfires, combined with declining water quality, have led to habitat destruction, threatening the sustainability of local communities and undermining social cohesion. In addition, the cultural and spiritual significance of peatland ecosystems to specific ethnic groups increases the potential for loss of cultural heritage and the traditional knowledge associated with resource management.

Addressing this circumstance requires a comprehensive and integrated approach that actively involves local communities, government agencies, and the private sector. Social and agricultural improvements, efficient fire management, and the pursuit of long-term economic viability should be given equal priority. Therefore, this research aims to assess the various issues facing peatland ecosystems in Indonesia, focusing on the social, economic, and cultural aspects. Using case studies in Sumatra and Kalimantan, the study demonstrates the profound impacts of the peatland crisis on the livelihoods of those who depend on these ecosystems. The lessons learned from this research will inform policymakers in formulating sustainable policies and practices that mitigate the crisis and protect peatland ecosystems and the communities that depend on them.

11.2 Methodology

This research employs a qualitative methodology using a case study approach. Data collection encompassed both desk research and field investigations from 2017 to 2022. The chosen unit of analysis for this study consists of multiple cases across various sites (Creswell 2013) within Sumatra and Kalimantan islands (Fig. 11.1), which include:

1. Pulang Pisau Regency and Kapuas Regency in Central Kalimantan Province.
2. West Tanjung Jabung Regency and East Tanjung Jabung Regency in Jambi Province.

The desk study drew on the results of previous research. Qualitative descriptive analysis was used with discourse analysis to examine the data from the desk and field studies. This discourse analysis was explicitly used to examine the various challenges facing peatland ecosystems in Indonesia, focusing on the social, economic, and cultural aspects. Four different approaches to discourse are distinguished: discourse as communication, discourse as text, discourse as frame, and discourse as social practice (Arts and Buizer 2009). This study uses a frame-based discourse approach and a social practice discourse.

11.3 Theoretical Background

Peat ecosystems, largely located in Sumatra and Kalimantan, are essential for global carbon storage, biodiversity conservation, and water regulation. Despite their importance, these ecosystems are currently facing several crises that endanger their sustainability and the well-being of local communities. The utilization of natural resources, particularly peatlands, is a contentious topic sparking ongoing discussion.

The expansion of agriculture, particularly palm oil plantations, has led to extensive deforestation and land conversion. Peatlands are often drained to make way for crops, which leads to the oxidation of peat soil and the release of significant amounts of carbon dioxide into the atmosphere. This process not only contributes to climate change but also degrades the land, making it more susceptible to fires.

The crisis facing Indonesia's peatland ecosystem is a significant challenge that requires collective attention and concerted action from all relevant stakeholders. This dilemma is inherently complex, requiring a delicate balance between meeting economic needs, protecting the environment, and empowering local communities in natural resource management. Collaborative efforts and participatory policy formulation are expected to reduce these problems. Therefore, restoration of damaged peatlands must be planned and implemented carefully.

This article reviews the various crises impacting Indonesia's peatlands and suggests strategies for sustainable management. This study of the peat ecosystem crisis was analyzed through an analytical discourse, incorporating findings from secondary and primary data sources. The analytical framework used in this study includes discourse as a frame and discourse as a social practice. Arts and Buizer (2009) provide the following insights into the concepts of discourse as a framework and discourse as social practice.

11.3.1 Discourse as Framework

Discourse as a frame draws upon society's historical and experiential context, whether consciously or unconsciously, and significantly shapes societal behaviors. Schön and Rein (1994) consider that uncovering diverse frames underpinning a problem can lead to a process of re-framing, which is essential for reaching a shared resolution. In essence, the frames employed by individuals provide the context for understanding the problem situation that needs to be identified.

In the context of this study, it becomes imperative to recognize the multitude of crises confronting peat ecosystems in Indonesia and elucidate the associated issues. When the crisis is framed as a predicament with both local and global origins, the ensuing policies required for addressing it will also differ. This research aims to delineate and pinpoint the challenges of managing

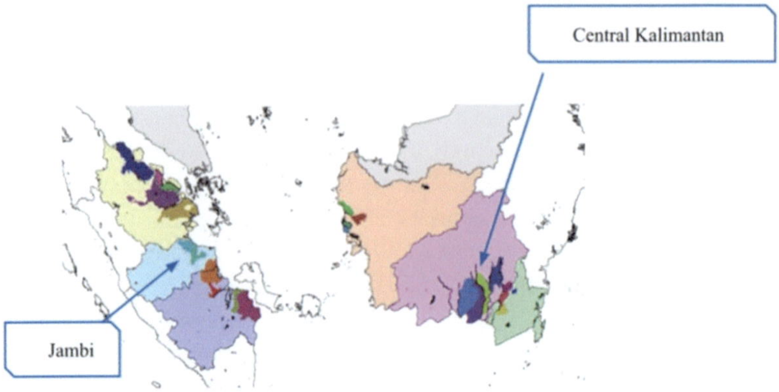

Fig. 11.1 Map of research locations. (*Source* Ministry of Environment and Forestry (2023))

peat ecosystems rooted in historical contexts. Furthermore, it assesses the efficacy of prevailing governance policies in addressing these issues and promoting sustainable management of peat ecosystems.

11.3.2 Discourse as Social Practice

The social practice approach to discourse finds its roots in the work of Michel Foucault (1984). According to Foucault, the relationship between discourse and power is profound, extending beyond the conventional notion of individuals wielding power through texts or interactions. Instead, power is inherently embedded within the discourse itself. Foucault's perspective on discourse suggests that it actively shapes society across various dimensions, including the construction of knowledge, the definition of social subjects, the formation of self-identities, the establishment of social relations, and the framing of perception. This approach underscores the interconnectedness of social discourse and practice, encompassing institutional structures and power dynamics (Foucault 1994).

In the context of this study, we critically examine the crisis in peat ecosystems and the pursuit of sustainable peatland management through three critical lenses:

1. We investigate the historical emergence of this discourse, exploring the underlying episteme and epistemic community that have shaped it over time.
2. From both social and economic perspectives, we assess how this discourse functions to discipline modes of thinking, speaking, and acting concerning the peat ecosystem crisis.
3. We analyze how this discourse normalizes specific behaviors, particularly concerning governance and the sustainable management of peatlands.

11.4 Results and Discussion

11.4.1 Peat Ecosystems in Indonesia

The peatland area in Indonesia is about 14.91 million ha, of which Sumatra accounts for 6.44 million ha (43%), Kalimantan for 4.78 million ha (32%), and Papua Islands for 3.69 million ha (25%), as shown in Table 11.1 (Anda et al. 2021; Surahman et al. 2018; Yuwati et al. 2021). Central Kalimantan Province has the largest peatland on the island of Kalimantan, with more than 2.5 million hectares. Likewise, Jambi Province is one of the provinces with the largest peat area on the island of Sumatra. The area of peatland forest in Jambi Province extends to 736,227.20 ha or around 14% of the province's total area spread over six districts: East Tanjung Jabung Regency (covering 311,992.10 ha), Muaro Jambi Regency (229,703.90 ha), West Tanjung Jabung Regency (covering an area of 154,598 ha), Sarolangun

Regency (covering an area of 33,294.2 ha), Merangin Regency covering an area of 5809.8 ha, and Tebo Regency (covering an area of 829.2 ha) (Yari 2017).

Table 11.1 Peatland area in Indonesia

	Area (Million ha)	Percentage (%)
Sumatera Islands	6.44	43
Kalimantan Islands	4.78	32
Papua Islands	3.69	25
Total	14.91	100

Sources Processed data, 2023

Given the extensive peatlands on these islands, inappropriate management poses an environmental, social, and economic threat that can lead to disasters. This vulnerability stems from the fragile and degradation-prone nature of peat ecosystems, which contrasts sharply with mineral soils. It is crucial to give careful consideration to the utilization of peat. Alongside efforts to restore and rehabilitate affected regions, there is an expectation that this will reduce the risk of forest and land fires. Hence, it is of utmost importance for both the government and the community to be vigilant and take actions that preserve and safeguard the sustainability of our forests.

The formal utilization of peat in Indonesia commenced during the 1960s. Inadequate management of peat has led to its degradation, prompting the government to introduce multiple regulations to enhance peatland governance. Following the extensive fire outbreaks of 2015, primarily on peatlands, the government implemented a policy prohibiting land clearing for burning and initiated ongoing efforts to prevent forest and land fires. One notable initiative in this regard was the establishment of the Peat Restoration Agency (BRG).

The issues within peat ecosystems include the following:

1. They are vulnerable to forest and land fires.
2. The productivity of peatlands relies heavily on management and human activities. Consequently, if peatland management neglects the specific characteristics of the peatland, it can result in decreased productivity due to soil fertility degradation and alterations in the physical and biological properties of the soil (Masganti et al. 2017).
3. Diverse methods of forest clearance are practiced within peat ecosystems, leading to the drainage of peat soil. This drainage results in significant carbon emissions, which escalate during periods of fires, particularly in El Niño years. These peatland fires also entail substantial economic and public health losses and frequently give rise to diplomatic tensions with neighboring nations (Giesen and Sari 2018).
4. The types of plants cultivated on peatlands struggle to thrive when their roots are submerged in the acidic, nutrient-poor waters of peatlands. Consequently, communities often resort to drainage to remove water from these peatlands. Mismanagement of water and unsuitable drainage techniques in peatlands can result in severe environmental issues for peatland ecosystems, including land subsidence, increased fire incidents, and elevated greenhouse gas emissions (Napitupulu and Mudiantoro 2015).

11.4.2 Economic, Socio-Cultural, and Ecological Advantages

Peatlands are the primary source of livelihood for local communities, albeit with some limitations associated with peat cultivation. Drainage practices associated with peat cultivation carry the risk of triggering fires. Social and economic interests often come into conflict with ecological concerns. Most of these peatlands are located near the coast, especially along the east coast of Sumatra and Kalimantan's southern and western coastal regions.

The experience from Jambi Province underscores the notion that community-managed assets in the form of peatlands can yield more economic benefits if used to grow crops consistent with the community's natural capabilities and knowledge. This approach contrasts with government policies that implement initiatives to restore peatland ecosystems. Community efforts to grow sengon (*parasarianthes falcataria*) or similar plants that are economically valuable to the community have not yet proven suitable for planting in peat swamp forests.

Most people living near forested regions are primarily engaged in agriculture. Therefore, peatlands converted into agricultural fields and plantations are the primary source of their livelihood. These people construct drainage ditches or canals, either manually or with mechanical aids, to establish drainage systems to regulate water and facilitate the transportation of agricultural and plantation products. Given the presence of these communities in peatlands, the Forestry Administration, Forest Management Unit (KPH) is considering using agroforestry techniques to restore the land. This approach involves creating plant cover that has economic and environmental value and provides employment opportunities and a sustainable source of income for the community. Agroforestry, characterized by the integration of agriculture and forestry, is a widespread practice among smallholder farmers in Indonesia and plays a critical role in promoting sustainable livelihoods, for example in agroforestry on peatland in Sido Mukti Village, Jambi Province (Surati 2021) (Fig. 11.2).

In Jambi Province, the population mostly uses the peatlands as residential farms, rice fields for rice cultivation, and gardens (Surati et al. 2019). Most of the inhabitants in the forested areas are engaged in agriculture, using peatlands as a primary resource for agricultural purposes and plantations to sustain their livelihoods. Agricultural use of peat soils is generally more complicated than mineral soils, requiring higher input use and a more complex water management system, which can have negative environmental impacts (Noor et al. 2014). In the West Tanjung Jabung region, people rely mainly on the cultivation of oil palm and areca nuts, as these crops can be quickly grown on peat soils despite the continuing decline in production yields.

The societal significance of peatlands in Sumatra and Kalimantan encompasses adopting sustainable peat cultivation practices aligned with local customs and culture. The Dayak tribe, for instance, has preserved its Indigenous knowledge over time, which includes the cultivation of peat-adapted species such as *belangeran* (*Shorea balangeran*), *meranti kuning* (*Shorea macrobalanos*), *jelutong* (*Dyera polyphylla*), and *galam* (*Melaleuca cajuputi*). These species craft various products, such as woven mats and household items. In addition, the *handep* cooperative system of planting paddy fields on peat soils is still practiced today, although at a decreasing rate. In Sumatra, the tradition of rice cultivation on *sonor* peat lands remains active.

Peatlands are prone to progressive degradation, characterized by changes in physical, biological, and chemical properties that lead to impaired function and ecological degradation. These changes have adverse effects on both the environment and socioeconomic development. The primary driver of peat degradation is usually land-use conversion from peat swamp forests to agriculture and other purposes. Nevertheless, peatlands also possess ecological functions.

Peatlands in Sumatra and Kalimantan fulfill numerous ecological, hydrological, and biochemical functions at local and regional scales. The remaining peatlands serve as sanctuaries for endangered species in declining lowland forest areas on mineral soils (Miettinen and Liew 2010).

11.4.3 The Crisis in Peat Ecosystems

Peatland forests in Indonesia amounting to approximately 13 million hectares have suffered extensive degradation. This can be attributed to various factors, including agricultural activities, logging, and land conversion for other purposes. During the Suharto era, Indonesia initiated the ambitious one-million-hectare Peat Rice Fields (PLG) project to convert vast peat areas into rice fields. However, as this project was in an area unsuitable for rice cultivation, it resulted in minimal rice production but led to the deforestation and draining of hundreds of thousands of hectares of peat forest, which releases millions of tons of CO_2 into the atmosphere annually. Once tropical peat soils, such as those in Indonesia, dry

Fig. 11.2 Agroforestry on peatland in Sido Mukti Village, Jambi. (*Source* Research data collection (2018))

out, they are irreversibly damaged. Peat ecosystems are further characterized by their susceptibility to subsidence, limited carrying capacity, low fertility, and limited number of microorganisms (Masganti et al. 2014).

To meet local and national food needs, a food program was initiated to convert forested land into rice fields. However, implementation of the project fell short of government expectations as soil conditions proved unsuitable for rice cultivation, leading many people to opt for alternative crops, including cassava.

Rice cultivation on peatlands is viable when the peat layer is shallow or not part of a peat dome. Some areas have relatively shallow peat, making them suitable for rice cultivation. Alternatively, farmers may consider growing floating rice, which grows more slowly and produces lower yields but requires less extensive fertilizer application.

Changing a community's deeply ingrained behaviors and practices is a complex undertaking that requires both time and sustained effort. This is particularly evident in the case of peatland residents who are asked to participate in the rice paddy cultivation program. The community's deeply rooted farming traditions cannot be changed quickly to adopt rice cultivation because each social class or group is differently receptive to innovation (Ekawati et al. 2021). Numerous factors influence the acceptance of new practices. Despite many years of adopting sustainable peat management techniques, community acceptance remains relatively low. Implementing the rice paddy cultivation project on peatlands has also created social tensions and conflicts within the community, as not all members were able to participate in the project. Communities or farmers participating in the initiative receive support through seeds, fertilizer, and pesticides. An overview of the peat ecosystem and its problems at the study location is presented in Table 11.2.

From a socio-cultural and economic perspective, there are social problems with peat ecosystems in Jambi and Central Kalimantan Provinces resulting from the introduction of government policies without appropriate guidance and assistance. In Jambi, socio-cultural problems are caused by an inappropriate government program (transmigration) that is moving people who are used to working on mineral soils to peatlands without properly equipping them with knowledge about peatlands. This results in people switching from farming to working as laborers in palm oil or other companies.

Table 11.2 The peat ecosystem and its problems

Description	Jambi	Central Kalimantan
Socio-cultural		
• Population composition	• Generally migrants (transmigrants) from Javanese, Bugis, Mining, Ekso, Acehnese	• Dayaknese, Banjarese, Bugis, and Javanese
• Social problems	• Inappropriate government program (transmigration) that moves people who are used to working on mineral soils to peatlands without equipping them with knowledge about peatlands	• Conflicts over ownership and rights to peatlands between Indigenous communities, farmers, and industrial companies • Erosion of traditional land management as a result of climate change and government policies
• Social impact	• Switching from farmers to laborers in palm oil companies or other laborers	• The loss of the traditional *handep* culture of *gotong royong* planting. Weakening of identity
• Troubleshooting option	• Capacity building of farmers for understanding the biophysical conditions of peatlands and their management	• The Food Estate Program's inclusive policies can continue with continued support
Economic		
• Livelihood or main profession	• Farmer, gardener, and laborer	• Farmer and laborer
• Commodities	• Palm oil and areca nut • Oil palm interspersed with areca nut and vegetables in agroforestry system	• Rubber, palm oil, rattan, meranti, lime, ramin
• Economic problems	• Unsuitable use of peatland in terms of its biophysical properties • Some species planted on peatlands fail to grow properly while their roots are inundated in acidic water, and they find fewer nutrients	• Over-reliance on a single economic activity, such as palm oil cultivation or timber harvesting • Unsustainable agricultural practices • Lack of community involvement in economic decision-making
• Economic impact	• A significant amount of land is sold, while the local government acquires land on mineral soils • A considerable portion of land remains unproductive due to a lack of cultivation, particularly in deep peat areas	• Bareland increases due to the non-burning policy for agricultural activities • Community food reserves are obtained by purchasing • Sale of unproductive agricultural land
• Troubleshooting option	• Choosing the appropriate plant species based on local conditions and their significant economic worth • Socialization of alternative types of commodities that can grow well on land and have high economic value • Agroforestry combination: rambutan—(*Nephelium lappaceum*)—purun (*Lepironia articulata*), kopi liberika (*Coffee liberica*)—pinang (*Areca catechu*), jelutung (*Dyera polyphylla*)—palawija (secondary crops), jelutung (*Dyera polyphylla*)—karet (*Hevea brasiliensis*)	• Promote livelihood diversification by supporting alternative income sources for communities dependent on peat ecosystems • Advocate for sustainable agricultural methods that are compatible with peatland conditions • Promote community-based enterprises and cooperatives that empower local communities

Sources Processed data results, 2023

Meanwhile, in Central Kalimantan, socio-cultural problems arise due to conflicts over ownership and rights to peatlands between Indigenous communities, farmers, and industrial companies. There is also erosion of traditional land management because of climate change and government policies. This has an impact on the loss of the traditional *handep* culture of *gotong royong* (cooperative system) planting, thereby weakening the identity of the community which is generally the Dayak tribe.

Economic problems in Jambi Province include the inappropriate use of peatlands in terms of their biophysical properties. Some species planted on peatlands do not grow properly when their roots are inundated with acidic water, and they find fewer nutrients. Meanwhile, in Central Kalimantan, there is over-reliance on a single economic activity, such as palm oil cultivation or timber harvesting, as well as unsustainable agricultural practices and a lack of community involvement in economic decision-making.

Of the socio-cultural and economic problems that occur in Jambi and Central Kalimantan, the solutions that must be implemented by the government in sustainable development efforts include:

1. Promoting livelihood diversification by supporting alternative income sources for communities dependent on peat ecosystems.
2. Advocating for sustainable agricultural methods that are compatible with peatland conditions.
3. Promoting community-based enterprises and cooperatives that empower local communities.
4. Choosing the appropriate plant species based on local conditions and their significant economic worth.
5. Socialization of alternative types of commodities that can grow well on land and have high economic value.
6. Agroforestry combination: rambutan (*Nephelium lappaceum*)—purun (*Lepironia articulata*), kopi liberika (*Coffee liberica*)—pinang (*Areca catechu*), jelutong (*Dyera polyphylla*)—palawija (secondary crops), karet (*Hevea brasiliensis*).

The impacts of the peat ecosystem crisis include increased vulnerability to forest and wildfire, diminished water quality, and habitat degradation, all of which pose significant threats to the community's long-term viability. For certain ethnic groups, peat ecosystems hold profound spiritual and traditional significance. The loss of these ecosystems threatens cultural heritage, erodes local knowledge of traditional practices, and undermines knowledge of sustainable natural resource management and use. In addition, this crisis can potentially undermine the socio-cultural identity of affected populations. Addressing these multifaceted challenges requires a comprehensive, integrated approach that actively engages local communities, government agencies, and the private sector. Such an approach aims to improve social and agricultural practices, mitigate fire-related risks, and safeguard economic sustainability.

11.4.4 Governance Improvements

Efforts to improve governance and promote sustainable development in Indonesia's forest sector, particularly with regard to peatland ecosystems, often have limited long-term impact. Reform efforts tend to focus on bringing about legal and policy changes in an environment where numerous influential stakeholders have individual and collective interests that oppose implementing these regulations and policies. Consequently, outcomes are usually limited to the allocation and use of land and forest resources (Berenschot et al. 2022).

In 2015, a series of devastating forest fires, the most severe in the last two decades, affected more than 2.6 million hectares of forest, mostly in peatlands, resulting in ecological damage (Giesen and Sari 2018; Wicaksono and Zainal 2022). The following year, the Indonesian government initiated a peat restoration policy, and the Peat Restoration Agency was established following Presidential Decree No. 1 of 2016. The Peatland Restoration Agency (BRG) is a non-structural institution reporting directly to the President and overseeing future peatland restoration and management efforts. This is of utmost importance due to the critical role of peatlands and their historical association with wildfire disasters. The BRG is expected to be pivotal in coordinating and harmonizing various peat restoration programs.

In the second phase of the restoration policy, the government revived the BRG and transformed it into the Peat and Mangrove Restoration Agency (BRGM) with Presidential Regulation Number 120 of 2020. The three BRGM entities play a pivotal role in accelerating peatland restoration initiatives, improving the welfare of communities in peatland restoration areas, and implementing accelerated mangrove rehabilitation in designated provinces.

Government-led rehabilitation efforts are already underway, yet the outcomes achieved to date remain suboptimal. Several challenges remain, including weak law enforcement against violations of restoration activities and a lack of synergy among parties charged with implementing restoration activities. Peatland restoration activities have principally succeeded in approximately a million hectares of non-concession areas, i.e., regions directly managed by the government, accounting for 87% of the overall restoration target (Wicaksono and Zainal 2022). However, the implementation of peatland restoration activities within company concessions has been less successful, as the government is limited to monitoring restoration efforts.

Peatland management in Indonesia has been improved through various regulatory tools, including introducing the 3Rs approach (rewetting, revegetation, and revitalizing local livelihoods). In addition, the institutional aspects of peatland management now emphasize sustainability more than solely prioritizing economic considerations. The process of rewetting degraded peatlands is primarily through channel blocking. However, although recognizing that channel blocking does not immediately restore the hydrological function of peat, it gradually raises the water table. To improve the effectiveness of canal blocking, careful attention must be paid to construction techniques and spacing between canal blockages (Yuwati et al. 2021).

Many efforts are required to restore the hydrological balance of peat ecosystems:

1. Initiatives to restore peat hydrology include channel blocking, peat rehydration, and replanting with non-timber forest product (NTFP) species well adapted to peat conditions. A critical challenge in maintaining such a system is to provide undisturbed access to facilitate NTFP harvest. In areas that are entirely inundated, there are three primary approaches to improving access: alternative channel blockages, footpaths, rail systems, and adjustments to harvest schedules.
2. Increase community awareness of suitable species with high economic value, mainly focusing on indigenous plant species and flora native to peat swamps with potential for paludiculture programs.
3. Enhancing peatland management by fostering synergy and collaboration among diverse agencies.

11.5 Recommendation and Conclusion

11.5.1 Recommendation

1. To make the sustainable peatland management program a success, the central government should implement clear sustainable and responsible management practices. This requires a comprehensive approach that recognizes the region's ecological importance and considers its social and economic dimensions. It also includes steps to prevent forest fires, encourage responsible land-use practices, restore degraded peatlands, and support local communities in adopting more sustainable livelihoods.
2. Overcoming the multi-crisis facing peat ecosystems in Indonesia requires a collective effort from government authorities, environmental organizations, and local communities. It also requires a commitment to safeguard these valuable ecosystems to benefit present and future generations while mitigating adverse environmental and societal impacts.

11.5.2 Conclusion

The findings of this research on peat ecosystems in Indonesia, particularly in Sumatra and Kalimantan, reveal a multifaceted crisis encompassing social, economic, and cultural dimensions. Peat ecosystems, characterized by their high carbon content, are vital to Indonesia's natural landscape. They serve as critical carbon reservoirs, water regulators, and habitats for diverse flora and fauna, including endangered species. Despite their substantial expanse, these ecosystems face a precarious situation, prone to environmental disasters and associated ecological, social, and economic upheavals when not managed judiciously. This fragility is primarily attributable to their unique characteristics, notably their extreme susceptibility and sensitivity to degradation, which set them apart from conventional mineral soils.

Several vital issues plague these invaluable peat ecosystems. First and foremost, their vulnerability to forests and wildfires presents a significant challenge. The dry conditions often associated with peatlands make them susceptible to ignition, leading to devastating fires that release massive amounts of stored carbon into the atmosphere, exacerbating climate change and endangering local communities.

Moreover, the productivity of peat soils is intricately intertwined with human activities and proper management practices. Unsustainable land-use practices, such as extensive agriculture and drainage for industrial development, disrupt the delicate balance of these ecosystems, resulting in soil degradation, subsidence, and a host of associated environmental problems.

Another critical concern arises from the various methods employed to clear forests in peat ecosystems, which leads to the drainage of peat soils. These practices further exacerbate environmental issues, compromising the ability of these lands to store carbon and manage water effectively.

Furthermore, the cultivation of plants on peatlands encounters significant challenges. The inundation of plant roots with acidic water and a shortage of organic matter and nutrients in the water can hinder the optimal growth of vegetation, impacting agricultural productivity and food security for communities relying on these lands.

Indonesia's peat ecosystems are at a pivotal moment. The various crises they confront require immediate and unified efforts to halt further degradation and secure their long-term viability. Embracing a comprehensive strategy that includes restoration, community engagement, strong policy measures, and continuous research can set Indonesia on a sustainable course, benefiting both the environment and local populations. Safeguarding these crucial ecosystems is essential not just for the nation but also globally in the battle against climate change.

References

Abdurrahim AY, Arya HD, Soeryo A, Herry Y, van Noordwijk M (2023) Relational and instrumental values of tropical peat landscapes: morality and political ecology in Indonesia. Curr Opin Environ Sustain 64:1–10. https://doi.org/10.1016/j.cosust.2023.101318

Anda M, Sofyan R, Erna S, Sukarman MH, Edi Y, Anny M, Rudi ES, Suratman H (2021) Revisiting tropical peatlands in Indonesia: semi-detailed mapping, extent and depth distribution assessment. Geoderma 402:1–14. https://doi.org/10.1016/j.geoderma.2021.115235

Arts B, Marleen B (2009) Forests, discourses, institutions: a discursive-institutional analysis of global forest governance. Forest Policy Econ 11(5–6):340–347. https://doi.org/10.1016/j.forpol.2008.10.004

Berenschot W, Edward A, Marcus C, Angus M (2022) Politik Kehutanan Di Indonesia Deforestasi Dan Perampasan Lahan

Creswell JW (2013) Qualitative inquiry research design: choosing among five approaches. Third Edition. SAGE Publications

Ekawati S, Sylviani S, Ramawati H, Prasetyo BD, Lugina M, Hidayat DC, Sumirat BG, Hardiansyah G, Mariani Y, Yusro F, Oktoria S, Kusrini N, Pranoto DY, Roslinda E (2021) Factors affecting communities in adopting sustainable peat cultivation techniques and strategies for implementation (a case study in Pulang Pisau, Central Kalimantan). In: Proceedings of International conference of Indonesia Forestry Researchers (INAFOR) 2021, pp 1–10

Foucault M (1994) Power. The New Press, New York

Giesen W, Sari ENN (2018) Tropical Peatland restoration report: Indonesian case. Jakarta. https://doi.org/10.13140/RG.2.2.30049.40808

Goldstein J (2016) Carbon Bomb: Indonesia's Failed Mega Rice Project. Arcadia 6

Masganti KA, Susanti MA (2017) Potensi Dan Pemanfaatan Lahan Gambut Dangkal Untuk Pertanian. Jurnal Sumberdaya Lahan 11(1):43–52

Masganti W, Ai Dariah N, Rachmiwati Y (2014) Characteristics and potential utilization of degraded Peatlands in Riau Province. Jurnal Sumberdaya Lahan 8(1):59–66

Miettinen J, Soo Chin L (2010) Status of Peatland degradation and development in Sumatra and Kalimantan. Ambio 39:394–401. https://doi.org/10.1007/s13280-010-0051-2

Napitupulu SM, Mudiantoro B (2015) Pengelolaan Sumber Daya Air Pada Lahan Gambut. In Annual Civil Engineering Seminar. Pekanbaru : Pusat Studi Infrastruktur, pp 1–8

Noor M, Nursamsi D, Alwi M, Fahmi A (2014) Prospek Pertanian Berkelanjutan Di Lahan Gambut : Dari Petani Ke Peneliti Dan Peneliti Ke Petani. Jurnal Sumberdaya Lahan 8(2):69–79

Schön D, Rein M (1994) Frame reflection: toward the resolution of intractable policy controversies. Basic Books, New York

Surahman A, Soni P, Shivakoti GP (2018a) Reducing CO_2 emissions and supporting food security in Central Kalimantan, Indonesia, with Improved Peatland Management. Land Use Policy 72:325–332. https://doi.org/10.1016/j.landusepol.2017.12.050

Surati (2021) Study on socio-economic of Peatland Community in Tanjung Jabung Timur Regency, Jambi. Jurnal Ilmu Kehutanan 15(2):147–159. https://doi.org/10.22146/jik.v15i1.1531

Surati SI, Di Charity H, Ariawan K, Kurniawan AS, Mulyadin RM (2019) Analysis of livelihoods in Peat Land: the case of Tanjung Jabung Barat regency. Jurnal Penelitian Sosial Dan Ekonomi Kehutanan 16(2):81–93

Wicaksono A, Zainal (2022) Peatlands restoration policies in Indonesia: success or failure? In: IOP conference series: earth and environmental science. vol. 995. IOP Publishing Ltd, pp 1–9

Yari D (2017) Peat characterization at peat hydrological unity of londerang protected forest East Tanjung Jabung District Jambi Province. Universitas Sriwijaya, Palembang

Yuwati TW, Dony R, Pratiwi MT, Yonky I, Hunggul YSHN, Muhammad AQ, Budi HN, Bondan W, Sri L, Purwanto BS, Rahardyan NA, Endang S, Pamungkas BP, Reni SW, Retno P, Wawan H, Besri N, Bastoni DM (2021) Restoration of degraded tropical Peatland in Indonesia: a review. Land 10(1170):1–33. https://doi.org/10.3390/land10111170

Part IV

Natural Resource Management from an Environmental Perspective

The Extinction of the Right to Water for Communities in the Sinking Area: A Case Study of Timbulsloko Hamlet, Sayung District, Demak Regency

12

Karmilah Mila, Khaeruddin Akhmad, Handriana Eka, and Madrah Muna Yastuti

Abstract

Access to clean water in Indonesia remains relatively low, primarily due to a lack of available clean water facilities and infrastructure. This conflicts with the Sustainable Development Goals (SDGs), which emphasize the importance of providing water, sanitation, and hygiene (WASH) services as part of Goal 6. Dukuh Timbulsloko, located in the Sayung Sub-district, has been grappling with erosion and relentless tidal waves since 2017, resulting in a shortage of clean water and proper sanitation. This research aims to investigate whether residents in this affected area are able to fulfill their basic rights to clean water and proper sanitation. The study employs a qualitative research methodology, utilizing ethnographic studies to describe the daily activities and events within the community, thus providing a comprehensive understanding of the situation in the research area. The findings of this study reveal that the Timbulsloko community employs various methods to meet their clean water needs. While they primarily rely on PAMSIMAS (Provision of Clean Water and Environmental Sanitation Program) and artesian wells, they also engage in rainwater harvesting. However, these solutions are not sustainable in the long term due to ongoing erosion, land subsidence, and frequent tidal surges in the region.

Keywords

Extinction · Right to water · Sinking area · Abrasion · *Rob*

K. Mila (✉) · K. Akhmad · H. Eka
Amerta Air Indonesia Foundation, Surabaya, Indonesia
e-mail: mila.k@unissula.ac.id

K. Mila · K. Akhmad · M. M. Yastuti
Faculty of Islamic Studies, Universitas Islam Sultan Agung, Semarang, Indonesia

12.1 Introduction

Some of the world's population who struggle to access their right to water reside in Timbulsloko Hamlet, Timbulsloko Village, Sayung District, and Demak Regency. According to the Environmental Health Risk Assessment (EHRA) study in the Buku Putih Demak Regency Sanitation, only 10% of the freshwater supply in Timbulsloko Village is met. Consequently, 90% of the Timbulsloko Village population, including Timbulsloko Hamlet residents, cannot access clean water, either through piped or non-piped services (Pemerintah Kabupaten Demak 2012). Timbulsloko Hamlet is one of four hamlets in Timbulsloko

H. Warsilah and A. H. Purnomo (eds.), *Approaching Natural Resources for Sustainable Development in Indonesia*, Sustainable Development Goals Series, https://doi.org/10.1007/978-981-96-6913-4_12

Village. The other three hamlets are Bogorame Hamlet, Wonorejopasir, and Karanggeneng. Bogorame and Wonorejopasir are settlements directly adjacent to the Java Sea, while Timbulsloko Hamlet and Karanggeneng are farther inland. However, Dukuh Timbulsloko is the settlement most severely affected by coastal erosion and flooding. There is no dry land left in this hamlet; all settlements have become inundated areas. Residents of Timbulsloko Hamlet live in houses with floors elevated on wooden planks. Similarly, the connecting roads between houses in the settlement are elevated with wood. To leave the village, the people of Timbulsloko rely on boats and dirt roads that require constant maintenance due to tidal flooding. Coastal erosion and flooding in Timbulsloko can be attributed to two main causes: climate change, which is linked to rising global temperatures, accelerating the melting of polar ice caps; and land subsidence resulting from the consolidation and compaction of alluvial soil structures, heavy infrastructure loading, and extensive deep groundwater extraction (Batubara and Wagner 2020).

12.2 Theoretical Background

Water is an essential human need. To lead a decent life, humans require clean water for drinking and hygiene on a daily basis. The recognition of this need for clean water forms the basis for the human right to access water. After a lengthy journey, on July 28, 2010, the UN General Assembly passed Resolution No. 64/292 (The Right to Safe and Clean Drinking Water and Sanitation as Human Rights, 2010), officially acknowledging that the right to safe and clean drinking water and sanitation is a fundamental human right. This resolution includes a call to action for countries and international organizations to provide funding, capacity building, and technology transfer through international assistance and cooperation, with a special focus on developing countries. The aim is to enhance efforts to provide safe, clean, easily accessible, and affordable drinking water (Taufik Basari 2012).

On the other hand, water is a finite natural resource, making it susceptible to scarcity and accessibility issues. The global use of water is increasingly complicating the realization of the right to water. According to statistics presented in the Right to Water Sheet #1, published by the Center for Socio-Economic Rights in 2010, less than 1% of the world's total water availability is freshwater, with the remainder being saltwater, locked in ice caps, or otherwise inaccessible. Alongside population growth, industrialization, and pollution, there has been a 58% decrease in the availability of renewable water sources. Currently, 460 million people face severe water shortages, and 1.1 billion people, equivalent to one-sixth of the world's population, lack access to safe water. These challenges are exacerbated by inadequate water supply systems. For instance, in Africa, water loss through distribution systems has reached 39%, in Asia it is 42%, and in North America, 15%. Additionally, as much as 90% of waste in developing countries is discharged directly into rivers without proper treatment, further contaminating clean water sources (Center for Socio-Economic Rights 2010).

Clean water is defined as water that is safe and suitable for drinking: it should be colorless, odorless, and possess a fresh taste (Suripin 2002). According to Regulation No. 416/Menkes/PER/IX/1990 of the Minister of Health of the Republic of Indonesia, clean water is water intended for daily use, including drinking after proper cooking. Clean water, in essence, is healthy water, meaning it must be free from disease-causing germs, enabling its use in various daily activities. The quality of clean water with regard to health is assessed through microbiological criteria, such as the presence or absence of *E. coli* and/or total coliform bacteria. Additionally, the concentrations of elements including arsenic, fluoride, cadmium, nitrite, cyanide, and selenium, are considered. Sanitation and hygiene, as defined by the Indonesian Ministry of Health, encompass actions aimed at enhancing hygiene and health by maintaining both individual and environmental factors that influence health. These actions are intended to protect individuals from the threats posed by disease-causing microorganisms. Sanitation primarily focuses on preventive health measures aimed at achieving a healthy human life, while hygiene is defined as the effort to preserve and safeguard individual cleanliness (Durasih 2011).

Access to water, sanitation, and hygiene (WASH) is recognized as a global human right, ensuring that no one is excluded based on geographical location, socioeconomic status, or individual characteristics. In Indonesia, this right is enshrined in the 45th Constitution (Article 33) and Law 17/2019 on Drinking Water. Furthermore, the right to WASH access is recognized from a public health perspective, specifically through Law 36/2009 (Right to Health and a Healthy Environment) and Law 32/2009 (Right to a Healthy Environment). WASH in Indonesia is also designated as a government responsibility and duty, as outlined in Law No. 23/2014, Law No. 17/2019, Presidential Regulation No. 59/2017, and Presidential Regulation No. 18/2020 (Sudrajat 2020). More fundamentally, the utilization of water in Indonesia is governed by the 45th Constitution's amendment in Article 33 Paragraph (3), affirming that water, including clean water, is a common good right that must be ensured by the state for all Indonesian citizens. Consequently, all citizens, regardless of their circumstances, should have equitable access to clean water and proper sanitation (Undang-Undang Dasar Negara Republik Indonesia Tahun 1945).

Access to water and sanitation is acknowledged by the United Nations (UN) as a fundamental human right, essential for the well-being, dignity, and prosperity of all individuals (United Nations Water, n.d.-b). Clean water and adequate sanitation, as basic human needs, are also enshrined in the Sustainable Development Goals (SDGs). Sanitation and clean water are specifically addressed in Goal 6, though it is important to note that these goals are interconnected (Kementerian PPN/Bappenas, n.d.). However, a 2014 report by WHO and UNICEF revealed that 748 million people lack access to clean water, and 2.5 billion people lack adequate sanitation. The latest UN-Water report indicates that "billions of people still live without safely managed sanitation", with 'safely managed' implying consistent and sustainable access. Thus, billions of people have yet to fully enjoy their right to life. These marginalized and vulnerable groups include women, children, refugees, Indigenous peoples, the disabled, and the impoverished, among others. UN-Water highlights that these marginalized individuals are often overlooked and actively discriminated against by those responsible for planning and managing water and sanitation services. Discrimination may also come from other service users (United Nations Water, n.d.-a).

Another fundamental aspect of water as a human right is the right to clean water, given its paramount role as the primary necessity for sustaining daily economic activities. The international average standard for daily water requirements is set at 20 L per person, with 4 L designated for consumption and the remainder allocated for various other activities. However, ensuring sufficient access to safe drinking water remains a critical challenge in Indonesia. In 2015, at the district level, only 49% of the population had access to clean water (Sukartini and Saleh 2016).

12.3 Methodology

Ethnography is a qualitative research methodology that focuses on uncovering sociological meaning through close-field observations of sociocultural phenomena. This method involves the detailed documentation of activities and events that unfold daily within a specific community, offering a comprehensive understanding of the research area. Ethnography is particularly useful for grasping the intricacies of the environment, the dynamics within communities, interactions among ethnic groups, power dynamics, communication patterns, and more (Spradley 1997). Timbulsloko Hamlet was selected as the research location due to its status as one of the most vulnerable areas affected by coastal erosion, flooding, and other climate-related changes. Data collection was carried out by immersing researchers in the daily life of Timbulsloko residents, involving first-hand experiences of seeing, hearing, and physically engaging in the community's activities, including how residents access water, sanitation, and hygiene (WASH) services. The analysis phase involved a thorough examination of theories related to the three primary objectives of this research. The final outcome of ethnographic research is a descriptive narrative that provides a comprehensive account, accompanied by interpretations of all aspects of life and a depiction of the complexities inherent in community life within the research location (Fig. 12.1).

12.4 Results and Findings

12.4.1 Abrasion and Tidal Flood (Rob) in Timbulsloko Hamlet

In the 1960s, the Timbulsloko settlement was surrounded by rice fields, with farming serving as the primary source of livelihood during that period. Significant changes occurred in the 1970s and 1990s when portions of the rice fields in Timbulsloko Hamlet were converted into aquaculture areas, a transformation that also extended to rice fields in the other three hamlets. The aquaculture ponds in Timbulsloko utilize a unique irrigation system that relies on tidal waves flowing back from the sea through two rivers, Kali Barijah and Kali Baru. Timbulsloko Village is traversed by five rivers, two of which pass through Timbulsloko Hamlet, with one of these rivers, Kali Bajirah, running through a residential area.

The number of residents employed as laborers is now the largest (Karmilah et al. 2023; BPS Kabupaten Demak 2021). In 2019, there were 15,000 laborers in Sayung, with less than half that number of farmers and fishermen. For residents who do not have ponds, their income from fishing is also smaller. The root of the problem is that the tidal floods are increasingly eroding the Sayung coast. As a result, many fishermen have become factory workers instead, in factories both in Sayung and in Semarang City (see Table 12.1).

The shift in land use from rice fields to aquaculture in Timbulsloko Village coincided with the emergence of tidal floods (locally known as *rob*), which initially affected the northern part of the village before encroaching further south, where Dukuh Timbulsloko is situated. The intrusion of *rob* into Timbulsloko Hamlet began in the early 2000s. The river within the heart of Timbulsloko could no longer accommodate the reverse flow of seawater, leading to overflow into the settlement. Between 2004 and 2005, tidal floods initially affected gardens and rice fields but subsequently began inundating house yards more frequently. Since then, the land in Timbulsloko Village has experienced significant coastal abrasion. The saltwater inundation wreaked havoc on the coconut trees in the front yards of houses in Timbulsloko Hamlet, and the gardens behind the houses, which once hosted various vegetables (including chili, eggplant, and white pumpkin), fruit trees such as coconut, and secondary crops, gradually succumbed to saltwater and became barren ground for plants. Similarly, the rice fields behind these gardens were gradually submerged in saltwater, leaving them fit only for pond management.

According to records from the National Land Agency (BPN) of Demak Regency in 2010, land conversion in Demak from 2002 to 2007 saw a remarkable increase. Most of this conversion was driven by the transformation of agricultural land into residential areas, industrial zones, and infrastructure development, particularly in flat topography areas. In 2002, data indicated that 83,363 square meters of agricultural land were converted into industrial, residential, and infrastructure areas. This figure witnessed a substantial rise in 2007, with agricultural land covering an area of 567,864 square meters converted into factories, residences, and infrastructure development. Besides exerting pressure on the subsoil due to the weight of heavy structures, the presence of industrial zones led to increased water pumping activities. The industrial sector's water demand significantly exceeded that of households, resulting in overpumping. A study by Suryanti and Marfai published in 2016 (Ni Md. Widya A. Suryanti and Muh A. Marfai 2016) revealed that the pumping for industrial water needs to reach (and reached) depths of 110–120 m. This excessive pumping created voids in the soil, causing

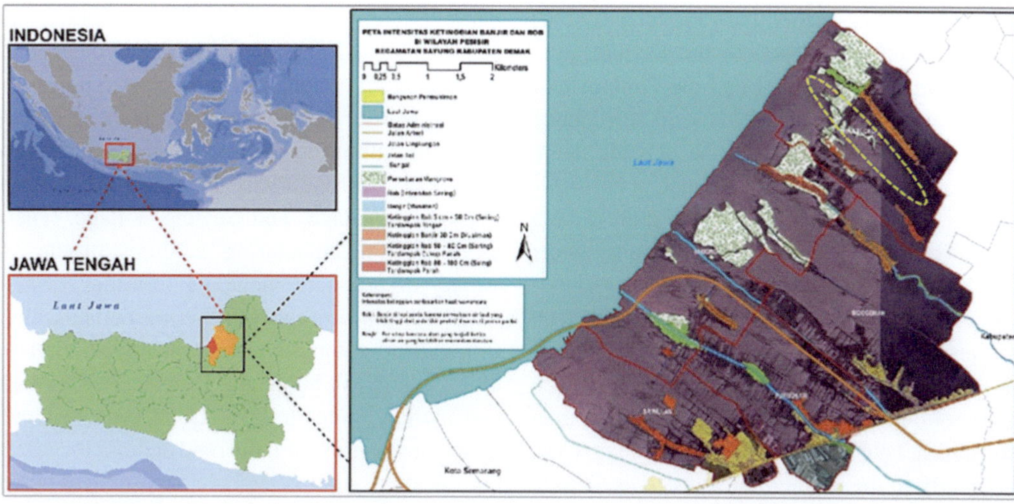

Fig. 12.1 Location of Timbulsloko Hamlet, Sayung, Demak. (*Source* Primary Survey 2022)

Table 12.1 Occupation types of Sayung residents, 2019

Desa	Petani	Buruh	Neleyan	Wiraswata	Pedagang	ABRI/PNS	Lain-Lain
Gemulak	670	1.710	292	15	228	38	258
Sriwulan	833	3.681	22	24	1.065	639	1.291
Sidogemah	941	2.316	46	16	708	119	287
Purwosari	630	1.853	43	21	829	138	408
Timbulsoko	545	2.291	206	4	276	35	302
Bedono	200	2.015	716	5	466	57	257
Surodadi	348	1.408	198	10	225	36	206
Jumlah	4.167	15.274	1.523	95	3.797	1.062	3.009

Sumber Kecamatan Sayung Dalam Angka 2019
Source Sayung Regency 2019

water to vanish from the soil pores. The combination of subsoil voids and the substantial weight of factory buildings contributed to the high rate of land subsidence in Sayung. According to Suryanti and Marfai's research, land subsidence ranged from 0.06 to 1.15 m per year in Sayung District, with the lowest rates occurring near the coast and the most severe subsidence observed in densely industrialized areas (Ni Md. Widya A. Suryanti and Muh A. Marfai 2016).

In 2008, the effects of coastal abrasion extended further, turning former rice fields into waterlogged areas. By 2009, tidal floods (*rob*) carrying sediment began causing damage to residential areas. In addition to being influenced by the shift from paddy fields to aquaculture, coastal abrasion and tidal floods in Timbulsloko Hamlet were exacerbated by land subsidence.

At a consortium meeting held in June 2021, attended by the Geological Agency of the Ministry of Energy and Mineral Resources (ESDM), the ESDM Office of Central Java Province, Diponegoro University Semarang, and Wetlands International (an organization focused on maintaining and restoring wetlands and their resources for communities and biodiversity, headquartered in Wageningen, the Netherlands), and the presence of the Demak District Government, it was reported that the coast of Sayung District is experiencing annual land subsidence ranging from 4 to 12 cm.

The meeting also reached the following conclusions: (1) land subsidence in the Demak area, including the Sayung coast and the Timbulsloko coast, is an integral part of the land subsidence occurring in Semarang; (2) the Sayung area, located on the north coast of Demak, is not sufficiently stable to support heavy structures, such as large buildings, leading to land subsidence; and (3) there are indications of excessive groundwater extraction from drilled wells and industrial facilities in Sayung, contributing to land subsidence. Groundwater extraction and the structural loading of buildings in the Sayung industrial area also impact land subsidence in Semarang City, particularly in the northern areas such as Genuk, Terboyo, and Tanjung Emas Port. Conversely, groundwater extraction and structural loading in the northern industrial areas of Semarang City affect land subsidence in the Sayung area. The port of Tanjung Emas has drilled wells to extract groundwater to meet the needs of large ships docking there. As the land surface continues to subside, tidal waves in the subsidence area rise, expanding into previously unaffected regions.

Based on the findings of Mila Karmilah et al. (2023), it is evident that land-use changes began as early as 1985, leading to land subsidence and rising water levels that have consequently made it challenging for residents to access clean drinking water. Figure 12.2 illustrates the land-use changes over approximately 35 years. These alterations are influenced by various factors, including industrial development and expansion in various areas surrounding the region and within the city of Semarang. This encompasses harbor expansion, residential development, and port construction (Asrofi et al. 2017; Damayanti 2019).

Meanwhile, Sayung District hosts the largest industrial zone in Demak Regency, with 30–40 factories established in the area. The industrial development initially began in Genuk District, Semarang City, which before 1976 was part of the Demak Regency. In response to the industrial expansion in Genuk, the Demak Regency government introduced Demak Regency Regional Regulation Number 19 of 1991, which defined the boundaries of Sayung District. This regulation specified that a 140-hectare area in the western part of Sayung District, jointly developed with the Municipality of Semarang, would be designated for industrial, commercial, and service purposes. Subsequently, a growing number of factories emerged within the designated industrial zones, with key concentrations in Sriwulan Village, Purwosari Village, and Sayung Village. These large-scale manufacturing facilities produce a variety of products, including furniture, printed materials, plastics, cartons, pharmaceuticals, electronics, and automotive components, all situated along Jalan Raya Semarang-Demak, a major trade distribution route in Java. Conversely, villages further away from the main road or those located in coastal areas, such as Timbulsloko, serve as buffer zones and are not designated for industrial use.

The most severe erosion in Timbulsloko Village took place between 2005 and 2010, resulting in the loss of over 107.31 hectares of land within the village. Despite some accretion, the addition of new deposits during this period was significantly smaller than the land eroded, amounting to just 0.362 hectares. This land erosion exacerbated the impact of tidal floods (*rob*), which inundated the villages in Timbulsloko. Additionally, the road infrastructure in Timbulsloko Village was eroded, causing disruptions in connectivity among different parts of the village (Table 12.2).

Table 12.2 Area of abrasion and accretion, 2000–2017

Year	Abrasion area (ha)	Accretion area (ha)
2000–2005	19,3872	–
2005–2010	107, 3174	0,3622
2010–2015	39,6483	3,016
2015–2017	0,9502	7,2646

Source Chris Antoni P. Purba, 2019

This situation has significantly impacted the lives and livelihoods of Dukuh Timbulsloko residents, including their access to fresh water. The physical condition of Dukuh Timbulsloko is deteriorating day by day, with entire houses becoming submerged in water, prompting residents to raise the floor levels of their homes. This approach

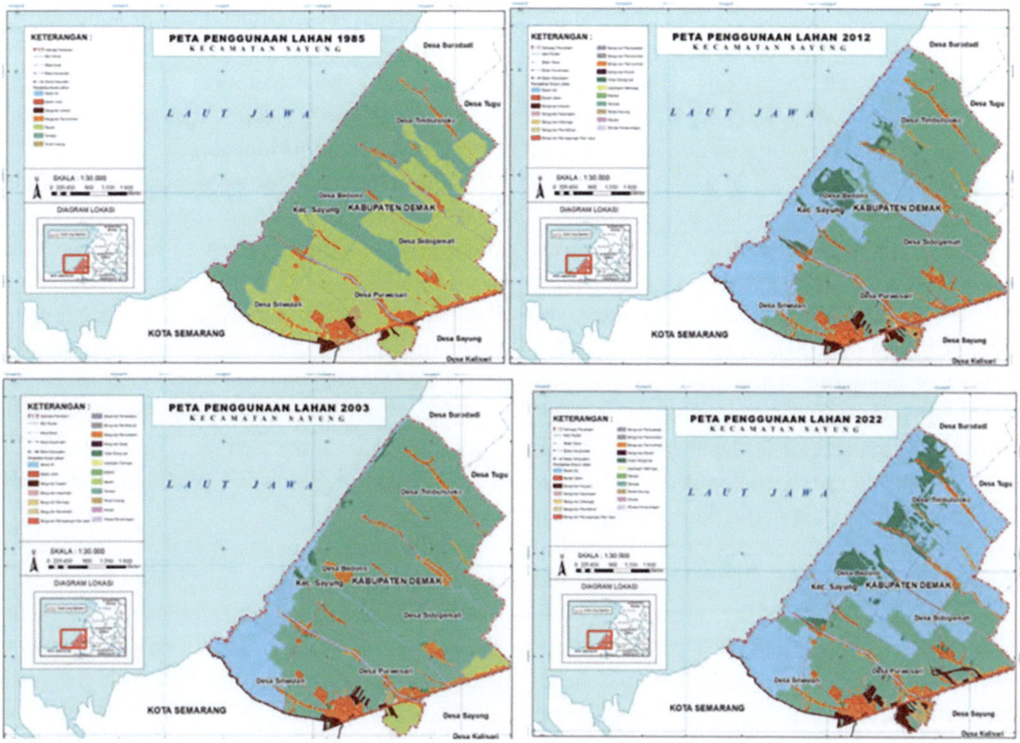

Fig. 12.2 Land-use change, 1985–2022. (*Source* Primary Survey 2022)

results in water channels, PDAM installations, bathrooms, latrines, and septic tanks remaining at their original levels, causing these crucial infrastructures to be submerged in seawater, damaged, and rendered non-functional. As a result, the people of Timbulsloko now grapple with the challenge of accessing clean water and maintaining healthy and hygienic sanitation facilities.

12.4.2 Access to Water in Timbulsloko

Following the disconnection of the Demak Regency PDAM in Timbulsloko Hamlet due to its submerged and damaged piping installation, residents attempted various means to access fresh water. In 2012, a seawater supply installation (desalination) was established in Timbulsloko Hamlet, reportedly provided as government assistance. The operation of this installation relies on electricity, funded through the sale of water to Dukuh Timbulsloko residents. Two plastic tanks, each with a capacity of 5000 L, were installed to store desalinated water. According to residents, the desalinated water from the purification machine is comparable to typical freshwater. The desalinated water is stored daily in these two plastic reservoirs, each with a capacity of 5000 L. However, the air produced by the desalination installation falls short of meeting the clean air needs of Dukuh Timbulsloko residents. Only half of one tank (2500 L) of fresh water can be produced from the two saltwater tanks (10,000 L), and this process consumes a substantial amount of electricity. Consequently, the revenue generated from selling water to residents does not cover the operational costs of the desalination machine. Furthermore, these installations are not resilient to saltwater damage, with both the engine and the guard having sustained damage. Presently, only two plastic tanks have been repurposed to store freshwater sourced from artesian wells.

At present, the population of Timbulsloko Hamlet, numbering approximately 350 people, relies on artesian wells (bore wells), locally known as *umbul,* as their primary source of freshwater. There are two wells, with depths of 135 m and 150 m, which are presently in use and managed under the Community-Based Water Supply and Sanitation (PAMSIMAS) model. These two artesian wells were constructed by the Department of Public Works and Spatial Planning (Dinputaru) of Demak Regency in 2014 and 2015, following the cessation of operation of the seawater desalination supply installation in Timbulsloko Hamlet. PAMSIMAS management faced similar challenges in covering operational costs as the desalination installation. The extraction of freshwater from the two artesian wells incurs monthly expenses of approximately IDR2.2 million and IDR1.5 million to cover the electricity used. Residents are charged IDR2500 per month for PAMSIMAS water usage. The total payments made by residents for using PAMSIMAS water are sufficient to cover electricity expenses, maintenance, and damage repair.

However, the problem arises from within the well, as the volume of freshwater extractable from both wells has been decreasing. This has resulted in a shortage of clean water for Dukuh Timbulsloko residents. PAMSIMAS management suspects a pipe leak in the ground as the cause of this reduction in freshwater discharge, which could prove difficult to repair. The PAMSIMAS management recommends that water installations not be buried underground to simplify maintenance. Nevertheless, it may well also be that the decreased water discharge from the two wells in Timbulsloko is a result of excessive groundwater pumping carried out by factories in the nearby Sayung industrial area, located just 2–3 km from Dukuh Timbulsloko.

The diminishing freshwater discharge prompted the initiation of a third well in mid-2022, although this drilling effort proved unsuccessful. The drill reached a depth of over 115 m, but no fresh water emerged. The first and third wells are positioned in a line parallel to former roads and rivers. Following the failure of the third drilling, a fourth drilling attempt was made in early October 2022, situated approximately 20 m south of the third hole. By October 10, 2022, the depth of the fourth drilling point had reached over 80 m, but there has been no indication that fresh water will be obtained from this location.

Because the need for clean water was not met by the two artesian wells, some residents of Timulsloko Hamlet purchased bottled water. There are at least three grocery stalls selling bottled water. Residents also often buy refill-water packs from the neighboring hamlet, which requires them to travel by ferry costing from Rp. 5000 to 10,000 each way, depending on their luggage. Stalls selling beverages like iced syrup and others also offer refillable water packs, priced at IDR12,000 per 19-L bottle.

Additionally, residents of Dukuh Timbulsloko collect rainwater through a method known as rainwater harvesting to fulfill their clean water needs. The harvested rainwater serves various purposes, including watering ornamental plants and vegetables, tending to chickens, washing clothes and dishes, bathing, cleaning after urinating and defecating, as well as cooking and drinking. Not all Timbulsloko residents or families use rainwater, nor do all use it for all these purposes. Rainwater harvesting for cooking and drinking typically does not begin with the first rainfall after the dry season. The initial rain is primarily used for watering plants, as it helps clean dirt from the roof. Following heavy rains, the collected rainwater is suitable for consumption, cooking, drinking, bathing, washing, and toilet activities. Despite the initial roof-cleaning rain containing less dirt than the first rain collected, the harvested rainwater undergoes a sedimentation process first. Typically, the collected rainwater is allowed to settle for about two hours to obtain clean water (Fig. 12.3). Clean water is stored in plastic bottles, with separate bottles for drinking and washing water.

At times, some residents directly consume rainwater stored in bottles. This decision is not made without consideration. One resident conducted an observation comparing refilled water with his rainwater collection in the same container. Within two weeks, the refilled water bottles developed mold, whereas the bottles of harvested rainwater

Fig. 12.3 Rainwater harvesting using bucket. (*Source* Primary Survey 2022)

showed no signs of mold after a month of storage. The benefits of consuming rainwater for the residents of Timbulsloko Hamlet include its accessibility during the rainy season, cost-effectiveness, as it does not entail a consumption-based cost, and the taste, which is comparable to refilled and bottled drinking water, and often fresher.

12.4.3 The Importance of the Right to Water

In arid regions with stable soil conditions, programs like PAMSIMAS can still effectively address the shortage of piped clean water services provided by the Regional Drinking Water Company (PDAM). As of 2021, 13 years since its launch in 2008, PAMSIMAS continues to operate in 33 provinces, 408 regencies/cities, and more than 35,000 villages. However, PAMSIMAS faces significant challenges in coastal areas susceptible to tidal waves and erosion, such as Timbulsloko Hamlet in the Sayung District.

The high cost of obtaining water in Dukuh Timbulsloko can lead to the impoverishment of residents. A study conducted in 311 villages across Indonesia with 13,469 household respondents, carried out by the Research and Development Section of the Ministry of Public Works and Public Housing, demonstrates a significant correlation between access to water and income levels, influencing poverty rates. Households lacking access to clean water facilities exhibit an income level 17.17% lower than households with such access. These households are also 1.29% more likely to fall into the category of impoverished households compared to those with access to clean water (Syah Putra et al. 2016). Furthermore, the Sustainable Development Goals, which are based on human rights and equality, place access to clean water and sanitation at the core of SDG6 Importantly, the SDGs emphasize the significance of access to clean water and sanitation in collaboration with other sectors to achieve several goals, including SDG1, which is to eradicate poverty (Director General of KIA Nutrition 2015). This is because water availability and poverty are closely intertwined, with the absence of freshwater exacerbating poverty levels. Therefore, access to sustainable water sources is pivotal for reducing poverty levels.

12.5 Conclusion and Recommendation

12.5.1 Conclusion

The situation in Timbulsloko Hamlet suggests that to achieve sustainable development in this area, the government must ensure the restoration of the residents' right to water. From the explanations above, it is evident that the loss of the right to water among Timbulsloko Hamlet's residents is due to a massive social and ecological crisis caused by: (1) the deterioration of the village's physical condition due to land subsidence triggered by industrialization (building loads and excessive groundwater extraction), compaction and consolidation of alluvial soil structures, and climate change; and (2) government policies that do not guarantee access to clean water in submerged areas like Dukuh Timbulsloko. Government initiatives, such as PDAM, PAMSIMAS, and the Healthy Latrine Program by BKM, aimed at ensuring access to WASH for Genuk and Sayung coastal residents, have proven ineffective. These government initiatives have also failed as solutions to the problem of lost WASH access. This is primarily due to the deteriorating physical conditions of the settlements, which make the implementation of these programs unfeasible. The settlement environments continuously experience subsidence, erosion from abrasion, frequent inundation from tidal floods, and even permanent submersion, rendering the entire settlement land inaccessible.

12.5.2 Recommendation

To achieve sustainable access to clean water, government policies need to intervene effectively. However, for these policies to be applicable to citizens at the grassroots level, they must be tailored to the specific circumstances of the affected population. Initiatives arising from the experiences of residents living in these areas should not be disregarded but instead, form the primary basis for policymaking.

References

Asrofi A, Hardoyo SR, Hadmoko DS (2017) Strategi Adaptasi Masyarakat Pesisir Dalam Penanganan Bencana Banjir Rob Dan Implikasinya Terhadap Ketahanan Wilayah (Studi Di Desa Bedono Kecamatan Sayung Kabupaten Demak Jawa Tengah). Jurnal Ketahanan Nasional 23(2):125–144

Batubara B, Wagner I (2020) Maleh dadi Segoro; socio-ecological crisis of the Semarang-Demak Coastal area. https://www.researchgate.net/publication/344804419

Center for Socio-Economic Rights (2010) H20 Fact Sheet #1. https://www.worldwatercouncil.org/fileadmin/wwc/Programs/Right_to_Water/Pdf_doct/Center_for_Economic_and_Social_Rights__RTW.pdf

Damayanti R (2019) Hilangnya Dua Kampung Pesisir Desa Bedono Kecamatan Sayung Kabupaten Demak Dalam Pusaran Abrasi dan Industrialisasi Tahun 1990–2010 [Skripsi, Universitas Diponegoro]. http://www.bbc.com/indonesia/berita_indonesia/2015/12/151212_indonesia_dem

Director General of KIA Nutrition (2015) Health Within the Framework of Sustainable Development Goals (SDGs)

Durasih IPA (2011) Gambaran Sanitasi Kantin Sekolah Dasar di Desa Taman Kecamatan Abiansemal, Kabupaten Badung. Politeknik Kesehatan Denpasar

Kementerian PPN/Bappenas (n.d.) Tujuan Pembangunan Berkelanjutan (SDGs). Retrieved September 12, 2022, from https://sdgs.bappenas.go.id/tujuan-6/

Mila K, Eka H, Syarifah A, Umdatin N (2023) Urip Dioyak-Oyak Banyu: Perjumpaan Manusia, Abrasi, Rob, dan Infrastruktur di Sayung (Bagas Yusuf Kausan, Ed.; 1st ed.). Mata Kata Inspirasi

Ni Md. Widya AS, Muh AM (2016) Multi-hazard analysis in the coastal area of Demak regency. Indones Earth J 5(2)

Pemerintah Kabupaten Demak (2012) Buku Putih Sanitasi Kabupaten Demak

Spradley JP (1997) Metode Etnografi. Tiara Wacana

Sudrajat A (2020) Pemenuhan Kebutuhan Air Minum Sanitasi dan Hygiene (WASH) di Indonesia. Kelompok Keahlian Teknologi Pengelolaan Lingkungan. https://www.itb.ac.id/berita/detail/57573/pemenuhan-kebutuhan-air-minum-sanitasi-dan-hygiene-wash-di-indonesia

Sukartini NM, Saleh S (2016) Akses Air Bersih di Indonesia. Jurnal Ekonomi Kuantitatif Terapan 9(2):89–98

Suripin (2002) Pelestarian Sumber Data Tanah dan Air (1st ed.). Andi

Syah Putra H, Rianto N, Pekerjaan Umum dan Perumahan Rakyat K (2016) Pengaruh Akses Air Bersih Terhadap Kemiskinan Di Indonesia: Pengujian Data Rumahtangga Impact Of Clean Water Access To Poverty In Indonesia: Micro Analysis At Household Level

Taufik B (2012) The right to clean and safe water. Directorate General of Human Rights, Ministry of Human Rights Law of the Republic of Indonesia

The right to safe and clean drinking water and sanitation as human rights, Pub. L. No. 64/292 (2010)

Undang-undang Dasar Negara Republik Indonesia Tahun 1945 (1945)

United Nations Water (n.d.-a) Human rights to water and sanitation. United Nations Water. Retrieved September 12, 2022, from https://www.unwater.org/water-facts/human-rights-water-and-sanitation

United Nations Water (n.d.-b) Water Facts. United Nations Water. Retrieved September 12, 2022, from https://www.unwater.org/water-facts

Protecting South Sumatra's Peat Swamp Forests: Economic Value and Conversion Challenges

13

Ari Nurlia, Nur Arifatul Ulya, Edwin Martin, and Mamat Rahmat

Abstract

This article examines peatland conversion in South Sumatra and its impact on environmental management and the economy. As unique ecosystems, peat swamp forests have significant economic and ecological value. However, the increasing conversion of land for other economic uses, such as oil palm plantations and forest plantations, has threatened the sustainability of this ecosystem. This research analyzes the economic value of various ecosystem services provided by peat swamp forests, including timber, fish, water, and other environmental aspects. We also consider the impact of this land conversion on biodiversity, carbon emissions, and the environment. The results show that peatland conversion significantly negatively impacts economic value and biodiversity. Therefore, we provide several policy recommendations to balance economic development and preserve this important peat swamp forest ecosystem. The chapter provides in-depth insight and a comprehensive analysis of the surrounding conversion of peatlands, emphasizing the significance of conserving these ecosystems within the framework of environmental sustainability. Quantitative methodology is used.

Keywords

Peatland conversion · Peat swamp forests · South Sumatra · Total economic value · Environmental quality

A. Nurlia (✉) · E. Martin
Research Centre for Society and Culture - National Research and Innovation Agency, Jakarta, Indonesia
e-mail: ari.nurlia@brin.go.id

N. A. Ulya
Research Center for Behavioral and Circular Economics - National Research and Innovation Agency, Jakarta, Indonesia

M. Rahmat
Directorate General of Social Forestry and Environmental Partnership, Ministry of Environment and Forestry, Jakarta, Indonesia

13.1 Introduction

Indonesia is a country that includes 47% of the world's tropical peatland area, with an area of 20.6 million hectares or around 10.8% of Indonesia's land area (Osaki et al. 2016; Page et al. 2011; Rahman and Yuliani 2018). Tropical peat swamp forests function as a dynamic bridge between land and water; they are zones of water flow exchange and nutrient cycling and are important to wetland ecosystems worldwide (Anamulai et al. 2019). Indonesian peatlands, which are included in the tropical peat category , naturally have land cover

H. Warsilah and A. H. Purnomo (eds.), *Approaching Natural Resources for Sustainable Development in Indonesia*, Sustainable Development Goals Series, https://doi.org/10.1007/978-981-96-6913-4_13

in the form of forests (Dohong et al. 2017). Peatlands in Indonesia are a concern due to forest fires, which in 2015 resulted in the destruction of peatlands and caused an exceptionally severe haze disaster with estimated losses reaching USD16.1 billion or the equivalent of IDR221 trillion (Pinem 2016; World Bank 2015), and climate change issues related to the use of peatlands for the development of oil palm plantations (Liu et al. 2020; Wijedasa et al. 2017).

Indonesian peatlands are being destroyed by land conversion (Maskun et al. 2021) which began in the 1970s. Logging roads, wooden rails, and drainage canals were built to earn foreign exchange from natural forest logging, including peat swamp forests (Dohong et al. 2017). Transmigration settlements led to the conversion of peatlands on the east coast of South Sumatra in the early 1970s (Yunardy et al. 2018), while on peatlands in Central Kalimantan one million hectares of rice fields were developed (Dohong et al. 2017; Ritzema et al. 2014). The latest is the conversion of peatlands for oil palm plantations (Dohong et al. 2017; Wijedasa et al. 2017). The mass expansion of oil palm plantations on peatlands has caused severe degradation of the peatland ecosystem (Dohong et al. 2017; Maskun et al. 2021). An investigation by the Madani Foundation found that Indonesian fires destroyed 1.6 million ha of forest and land in 2019, 44% of which was peatland (Maskun et al. 2021). Approximately 20% of oil palm plantations in Indonesia are located on peatlands.

Forests can be converted for non-forestry purposes under current rules. Area utilization and conversion policies must be based on a comprehensive study to include comparison of financial benefits and technical aspects. Ecological concerns, legislation, and reality often conflict when determining a forest area's function without considering its advantages (Bann 1998; Ulya et al. 2015). Assessing the economic value of forest resource benefits can help solve this problem (Bann 1998; Munasinghe 1992). In the case of forest areas facing conversion, the economic value must be used to compare natural forest advantages to alternative uses (Gregersen and Contreras 1979). The assessment should include natural forests and monoculture plantation forests or oil palm plantations. A comprehensive assessment of the benefits of each method of management represented in the total economic value can help determine peat swamp forest management decisions.

The question arises as to whether peatland conversion is a sustainable solution or a threat to environmental management in South Sumatra. This chapter aims to explain and analyze the impact of peatland conversion on the economic value and sustainability of peat swamp forest ecosystems in this region. By combining empirical data and a scientific approach, we will identify the economic value of the various ecosystem services provided by peat swamp forests, including timber, fish, water, and several other benefits. We will also investigate the impact of this land conversion on biodiversity, carbon emissions, and the environment as a whole.

13.2 Theoretical Framework

Forests are a scarce economic resource and are needed in the economy. National economic development uses forests to supply goods and the environment (Costanza et al. 1997, 2014, 2011). However, forests not only produce commodities as input for various industries but also produce various environmental services that originate from ecosystem functions that are important for economic sustainability.

With proper resource allocation, scarce forest resources can yield their full benefits. Market prices must be identified to distribute scarce resources optimally. For natural resources (including forests), Hanley et al. (2001) suggest using monetary value as a measure of resource scarcity, which is reflected by the economic value of the resource. Efficient allocation of forest resources requires estimating the value of benefits generated by forest resources in monetary terms. The value of these benefits is assessed using the same criteria, namely economic benefits. The sum of the economic values of all benefits is called total economic value (TEV) (Munasinghe 1992). TEV of forest resources is the sum of direct use value (DUV), indirect use value (IUV), option value (OP), and bequest value (BV).

Economic development exposes peatland forests to conversion for forestry and other purposes. Whether an area is maintained in its natural condition or converted for other uses must be based on the comparison criteria between the TEV provided by the forest and the TEV if the forest area is used for other purposes. When activities are undertaken to convert natural forest areas to be managed in different forms, like development activities in general, benefits commensurate with the costs incurred are expected. The benefits and costs must be assessed within an economic analysis framework.

13.3 Methodology

The research was conducted in the Merang Kepayang Peat Swamp Forest Area in Bayung Lencir Sub-District, Musi Banyuasin District, South Sumatra Province. The Merang Kepayang Peat Swamp Forest Area has an area of 138,200 hectares located between 01° 45′ and 02 03′ South Latitude and 103° 51′–104° 17′ North Latitude (Fig. 13.1).

The research was conducted using the quantitative method (Cresswell 2016). Quantitative research data on the economic value of peatlands was obtained by interviewing respondents, searching documents from government agencies and companies that obtained business permits at the research location, and a literature study. A structured questionnaire with open questions was used as an interview guide for respondents living around peat swamp forests.

Data analysis using total economic value (TEV) calculations was carried out to establish the economic value and conversion of peat swamp forests. The TEV concept measures the utilitarian value of ecosystems to support efficient resource allocation. According to Munasinghe (1992), the TEV of the peat swamp forest area is calculated as:

$$\begin{aligned} \text{TEV} = &\text{Vt} + \text{Vf} + \text{Vwh} + \text{Vwt} \\ &+ \text{Vp} + \text{Vc} + \text{Vff} + \text{Vbio} + \text{Ve} \end{aligned}$$

where:

Vt = value of timber.
Vc = value of carbon.
Vf = value of fish.
Vff = value of flora and fauna.
Vwh = value of water for households.
Vbio = value of biodiversity.
Ve = value of existence.
Vp = value of protection of environment.
Vwt = value of water for transportation.

Data analysis was carried out using market price, cost of illness, benefits transfer, and contingent valuation methods (CVM) (Adger et al. 1995; Baral et al. 2016; Groot and van der Meer 2010; Kakuru et al. 2013). These methods estimate the economic value of various environmental aspects of peat swamp forests and their conversion.

13.4 Result and Discussion

13.4.1 Economic Value of Peat Swamp Forests and Their Conversion

The total economic value is calculated based on nine variables: (1) the economic value of timber; (2) the economic value of fish; (3) the economic value of household water; (4) economic value of water for transportation; (5) economic value of environmental quality protection; (6) economic value of carbon; (7) value of choice; (8) value of biodiversity; and (9) economic value of existence.

13.4.1.1 Economic Value of Timber

The economic value of timber in peat swamp forests is the sum of the economic value of logs and small logs. A log is wood with a diameter of more than 30 cm, while the small log is wood with a diameter between 10 and 30 cm. The estimated economic value of stands in the Merang Kepayang Peat Swamp Forest are presented in Table 13.1. The economic worth is calculated by multiplying the stumpage sales price (SSP) by its potential.

Natural forest timber in the Merang Kepayang Peat Swamp Forest has a high economic value, showing that natural management may produce a lot of timber. Due to the timber potency, a

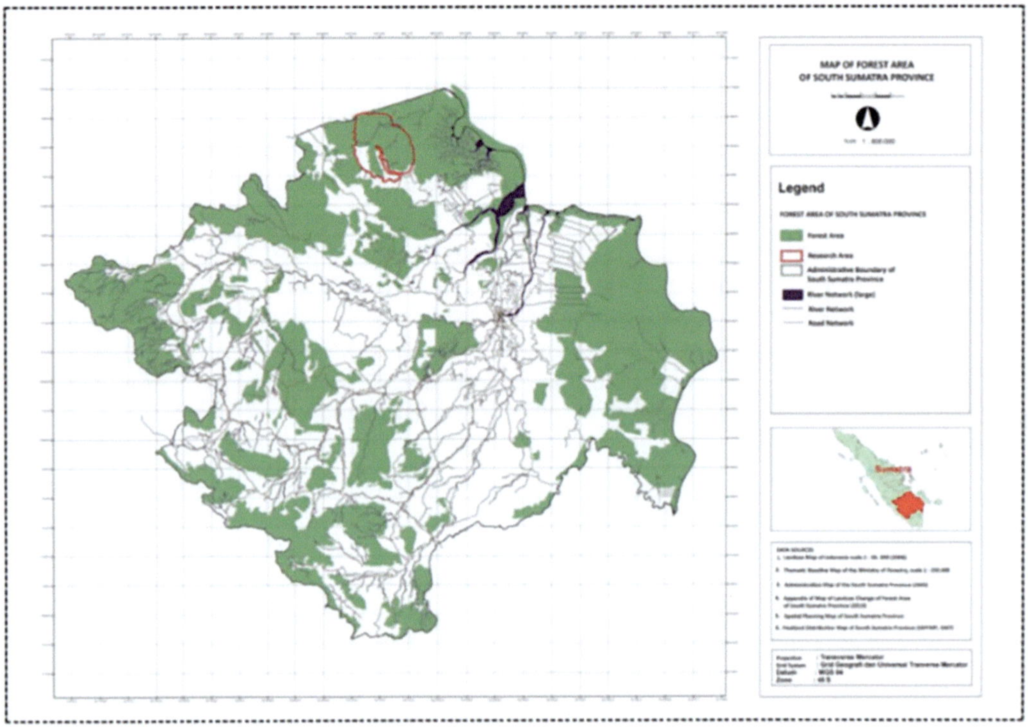

Fig. 13.1 Merang Kepayang Peat Swamp Forest Area. (*Source* Ulya (2015))

Table 13.1 Timber potential in the Merang Kepayang Peat Swamp Forest

Timber size	Volume (m^3 Ha^{-1})	Value (USD Ha^{-1})
Log	19.67	1371.83
Small log	71.38	1537.64
Amount		2909.47

Source Data analyzed

Table 13.2 Estimated timber value in peat swamp conversion plantation forests

Timber classification	Volume (m^3 Ha^{-1})	Value (USD Ha^{-1})
Main plantation	25.00	539.99
Livelihood plantation	8.69	187.63
Amount		727.62

Source Data analyzed

conversion that begins with clear-cutting is unsuitable for Merang Kepayang Peat Swamp Forest management.

Acacia crassicarpa, A. *mangium*, and *Eucalyptus pellita* are planted in peat swamp natural forest conversion plantations. The plant cycle is six years. According to a study on permanent sample plots owned by forest corporations, the first-cycle plants were 25 m^3 ha^{-1} Yr^{-1}, and the subsequent cycles were 30 m^3 ha^{-1} yr^{-1} (Ulya 2015). Table 13.2 shows the SSP-estimated economic value of timber in converted peat swamp forest plantings.

Meanwhile, in oil palm plantation conversion, development begins with clear-cutting of natural forests. Clear-cutting of natural forests is assumed to cause the loss of all natural forest timber value. The economic value of oil palms built by converting natural forests is estimated using the economic value of oil palms as reflected by the economic net present value (ENPV) of oil palm plantations (Yaron 2001). In the exploitation of peat swamp forest conversion to oil palm plantations, with an exploitation period of 25 years, it is estimated that the value of wood will be lost at USD 9,647,422.02 Ha^{-1}.

13.4.1.2 Economic Value of Fish

Estimating the economic value of fish originating from peat swamp forests is carried out by estimating a willingness to pay (WTP) to be able to consume fish. The value of consumers' willingness to pay to consume fish is obtained from a regression analysis of the relationship between fish consumption (Y) and socio-economic variables that are considered to influence fish consumption, which are fish procurement costs (X1), age (X2), duration living in the village (X3), distance from house to forest (X4), education (X5), total income (X6), and number of family members (X7). The results of the regression are shown in the equation below.

$$\ln Y = 17,1 - 0,985 \ \ln X1 - 0,746 \ \ln X4$$

where: Y = fish consumption (kg $person^{-1}$ yr^{-1})

X1 = fish procurement costs (USD Kg^{-1}).

X4 = distance from house to forest (km).

The regression equation above is then used to estimate the economic value of fish originating from the Merang Kepayang Peat Swamp Forest. Based on the equation, the economic value of fish for communities around the forest is around USD277.70 ha^{-1} yr^{-1} (Ulya 2015).

Even though environmental degradation reduces fish harvests in peat swamp forest conversion plantations, community access to fish remains. The Merang River, the research location, saw a 70% reduction in fish output from 1975 to 2004 (Wardoyo 2006). According to Wardoyo (2006), peat swamp forest plantation fish have a total economic value of USD83.31 ha^{-1} yr^{-1}.

Meanwhile, the conversion of peat swamp forests to oil palm plantations eliminates the economic value of fish. Based on field observations, in the oil palm plantation area, no people are fishing, and no fish can live in ditches and canals due to poor water quality. Inputs for cultivation in the form of chemical fertilizers and pesticides cause pollution and a decrease in water quality.

13.4.1.3 Economic Value of Household Water

The value of WTP to consume household water is obtained from a regression analysis of the relationship between household water consumption (Y) and socio-economic variables, which are thought to influence household water consumption. These socio-economic variables are the cost of providing water for the household (X1), age (X2), duration living in the village (X3), distance from house to forest (X4), education (X5), total income (X6), and number of family members (X7). The results of the equation from the regression for the economic value of household water for communities around the peat swamp forest are:

$$\ln Y = 6,16 - 0,431 \ \ln X1 + 0,261 \ \ln X7$$

where: Y = household water consumption (USD $person^{-1}$ yr^{-1})

X1 = cost of providing water for the household (USD per m^3).

X7 = number of family members (person).

Based on this equation, the economic value of household water for communities around peat swamp forests is 191.77USD ha^{-1} yr^{-1} (Ulya et al. 2014).

In plantation forests, water availability for household and transportation purposes is assumed not to change consequent upon changes from natural forest cover to plantation forests. So the economic value of household and transportation water in plantation forests is considered the same as that of natural forests.

The existence of oil palm plantations converted from peat swamp forests causes a 13% decrease in regional water availability compared to when the area was still peat swamp forest (Widodo 2011). Thus, it is assumed that water availability for households and transportation will also decrease by 13% when compared to the value of water in the economic value of natural peat swamp forests.

13.4.1.4 Economic Value of Water for Transportation

The value of willingness to pay for water consumption for transportation is obtained from a regression analysis of the relationship between water consumption for transportation (Y) and socio-economic variables that are thought to influence water consumption for transportation, including the cost of procuring water for transportation (X1), age (X2), duration living in the village (X3), distance from house to forest (X4), education (X5), total income (X6), number of family members (X7), and frequency of use of water transportation in a year (X8). The results of the best regression equation for the economic value of water for transportation in peat swamp forests are:

$$\ln Y = 14,9 - 0,876\ \ln X1 - 1,01\ \ln X2 - 0,542\ \ln X5 + 0,919\ \ln X8$$

where: Y = water consumption for transportation (km yr^{-1})

X1 = cost of procuring water for transportation (USD per m^3).

X2 = age (years).

X5 = education (years).

X8 = frequency of use of water transportation in a year (trip yr^{-1}).

Based on this equation, the economic value of water for transportation for communities around peat swamp forests is USD618.69 ha^{-1} yr^{-1} (Ulya et al. 2019, 2014). The economic value of water for transportation in plantation forests is considered to be the same as the total economic value of natural forests because there is no very significant change when there is a change due to natural forest cover becoming plantation forest. Meanwhile, when peat swamp forests are converted into oil palm plantations, the economic value of water decreases by 13% when compared to the water value of natural peat swamp forests.

13.4.1.5 Economic Value of Environmental Quality

Environmental protection in peat swamp forests is estimated by drought, flood, and haze-related medical expenditures and opportunity costs. Peat swamp forests defend against floods, droughts, and forest and land fire haze. The community's losses from peat swamp forest disturbances approximate the entire worth of protecting the area's environmental quality at USD6.85 ha^{-1} yr^{-1}. Natural forests lose their environmental significance when peat swamp woods are converted into plantation forests and oil palm plantations.

13.4.1.6 Economic Value of Carbon

Merang Kepayang has 138,200 hectares of peat swamp forest (Ballhorn et al. 2007). Peat swamp forests have carbon in the vegetation above and below the ground. The below-ground carbon content was calculated using a three-dimensional bedrock model and 80 peat drilling locations. The economic value of carbon is estimated using the benefit transfer method. The Ulu Masen carbon price was used as reference for benefit transfer because this scheme was already running during data collection. Based on the Ulu Masen, the carbon price is USD4 per ton of CO_2 emissions, so the total carbon value is USD11,412.54 ha^{-1} yr^{-1}.

The carbon value of the plantation forest was estimated by the biomass value of A. *crassicarpa.* This approach was chosen because A. *crassicarpa* was developed by industrial forest plantations on peatlands in South Sumatra (Ulya et al. 2014). The above-ground carbon value of natural peat land forest at the research location is lower than that of plantation forest stands, which is thought to occur because not all the land cover is tree. There are 45,040 hectares (33%) of land cover in the form of water shrubs, agricultural land, plantations, and mining with low or even zero CO_2 uptake. The carbon value of plantation forests is the sum of the economic value of above-ground and below-ground biomass of USD13,227 ha^{-1} yr^{-1} (Ulya et al. 2019).

Oil palm plants can store more than 80 tons of carbon per hectare after 10–15 years of growth, resulting in an average of 60.4 tons per hectare, or 2.44 tons of carbon ha^{-1} yr^{1-}, or 8.95 tons of CO_2. The sum of the carbon economic value of oil palm plantations and below-ground biomass is USD10,645.93 ha^{-1} yr^{-1} (Ulya et al. 2015). This value is lower than the economic value of carbon of plantation forest and natural peat swamp forests.

13.4.1.7 Option Value

Option value is the potential of natural resources that have no economic value but could benefit society in the future (Munasinghe 1992). The flora and fauna in peat swamp forests are often considered by the community to have no benefits, but have the potential to have great economic value in the future. Calculation of peat swamp forest options using the CVM suggest a value of USD0.30 ha^{-1} yr^{-1}. The community around the peat swamp forest's WTP for future advantages that have no economic worth determines this value. Meanwhile, the conversion of peat swamp forests into plantation forests and oil palm plantations has no option value because it eliminates natural resource potential (Ulya et al. 2015).

13.4.1.8 Economic Value of Biodiversity

The biodiversity value of peat swamp forests is estimated using the benefit transfer method (Richardson et al. 2015). The reference data is the biodiversity value of the Berbak-Sembilang National Park (BSNP). BSNP has similar ecological conditions and is located close to the research location. The BSNP biodiversity value is USD29.72 ha^{-1} yr^{-1}. Meanwhile, the biodiversity value of plantation forests is USD21 ha^{-1} $yr^{-}1$, or around 70% of the biodiversity value of natural forests (Ulya 2015; Yaron 2001).

Conversion of natural forests into oil palm plantations causes a decline in plant biodiversity by 80% and 80–90% for mammals, birds, and reptiles. In this study, the figures proposed by Fairhurst and McLaughlin (2009) were used to assess the biodiversity remaining due to the conversion of oil palm plantations to peat swamp forests. Thus, the biodiversity value of oil palm plantations converted to peat swamp forests is USD4.46 ha^{-1} yr^{-1}.

13.4.1.9 Economic Value of Existence

Existence value is the value given by the community around the peat swamp forest to its spiritual, cultural, and aesthetic benefits (Munasinghe 1992). Estimation of the value of the existence of peat swamp forests was carried out using the CVM with open questions. The existence value of peat swamp forest is USD0.5 ha^{-1} $year^{-1}$. Meanwhile, the value of plantation forests and oil palm plantations is lost due to clear-cutting of natural forests.

The total economic value of Merang Kepayang Peat Swamp Forest is estimated by adding up the economic value of all the variables described above. The resulting calculation shows that the total economic value of the peat swamp forest is around USD15,447.52 per hectare per year. On the other hand, the total economic value of peat swamp forest conversion plantations is around USD14,019.96 per hectare per year, while the total economic value of peat swamp forest conversion oil palm plantations is around USD-9,636,066.53 per hectare per year.

13.4.2 Economic Value for Sustainable Forest Management

The results of the analysis show that the Merang Kepayang Peat Swamp Forest has very high economic value, especially if managed sustainably. On the other hand, the conversion of peat swamp forests into plantation forests and oil palm plantations can eliminate much of the economic value of peat swamp forests. Conversion to oil palm plantations in particular results in the loss or degradation of the economic value of timber, fish, water for households, water transportation, environmental protection, biodiversity, and the value of existence. The total economic value of oil palm plantations resulting from peat swamp forest conversion is much lower than that of sustainably managed peat swamp forests.

This analysis emphasizes the need to preserve and maintain peat swamp forests. Besides providing economic benefits, peat swamp forests help maintain environmental quality, combat climate change, and preserve biodiversity. To safeguard natural resources and offer sustained economic advantages for the community, Merang Kepayang Peat Swamp Forest regulations must encourage conservation and sustainable management.

Plantation forests have relatively high carbon stores compared to natural forests, because the research location is logged over forest that is not completely covered by trees, while the plantation forest area is entirely planted. Planted forests in peat swamp areas can be sources and sinks of carbon, depending on age, operation, and environmental boundary conditions such as location and climate (Scholz 2011). Therefore, emissions trading mechanisms can be considered an important new incentive for investment in plantation forests in the tropics (Ulya et al. 2019).

Natural forest fish are more valuable than plantation forest and oil palm plantation fish. Fish have no economic value in oil palm fields since they cannot be caught. Changes in vegetation cover impair fish habitat, reducing production and threatening sustainability. Environmental protection, plantation forests, and oil palm plantations are zero. Clear-cutting at the start of plantation forest and oil palm plantation development destroys native forests and their functions. Forest ecosystems lose economic value due to degradation (Baral et al. 2016; Groot and van der Meer 2010).

The biodiversity value is highest in natural forests, followed by plantation forests and oil palm plantations. The types of plants in monoculture plantations and oil palm plantations result in a reduction in biodiversity. Conversion of peat swamp forests, causing the extinction of endemic peat swamp forest species, increases global carbon emissions and climate change (Adila et al. 2017; Hawa et al. 2016; Miettinen et al. 2012; Sasidhran et al. 2016).

The research found that plantation forest and oil palm plantation development projects in peat swamp forests generate more negative externalities than direct benefits, including economic value loss and carbon emissions. However, for the community at national level, who are the owners of the resources, the development of plantation forests and oil palm plantations from the conversion of peat swamp forests cannot provide economic benefits to the community because the costs sacrificed are greater than the benefits received by the community. This study illustrates that a natural forest can benefit the community economically more than cultivation of plantation forests and oil palm plantations by converting peat swamp forests. Forests protect people from disasters and provide food and water, which are essential to life. Forests affect human lives and economic sustainability. Forests bring spiritual worth, psychological comfort, and cultural influence. Pearce and Moran (1994) even remind us that the formula above does not accurately calculate the overall economic value, which is higher. Many ecologists say the overall economic value cannot be computed using a simple formula because multiple basic ecological functions are synergistic, making it much higher. Forests also provide timber, regulate water, shelter species, produce food, provide environmental services, absorb CO_2, and attract tourists. All experts agree that these functions interact dynamically, making it impossible to establish their boundaries.

Peat swamp forests in natural conditions are able to provide a higher TEV than that of plantation forest exploitation or oil palm plantations. Thus, it can be stated that the Peat Swamp Forest area is more suitable to be managed as a natural peat swamp forest. The management of peat swamp forest areas located in production forests can be focused on natural forest timber production using selective logging methods. This approach does not involve changes to the peat swamp ecosystem, such as peat drainage, and the potential for natural forest wood production in the Merang Kepayang Peat Swamp Forest is quite high. A peat swamp forest areas designated as an 'other use area' is currently managed as an oil palm plantation and mixed agriculture. To minimize the negative impact of conversion of natural peat swamp forests, a mixed planting pattern approach can be implemented by including natural peat swamp plant species without intensive peat drainage. The hope is that the natural plant species introduced will not only produce wood, but also produce non-wood products that can provide an additional source of income for the community.

13.5 Conclusion and Recommendation

13.5.1 Conclusion

This study produced essential findings regarding the economic value of peat swamp forests, especially in the context of sustainable management versus conversion to plantation forests or oil palm plantations. The analysis results show that the Merang Kepayang Peat Swamp Forest has very high economic value when managed sustainably. This economic value covers various aspects, including the value of wood, fish, water for households, water transportation, environmental protection, carbon absorption, choice-value, biodiversity, and existence value.

However, converting peat swamp forests into plantation forests or oil palm plantations can eliminate much of the economic value found in peat swamp forest ecosystems. This conversion results in the loss of the economic value of wood, fish, water for households, water transportation, environmental protection, biodiversity, and the value of existence. In addition, the total economic value of converted oil palm plantations is much lower than that of sustainably managed peat swamp forests.

13.5.2 Recommendation

Based on our analysis and findings, we recommend the following actions:

1. *Preserving Peat Swamp Forests*: It is important to prioritize conservation and sustainable management of peat swamp forests. Management policies and practices must be based on comprehensive considerations, including economic and environmental values.
2. *Sustainable Plantation Forest Development*: The development of plantation woods must incorporate peat swamp ecosystem sustainability and preservation considerations, prioritizing ecosystem-friendly and carbon-sequestration management.
3. *Mixed Cropping Patterns in Other Use Areas*: If peat swamp forests are used for mixed agriculture or oil palm plantations, a mixed cropping pattern approach that includes natural peat swamp plant species without intensive peat drainage can reduce negative impacts. This can also provide an additional source of income for the community.
4. *Introduction of a Carbon Trading Mechanism*: A carbon trading mechanism can be an alternative incentive for peat swamp forest conservation efforts compared to converting it to other uses.
5. *More Comprehensive Economic Value Assessment*: A more comprehensive economic value evaluation that includes ecological and social advantages of natural forests requires more investigation. This will show natural resource values more accurately.

These conclusions and recommendations should help policymakers make better peatland management decisions. Protecting the environment and promoting community welfare requires prioritizing natural resource conservation and sustainable economic advantages.

References

Adger WN, Brown K, Raffaello C, Moran D (1995) Total economic value of forests in Mexico. Ambio 24(5):286–296

Adila N, Sasidhran S, Kamarudin N, Puan CL, Azhar B, Lindenmayer DB (2017) Effects of peat swamp logging and agricultural expansion on species richness of native mammals in Peninsular Malaysia. Basic Appl Ecol 22:1–10. https://doi.org/10.1016/j.baae.2017.04.002

Anamulai S, Sanusi R, Zubaid A, Lechner AM, Ashton-Butt A, Azhar B (2019) Land use conversion from peat swamp forest to oil palm agriculture greatly modifies microclimate and soil conditions. PeerJ 7:e7656. https://doi.org/10.7717/peerj.7656

Ballhorn U, Mott C, Siegert F (2007) Peat dome mapping and analysis. S. S. F. F. M. Project

Bann C (1998) The economic valuation of tropical forest land use option: a manual for researchers. Economy and Environment Program For Southeast Asia

Baral H, Guariguata MR, Keenan RJ (2016) A proposed framework for assessing ecosystem goods and services from planted forests. Ecosyst Serv 22:260–268. https://doi.org/10.1016/j.ecoser.2016.10.002

Costanza R, d'Arge R, de Groot R, Farber S, Grasso M, Hannon B, Limburg K, Naeem S, O'Neill RV, Paruelo J, Raskin RG, Sutton P, van den Belt M (1997) The value of the world's ecosystem services and natural capital. Nature 387(6630):253–260. https://doi.org/10.1038/387253a0

Costanza R, de Groot R, Sutton P, van der Ploeg S, Anderson SJ, Kubiszewski I, Farber S, Turner RK (2014) Changes in the global value of ecosystem services. Global Environ Change 26:152–158

Costanza R, Kubiszewski I, Ervin D, Bluffstone R, Boyd J, Brown D, Chang H, Dujon V, Granek E, Polasky S, Shandas V, Yeakley A (2011) Valuing ecological systems and services. F1000 Biol Rep 3:14. https://doi.org/10.3410/B3-14
Cresswell JW (2016) Research Design Pendekatan Metode Kualitatif, Kuantitatif dan Campuran (Edisi Keem ed.). Pustaka Pelajar
Dohong A, Aziz AA, Dargusch P (2017) A review of the drivers of tropical peatland degradation in South-East Asia. Land Use Policy 69:349–360. https://doi.org/10.1016/j.landusepol.2017.09.035
Fairhurst T, McLaughlin D (2009) Sustainable oil palm developmenton degraded Land in Kalimantan. In: WWF (ed) Gland, Switzerland
Gregersen HM, Contreras AH (1979) Economic analysis of forestry projects. Food and Agriculture Organization of The United Nation
Groot R, van der Meer P (2010) Quantifying and valuing goods and services provided by plantation forests. Ecosyst Goods Serv Plant for. https://doi.org/10.4324/9781849776417
Hanley N, Shogren J, White B (2001) Introduction to environmental economics. Oxford University Press
Hawa A, Azhar B, Top MM, Zubaid A (2016) Depauperate avifauna in tropical peat swamp forests following logging and conversion to oil palm agriculture: evidence from mist-netting data. Wetlands 36(5):899–908. https://doi.org/10.1007/s13157-016-0802-3
Kakuru W, Turyahabwe N, Mugisha J (2013) Total economic value of wetlands products and services in Uganda. Sci World J 2013:192656. https://doi.org/10.1155/2013/192656
Liu FHM, Ganesan V, Smith TEL (2020) Contrasting communications of sustainability science in the media coverage of palm oil agriculture on tropical peatlands in Indonesia, Malaysia and Singapore. Environ Sci Policy 114:162–169. https://doi.org/10.1016/j.envsci.2020.07.004
Maskun, Achmad, Naswar, Assidiq H, Bachril SN (2021) Palm oil cultivation on peatlands and its impact on increasing Indonesia's greenhouse gas emissions. IOP Conf Ser: Earth Environ Sci 724(1):012092. https://doi.org/10.1088/1755-1315/724/1/012092
Miettinen J, Shi C, Liew SC (2012) Two decades of destruction in Southeast Asia's peat swamp forests. Front Ecol Environ 10(3):124–128. https://doi.org/10.1890/100236
Munasinghe M (1992) Environmental economics and valuation in development decision making
Osaki M, Nursyamsi D, Noor M, Wahyunto, Segah H (2016) Peatland in Indonesia. In: Osaki M, Tsuji N (eds) Tropical Peatland ecosystems. Springer Japan, pp 49–58
Page SE, Rieley JO, Banks CJ (2011) Global and regional importance of the tropical peatland carbon pool. Global Change Biol 17(2):798–818. https://doi.org/10.1111/j.1365-2486.2010.02279.x
Pearce D, Moran D (1994) The economic value of biodiversity. Earthscan Publications
Pinem T (2016) Kebakaran hutan dan lahan gambut. Jurnal Gema Teologika 1(2):61–68. https://doi.org/10.21460/gema.2016.12.219
Rahman A, Yuliani F (2018) Mitigasi Bencana Kebakaran Lahan Gambut Dan Pemberdayaan Masyarakat Melalui Metode Restorasi. Sosio Informa. https://doi.org/10.33007/inf.v4i2.1460
Richardson L, Loomis J, Kroeger T, Casey F (2015) The role of benefit transfer in ecosystem service valuation. Ecol Econ 115:51–58. https://doi.org/10.1016/j.ecolecon.2014.02.018
Ritzema H, Limin S, Kusin K, Jauhiainen J, Wösten H (2014) Canal blocking strategies for hydrological restoration of degraded tropical peatlands in Central Kalimantan, Indonesia. CATENA 114:11–20. https://doi.org/10.1016/j.catena.2013.10.009
Sasidhran S, Adila N, Hamdan MS, Samantha LD, Aziz N, Kamarudin N, Puan CL, Turner E, Azhar B (2016) Habitat occupancy patterns and activity rate of native mammals in tropical fragmented peat swamp reserves in Peninsular Malaysia. For Ecol Manag 363:140–148. https://doi.org/10.1016/j.foreco.2015.12.037
Scholz M (2011) Carbon storage and fluxes within wetland systems. Green Energy Technol. https://doi.org/10.1007/978-1-84996-459-3_3
Ulya NA (2015) Penilaian kawasan hutan rawa gambut berdasarkan konsep nilai ekonomi total (studi kasus di Merang Kepayang Sumatera Selatan) Gadjah Mada University]. Yogyakarta
Ulya NA, Nurlia A, Kunarso A, Martin E, Waluyo EA (2019) Valuation of goods and services derived from plantation forest in peat swamp forest area: the case of South Sumatra Province. IOP Conf Ser: Earth Environ Sci 308(1):012047. https://doi.org/10.1088/1755-1315/308/1/012047
Ulya NA, Warsito PS, Andayani W, Gunawan T (2014) Economic value of water for domestric and transportation—case study in villages around Merang Kepayang Peat Swamp Forest, South Sumatera Province. Jurnal Manusia Dan Lingkungan 21(2):232–238
Ulya NA, Warsito PS, Andayani W, Gunawan T (2015) Economic value of carbon of Merang Kepayang Peat Swamp Forest, South Sumatera Province. Jurnal Manusia Dan Lingkungan 22(1):52–58
Wardoyo SA (2006) Pengaruh Kerusakan Penutupan Vegetasi Rawa Gambut terhadap Komunitas Ikan di Sungai Merang Kabupaten Musi banyuasin, Sumatera Selatan [IPB University]. Bogor
Widodo IT (2011) Estimasi Nilai Lingkungan Perkebunan Kelapa Sawit Ditinjau dari Neraca Air Tanaman Sawit (Studi Kasus Perkebunan Kelapa Sawit di Kecamatan Dayun, Kabupaten Siak, Propinsi Riau) [IPB University]. Bogor
Wijedasa LS, Jauhiainen J, Kononen M, Lampela M, Vasander H, Leblanc MC, Evers S, Smith TE, Yule CM, Varkkey H, Lupascu M, Parish F, Singleton I, Clements GR, Aziz SA, Harrison ME, Cheyne S, Anshari GZ, Meijaard E, Andersen R (2017) Denial of long-term issues with agriculture on tropical peatlands

will have devastating consequences. Glob Chang Biol 23(3):977–982. https://doi.org/10.1111/gcb.13516
World Bank (2015) The World Bank Annual Report 2015 [Annual Report]. The World Bank
Yaron G (2001) Forest, plantation crops or small-scale agriculture? An economic analysis of alternative land use options in the Mount Cameroon Area. J Environ Planning Manag 44:85–108. https://doi.org/10.1080/09640560123194
Yunardy S, Ulya NA, Junaedy A, Kunarso A, Harbi J, Milantara N (2018). Hutan OKI Kembali Lestari Pembelajaran dari Pengelolaan Hutan di Kabupaten OKI Sumatera Selatan. UNSRI Press

14 Evergreen Amid Deforestation: A Case Study of Forest Protection, Livelihoods, and Fostering Harmony in Rantau Kermas Village, Jambi

Dicky Rachmawan, Robert Siburian, Herman Hidayat, and Angga Kurniawansyah

Abstract

The deforestation rate across the globe is expected to increase in 2022 as a result of human activities, particularly in areas managed by Indigenous peoples. Indonesia, which has the largest number of Indigenous peoples in Southeast Asia, is likely to be the most affected. This can, however, be contained in woodlands managed by customary law communities such as the Marga Sarampas Customary Law Community (MS CLC) in Rantau Kermas Village. This chapter explores the concept of collaborative action to promote the harmonization of forest protection with socioeconomic goals and objectives, within the MS CLC's integrative framework for collaborative governance (IFCG). The chapter is based on qualitative case studies, and its outcomes are innovative in demonstrating the internal collaboration dynamics among community members. The values of local wisdom in protecting the customary forest are upheld through collaborative efforts such as mutual reminders, forest patrol activities, the equitable enforcement of customary law, and *Kenduri Sako*, an advisory ceremony. In a slightly different result from that seen in previous research, the chapter shows that the environmental benefit has a substantial impact on increasing member awareness of forest protection (through shared motivation and principled engagement), which is further reinforced by the leader acting as local law enforcer.

Keywords

Deforestation · Customary forest · Local wisdom · Collaborative governance regimes

14.1 Introduction

The deforestation rate worldwide resulting from human activities continues to rise, and it remains a cause for concern in the relationship between humanity and the environment (Nuswantoro 2023; Reuters 2023; Rude et al. 2021; Silva and Rodgers 2018). Despite a decrease in the last decade, the global deforestation rate is still reaching 4.1 million hectares annually, primarily in indigenous forests, particularly in tropical regions (Nuswantoro 2023; Ritchie and Roser 2021; Rude et al. 2021). Indonesia boasts the ASEAN's most extensive tropical rainforest which is home to between 700 and 2240 Indige-

D. Rachmawan (✉) · R. Siburian · H. Hidayat
Research Centre for Society and Culture - National Research and Innovation Agency, Jakarta, Indonesia
e-mail: dickybrin@gmail.com

A. Kurniawansyah
Department of Geography, University of Indonesia, Jakarta, Indonesia

H. Warsilah and A. H. Purnomo (eds.), *Approaching Natural Resources for Sustainable Development in Indonesia*, Sustainable Development Goals Series, https://doi.org/10.1007/978-981-96-6913-4_14

nous groups (AMAN 2007; Asteria 2021; Errico 2017; Setra 2013). The country is also facing significant deforestation challenges (KLHK 2022; Ritchie and Roser 2021). Although Indonesia's deforestation rate has decreased significantly from 650,000 hectares per year in 2015 (Ritchie and Roser 2021) to 107,000 hectares per year in 2022 (Forestinsights.id 2023), the country remains in fourth place for deforestation globally, after Brazil, the Congo, and Bolivia (Nuswantoro 2023). Numerous studies from various parts of the world have highlighted the vital role of Indigenous communities in the protection and preservation of forests (Bullock et al. 2020; Hotte et al. 2021; Maru et al. 2020; Susanto et al. 2020; Timko and Webbe 2020).

In the Indonesian context, Jambi Province recognizes 48 customary forests, which constitute 40% of the traditional forests in the country legally recognized by the Ministry of Environment and Forestry (MoEF) (MoEF 2023). However, approximately 1.26 million hectares (60%) of the province's forested region have been encroached upon by conversion (Mairiadi 2022; Ramadhan 2022), resulting in only 882,272 hectares of forest remaining in protected and conservation areas (Solmi 2020) such as Taman Nasional Kerinci Seblat (TNKS). TNKS is a national park located in Jambi, West Sumatera, Bengkulu, and South Sumatera Provinces. It boasts high biodiversity and an abundance of germplasm that aligns with the culture. TNKS was recognized as a UNESCO World Heritage Site in 2004 due to its rich natural biodiversity and cultural harmony (Karyadi et al. 2018).

Unfortunately, deforestation continues to affect forested regions, with a primary impact in three districts: Tebo, Muara Jambi, and Merangin districts (Baitri 2016). For instance, 14,000–15,000 hectares of encroachment area in TNKS have been converted, mostly into coffee farming. This farming provides income for 13,000–15,000 Southern migrants in seven sub-districts within the Merangin District (Diana 2017b, 2017c; Kompas.com 2010). Natural disasters increased significantly in 2020 compared to the previous five years, with 36 flood disasters (BNPB 2023) causing damage to 2703 houses and suffering for 62,882 people.

Many Southern migrants have settled in various villages located in the Lembah Masurai sub-district; Sungai Lalang village is the closest to the Jangkat sub-district (Diana 2017c). Unfamiliar with the customary boundaries between the people of the Jangkat sub-district, especially with regard to CLCs, a number of migrants have attempted to clear forest areas and expand coffee cultivation within the customary territory of the Marga Sarampas (MS) (Diana 2017b; R. Saputra 2017). The MS community has always followed the traditional rules in their area, although this is still not recognized by the government. Several damaging conflicts between migrants and the MS CLC occurred in the struggle for control of the forest areas (Asikin 2016; Beritajambi.co 2018; jambupdate.co 2017; R. Saputra 2017; Solihin 2017). However, the MS CLC was able to protect their customary forest territory from deforestation expansion (Diana 2017c; D. Saputra 2019). This chapter explores how the MS CLC, through its internal group dynamics, enhances harmony in forest protection and promotes the socioeconomic goals of the integrative framework for collaborative governance (IFCG).

14.2 Methods

This study uses a qualitative research approach to understand complex situations. Research activities were conducted from June 16–30, 2022 at MS CLC, specifically in Rantau Kermas Village, Merangin Regency, Jambi Province, Indonesia. We interviewed seven key informants (representing government elements, community activists, and five community leaders in MS CLC) and 33 informants from members of the MS CLC group. The qualitative data were then processed using NVIVO 12 plus, coded, and aggregated to reflect the depth of the information obtained.

First, however, we describe the physical aspects and context of deforestation in the Kerinci Seblat Nation Park (TNKS), comparing the forest cover of the customary forests managed by MS

CLC and the land cover in TNKS in the Jangkat Sub-district. We use remote sensing with PlanetScope through the Dove-R sensor in the https://www.planet.com/ website, which uses filtering in the form of area coverage of 90% and initial coverage of 5–10%. The satellite image recordings were coordinated, adjusted, digitized, and delineated (Luthfina et al. 2019; Wardani et al. 2016) with MoEF decree number 5533/2017 in Pulau Tengah Village using ArcGIS software, version 10.4. At the conclusion of the process, we tried to describe the loss of native forest cover by calculating the geometry with a red line on a hectare scale.

14.3 Literature Review

Community participation in resource management, such as forest management, is an integral element of empowering and implementing government policies (Barrett et al. 2021; Fatem 2019; Pane et al. 2021; Sitienei et al. 2021; Vihma and Toikka 2021). In the global context of protected or conservation area management, such as forests, it is essential to include the Indigenous people who are estimated to manage at least 40% of protected areas worldwide (Bullock et al. 2020). Active community involvement in conservation activities is a significant challenge that takes time to achieve (Quazi et al. 2008). At the heart of regulating community involvement or participation based on agreement is the concept of collaborative governance (hereafter referred to as CG) (Michel 2017), and the most cited CG concept is that proposed by Ansel and Gash (Batory and Svensson 2020). Along the way, this concept was developed by Emerson and Nabatchi (2015) who introduced the term 'integrative framework for collaborative governance' (IFCG). The IFCG integrates theoretical, empirical, and practical approaches to recognize behaviors, relationships, and structural elements that help describe forms of collaboration (where, why, how, and to what effect) (Emerson and Nabatchi 2015). One-third of previous IFCG studies examined one or two predispositions from IFCGs (Emerson and Anh 2022). Furthermore, Emerson and Anh (2022) argue that it is essential to look at the dynamics of collaboration in non-Western and developing country contexts.

Many previous studies examining the CG process as it concerns Indigenous peoples and natural resource management have explored interactions between groups. Building mutual understanding between groups in such a way as to accommodate various interests is a positive factor mentioned in most previous studies (Bettis et al. 2020; Mattor et al. 2020; Ming'ate et al. 2019; Roengtam and Agustiyara 2022; Vihma and Toikka 2021). The next positive factor is the importance of mutual trust and a sense of responsibility (Bettis et al. 2020; Koning et al. 2017; Morley and Cashell 2017; Ran and Qi 2018) and accountability or transparency between parties (Amin and Marta 2021; Koning et al. 2017; Ming'ate et al. 2019; Quazi et al. 2008). A review of negative factors in relation to CG found conflicts of interest (Lindgren et al. 2021; Morley and Cashell 2017; Sitienei et al. 2021) such as the occurrence of elite capture (Tadaki et al. 2020), lack of institutional readiness (Quazi et al. 2008), unclear roles between groups (Sitienei et al. 2021), poor communication (Morley and Cashell 2017), and local power struggles (Quazi et al. 2008). In the more specific context of the characteristics of Indigenous peoples, a community or a group at site level, group participation, and cohesion are crucial aspects (Johansson 2018; Tonelli et al. 2018; Wanna 2008).

A number of scholars suggest deeper study of internal group collaboration in the context of CG theory, down to the individual or member level of a group (Batory and Svensson 2019; Bianchi et al. 2021; Hotte et al. 2021; Mattor et al. 2020; Robitaille et al. 2017). Because the CG concept tends to examine interactions between groups and sees small groups (such as Indigenous peoples) as homogeneous groups (Premauer and Berkes 2015), uncertainty can arise at various levels, so it is essential to consolidate the group internally first to achieve effective collaboration at the site level (Ulibarri 2019).

Most of the early studies related to the implementation of social forestry in Indonesia from the IFCG concept tended to look at collaborative governance regimes (CGRs) mainly from the perspective of joint capacity (JC) (Budi et al. 2021; Chandra et al. 2022; Husein et al. 2021; Massiri et al. 2020; Nurrochmat et al. 2021; Situmorang and Yen 2022; Ulya et al. 2022). Collaborative actions, on the other hand, are the least studied in previous similar studies (Fisher et al. 2020; Hiratsuka et al. 2019; Iriyani et al. 2020; Sahide et al. 2020). Even though collaborative action is an element that reflects a dynamic collaborative process (principled engagement (PE), shared motivation, joint capacity), it is indeed the one least paid attention to by scientists and is often left unspecified (Emerson and Nabatchi 2015: 81–83). Emerson and Nabatchi (2015) further state that effective collaborative actions come back to the dynamic collaboration process, leading to mutually agreed changes. This chapter aims to fill a niche that has received less attention from previous studies, namely to examine the collaborative actions associated with the harmonious goals of protecting ecosystems and socio-economics in MS CLC.

14.4 Result and Discussion

14.4.1 MS CLC Customary Forest: Evergreen Amid Deforestation

The exclusion of Indigenous peoples from their customary forests due to deforestation, as has occurred in the Amazon and Kalimantan, is tantamount to destroying their cultural values (Peluso 1992; Suhrke 1993). In the context of MS CLC, their customary forest is a traditional forest surrounded by the TNKS. 65% of the forest clearing and encroachment in the TNKS that has been occurring since the 2000s is based on coffee plantations (Diana 2017a; Santoso 2018; D. Saputra 2019). It is estimated that the enlarged TNKS area has reached 14,000–15,000 hectares which will become a source of livelihood for 13,000–15,000 farmers in seven sub-districts (Diana 2017b, 2017c). The number of migrants is now thought to have reached between 30,000 and 90,000 (DHR and RDD informants). Since 2014 alone, around 5000 hectares of TNKS have been encroached on in two sub-districts, Lembah Madurai and Jangkat (Herupitra 2015), with informants mentioning that the land in the Jangkat area and the Madurai Valley was fertile and suitable for coffee plantations (Infoman NRN and DSI). DSI stated that by only managing 4 hectares, his family could harvest coffee three times a year without using fertilizer, with the estimated production reaching 6 tons/year or 2 tons/harvest.

We examined the forest cover, comparing the MS CLC customary forest with the forest cover in the TNKS in the Jangkat District, with remote sensing centered on the location of the MS CLC ordinary forest. Based on a study of PlanetScope satellite imagery through the Dove-R sensor, we found around 2192.5 hectares of encroached land, the most prominent area, around 621.2 hectares, being close to Lake Pauh. This is a huge amount compared to the customary forest in MS CLC, which is only ± 130 hectares. Although the conditions are very different, the forest in MS CLC tends to be sustainable, even though it is small in area. The sustainability of customary forests in MS CLC can be seen on the map in Fig. 14.1.

Internally, the MS CLC group has five principal activities (collaborative actions) related to customary forest management: (1) customary forest management planning, which is carried out every 3 months; (2) monthly patrols; (3) delivery of routine information to maintain customary forests once a week or every two weeks; (4) advice given by traditional leaders to the next generation (including values related to customary forest management), called *Kenduri Sako*; and (5) mutual reminders between members.

The most significant activity carried out by MS CLC as a group, as viewed by the members, is supervision of each other and mutual reminders (17 aggregates). Group members at MS CLC supervise each other and are fearless in reporting to the customary forest manager if someone destroys the forest. Supervising and reminding one another is a form of support and a reflection of the cohesiveness of collaborative actions in MS CLC.

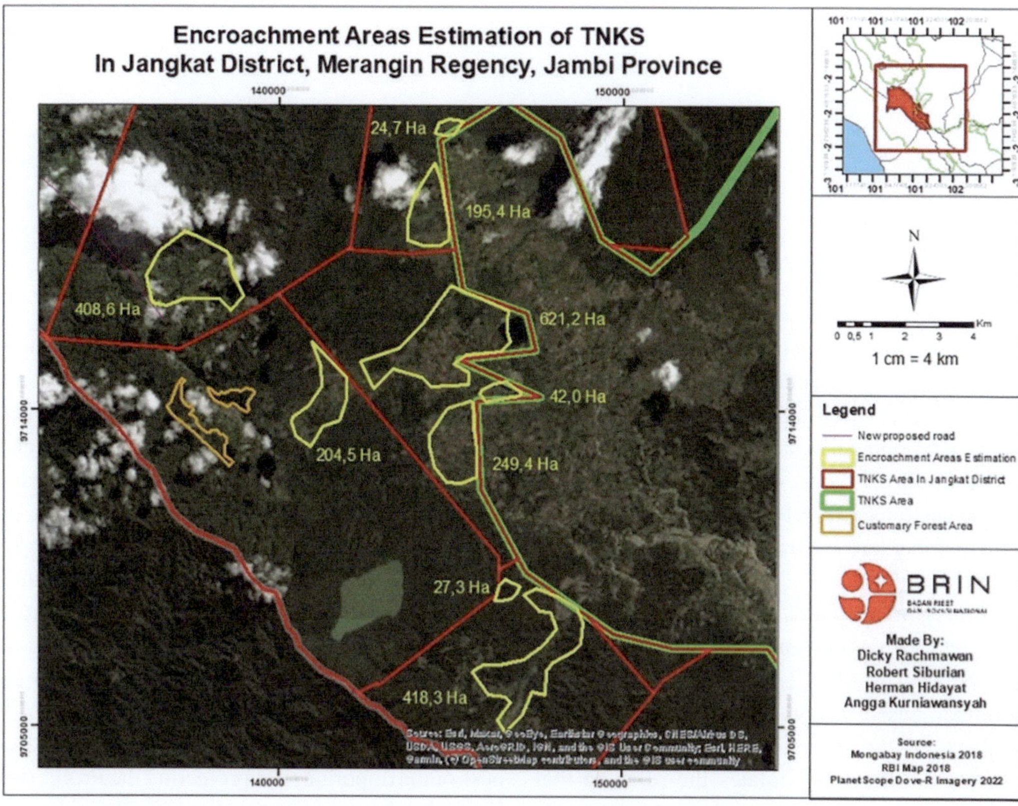

Fig. 14.1 Map of Estimated Encroachment in the TNKS Area of Jangkat District. (*Source* data processed by authors (2023))

The second collaborative action activity is a monthly patrol (15 aggregates). Most informants knew that the customary forest manager at MS CLC carried out a routine monthly patrol. Some of them also take part in patrol activities as a contribution to preserving customary forests. Some informants thought that this patrol activity was only carried out twice a year.

The third collaborative action activity is a three-monthly discussion of customary forest management planning (10 aggregates). At this point, the planning discussions conducted by MS CLC are two-way discussions between MS CLC customary forest managers and members to determine targets or activities to be carried out as a group. The two-way discussion made group members gradually understand more and become aware of the benefits of having customary forests.

The remainder with the same aggregate, namely eight aggregates, is filled with routine information once a week or every two weeks, and advice given in the *Kenduri Sako* ceremony, which includes regulations related to customary forests. In the MS CLC context, internet access is still challenging to obtain, but information related to preserving forests can still be conveyed and disseminated effectively through announcements at mosques delivered after Friday prayers. Then the distribution of customary rules in general, including protecting the forest, is again conveyed once a year during the *Kenduri Sako* ceremony, usually held after the *Idul Fitri* holiday. This advice is given by traditional leaders, village administrators, to figures who are members of customary forest management (Fig. 14.2). The following is a graph of activity significance based on the aggregate data obtained:

In line with Emerson and Nabatchi's (2015) argument, which is more specific on collaborative actions at the internal group level, it is found that

Fig. 14.2 The significance of collaborative actions at MS CLC in indigenous forest management

practical collaborative actions in internal MS CLC also further strengthen internal principled engagement (PE). The five collaborative actions combine to create a deliberative process that strengthens the definition of shared values about the importance of customary forest (determination) by each group member through monitoring and reminding one another.

With regard to shared motivation (SM), monitoring and reminding one another down to the member level in MS CLC is a form of internal legitimacy based on mutual understanding. The members of the MS CLC are committed to reporting anyone who violates the rules so that customary sanctions on the violator can be stipulated based on deliberations and agreements with traditional leaders.

Finally, the leadership of traditional leaders, together with the cohesiveness down to the level of group members and the adherence to customary rules (procedural and institutional arrangements), further strengthened forest protection in MS CLC. Leaders also have an essential role in making customary rules and upholding them fairly as "enforcers of local law", which is not explained by Emerson and Nabatchi (2015). The following section will explain the harmony between the protection of customary forests and aspects of socioeconomic community life in MS CLC.

14.4.2 The Value of MS CLC and Benefits of Conserving Customary Forest

Long before the recognition and establishment of the MS CLC customary forest in 2016, the MS CLC community had guarded the customary forest, which they called the Prohibited Forest. A Prohibited Forest is a forest area on an extreme slope upstream of a river or *ulu aek*, which functions for household needs (consumption) and maintains water resources to irrigate the rice fields of MS CLC members, who tend to work as farmers. The MS CLC customary forest shades 11 upstream rivers. Our MS CLC member informants consider customary forests are very important to protect, and their most significant meaning is as a source of water (21 aggregates), followed by preventing natural disasters, namely landslides (9 aggregates), and drought (7 aggregates). Awareness of the importance of protecting the Prohibited Forest has been instilled in the MS CLC community since the bitter experience of flash floods due to forest destruction in the upstream area in 1977 (Nurdin 2019). The fear of landslides has become a concern because the MS CLC community settlement is located in the valley. The location of the MS CLC settlement between two valleys is shown in Fig. 14.3.

In addition to the importance of customary forests as a water resource and for disaster prevention, the members of the MS CLC community also understand their role in supporting the preservation of nature and life (5 aggregates).

A further benefit that is felt directly by MS CLC members is the supply of electricity from a micro-hydro power plant (PLTMH) (7 aggregates), the construction of which was initiated by Warsi and WWF, who received funding from the Green Prosperity Millennium Challenge Account (MCA) Indonesia. The plant has a capacity of 41 KW valued at 3.4 billion rupiah. Thanks to this PLTMH, the MS CLC community has 24-h access to electricity at a relatively affordable price.

Fig. 14.3 The MS CLC settlement located between two customary forest valleys. (*Source* data collected from informants)

Finally, the customary forests benefit from additional funds from third parties through the foster tree program, assisted by the Warsi network. This program has been running since 2014, and funding from foster trees currently stands at IDR200 million/year. 60% of the income from the foster tree program is allocated to the development of village business entities (BUMDes), of which 10% is for social activities such as scholarships and 5% for customary activities. The remaining 40% is retained for customary forest management. The benefits are felt both in the form of environmental services and the flow of funds, making the objectives of protecting MS CLC's customary forests more harmonious. Figure 14.4 illustrates the importance of protecting the forest and fostering trees at MS CLC in order to maintain water discharge so that electricity can be supplied through a PLTMH generator.

14.5 Conclusion and Recommendation

14.5.1 Recommendation

This case study is a success story that demonstrates the important role of Indigenous peoples as communities protecting forests from widespread deforestation in the TNKS area. We suggest that the Indonesian government consider proposals from Indigenous groups, especially proposals for areas that can be protected from the threat of deforestation by encroachers. Although the internal and external dynamics of other Indigenous communities may be different from those of MS CLC, the government can support and facilitate them as Indigenous groups who deserve to manage their own customary forests. Finally, we suggest similar research in the future, related to the implementation of social forestry in Indonesia, to examine the internal dynamics of the community and the external dynamics between the community and its stakeholders.

14.5.2 Conclusion

The MS CLC is a collective protecting traditional forests in community-based forest management. To minimize the socio-ecological crisis of forest resources, it is important to encourage and strengthen collective groups such as the MS CLC, whose activities meet their livelihood needs in terms of environmental services (water, electricity, and prevention of natural disasters) and economic benefits (additional income from these trees through the nursery program). However, collaborative processes in groups will only run well if there is a balance of roles between group members. Traditional leaders can act as enforcers of existing regulations, and there are roles for initiators, sponsors, facilitators, participants, scientists, technical experts, and policymakers.

Fig. 14.4 MS CLC foster tree (top left), micro-hydro power generator (top right), and billboard promoting protection of forests to keep electricity on (bottom). (*Source* Authors' Documentation)

References

AMAN (2007) Request for consideration of the situation of indigenous peoples in Kalimantan, Indonesia, under the United Nations Committee on the Elimination of Racial Discrimination's Urgent Action and Early Warning Procedures Committee on the Elimination of Raci (Issue July). https://www.forestpeoples.org/sites/default/files/documents/Early Warning Urgent Action Procedure CERD submission Indonesia.pdf

Amin RM, Marta A (2021) Towards the collaborative governance in maintaining indigenous history of Pejangki Village in Indragiri Hulu, Indonesia. Otoritas: Jurnal Ilmu Pemerintahan 11(1):45–53. https://doi.org/10.26618/ojip.v11i1.3002

Asikin MN (2016) Mencekam, Penghancuran Pondok Perambah Hutan, Warga Renah Alai Siaga. JawaPos.Com, pp 6–8. https://www.jawapos.com/nasional/hukum-kriminal/08/11/2016/mencekam-penghancuran-pondok-perambah-hutan-warga-renah-alai-siaga/Close

Asteria D (2021) Contribution of customary law in sustainable forest management for supporting climate action. IOP Conf Ser: Earth Environ Sci 940(1):012080. https://doi.org/10.1088/1755-1315/940/1/012080

Baitri JH (2016) Tiga Kabupaten Ini Paling Banyak Deforestasi. TribunJambi.Com. https://jambi.tribunnews.com/2016/04/08/tiga-kabupaten-ini-paling-banyak-deforestasi

Barrett KJ, Cannon JB, Schuetter AM, Cheng AS (2021) Effects of collaborative monitoring and adaptive management on restoration outcomes in dry conifer forests. For Ecol Manag 488(January):119018. https://doi.org/10.1016/j.foreco.2021.119018

Batory A, Svensson S (2019) The fuzzy concept of collaborative governance: a systematic review of the state of the art. Central Eur J Public Policy 13(2):28–39. https://doi.org/10.2478/cejpp-2019-0008

Batory A, Svensson S (2020) Regulating collaboration: the legal framework of collaborative governance in Ten European Countries. Int J Public Adm 43(9):780–789. https://doi.org/10.1080/01900692.2019.1658771

Beritajambi.co (2018) Ini Penyebab Ratusan Petani dari Desa Datang ke Bangko. Beritajambi.Co, pp 30–31. www.beritajambi.co/read/2018/01/31/3170/ini-penyebab-ratusan-petani-dari-desa-datang-ke-bangko

Bettis A, Schoon M, Blanchette G (2020) Enabling regional collaborative governance for sustainable recreation on public lands: the Verde Front. J Environ Planning Manag. https://doi.org/10.1080/09640568.2020.1753178

Bianchi C, Nasi G, Rivenbark WC (2021) Implementing collaborative governance: models, experiences, and challenges. Public Manag Rev 23(11):1581–1589. https://doi.org/10.1080/14719037.2021.1878777

BNPB (2023) Data Informasi Bencana Indonesia (DIBI). Badan Nasional Penanggulangan Bencana. https://dibi.bnpb.go.id/home/index2

Budi KH, Nugroho B, Mardiana R (2021) Implementation of social forestry policy: analysis of community access. For Soc 5(1):60–74. https://doi.org/10.24259/fs.v5i1.9859

Bullock R, Zurba M, Reed MG, McCarthy D (2020) Strategic options for more effective indigenous participation in collaborative environmental governance. J

Plan Educ Res. https://doi.org/10.1177/0739456X20920913
Chandra A, Fauzi D, Khatimah FH, Wicaksono SA (2022) Assessing drivers of forest conservation in Simancuang village Forest, West Sumatra. Small-Scale for 21(1):93–118. https://doi.org/10.1007/s11842-021-09489-2
Diana E (2017a) Berebut Lahan di Lembah Masurai (Bagian 1). www.mongabay.co.id, February. https://www.mongabay.co.id/2017/02/28/berebut-lahan-di-lembah-masurai-bagian-1/
Diana E (2017b) Cerita Petani Kopi dari Lembah Masurai. www.mongabay.co.id, March. https://www.mongabay.co.id/2017/03/11/cerita-petani-kopi-dari-lembah-masurai-bagian-2/
Diana E (2017c) Konflik Lahan Berlarut di Lembah Masurai (Bagian 3). www.mongabay.co.id, April. https://www.mongabay.co.id/2017/04/10/konflik-lahan-berlarut-di-lembah-masurai-bagian-3/
Emerson K, Anh MW (2022) Informing practice through research. April
Emerson K, Nabatchi T (2015) Collaborative governance regimes. Georgetown University Press
Errico S (2017) The rights of indigenous peoples in Asia. Human rights-based overview of national legal and policy frameworks against. https://www.ilo.org/wcmsp5/groups/public/---dgreports/---gender/documents/publication/wcms_545484.pdf
Fatem SM (2019) Connecting social forestry to conservation policies in Tanah Papua. Forest Soc 3(1):141–147. https://doi.org/10.24259/fs.v3i1.5865
Fisher MR, Workman T, Mulyana A, Institute B, Moeliono M, Yuliani EL, Colfer CJP, Adam UEFB (2020) Striving for PAR excellence in land use planning: Multi-stakeholder collaboration on customary forest recognition in Bulukumba, South Sulawesi. Land Use Policy. https://doi.org/10.1016/j.landusepol.2017.09.057
Forestinsights.id (2023) Laju Deforestasi Indonesia Tahun 2022, Menteri LHK: Angkanya Terus Menurun. Forestinsights.Id. https://forestinsights.id/laju-deforestasi-indonesia-tahun-2022-menteri-lhk-angkanya-terus-menurun/
Herupitra (2015) Di 2014 , 5 . 000 Hektare Kawasan TNKS. TribunJambi.Com. https://jambi.tribunnews.com/2015/04/25/di-2014-5000-hektare-kawasan-tnks-jadi-kebun-kop
Hiratsuka M, Nakama E, Satriadi T, Fauzi H, Aryadi M, Morikawa Y (2019) An approach to achieve sustainable development goals through participatory land and forest conservation: a case study in South Kalimantan Province, Indonesia. J Sustain for 38(6):558–571. https://doi.org/10.1080/10549811.2019.1598440
Hotte N, Kozak R, Wyatt S, Wilkes R (2021) Discrimination impacts trust during collaborative natural resource governance involving indigenous communities. Soc Nat Resour. https://doi.org/10.1080/08941920.2021.1950883
Husein S, Herdiansyah H, Putri LS (2021) An ecofeminism perspective: a gendered approach in reducing poverty by implementing sustainable development practices in Indonesia. J Int Women's Stud 22(5):210–228
Iriyani S, Humam Hamid A, Setyarso A, Basri H (2020) Towards the implementation of social forestry policy in Aceh Province, Indonesia: process and institutional assessment. Pertanika J Social Sci Human 28(4):2689–2710. https://doi.org/10.47836/PJSSH.28.4.11
jambupdate.co (2017) Perambah Hutan Kian Meraja Lela, Jangkat Kembali Memanas. Jambupdate.Co. https://jambiupdate.co/artikel-perambah-hutan-kian-meraja-lela-jangkat-kembali-memanas.htm
Johansson J (2018) Collaborative governance for sustainable forestry in the emerging bio-based economy in Europe. Curr Opin Environ Sustain 32:9–16. https://doi.org/10.1016/j.cosust.2018.01.009
KLHK (2022) The State of Indonesia's Forests 2022: Toward FOLU Net Sink 2030. Ministry of Environment and Forestry Republic of Indonesia
Kompas.com (2010) Taman Nasional Terus Dirambah Pendatang. Kompas.Com2. https://nasional.kompas.com/read/2010/12/20/04042168/index.html
Koning M, de Nguyen T, Lockwoodc M, Sengchanthavong S, Phommasanee S (2017) Collaborative governance of protected areas: success factors and prospects for Hin Nam No national protected area, Central Laos. Conserv Soc 15(1):87–99. https://doi.org/10.4103/0972-4923.201396
Lindgren AY, Reed MG, Robson JP (2021) Process makes perfect: perceptions of effectiveness in collaborative environmental governance. Environ Manag. https://doi.org/10.1007/s00267-020-01402-5
Luthfina MAW, Sudarsono B, Suprayogi A (2019) Analisis Kesesuaian Penggunaan Lahan Terhadap Rencana Tata Ruang Wilayah Tahun 2010–2030 Menggunakan Sistem Informasi Geografis di Kecamatan Pati. Jurnal Geodesi Undip 8(1):74–82
Mairiadi N (2022) Walhi memperkirakan 60 persen hutan Jambi sudah dirambah atau rusak. AntaraJambi. https://jambi.antaranews.com/berita/507917/walhi-memperkirakan-60-persen-hutan-jambi-sudah-dirambah-atau-rusak
Maru Y, Gebrekirstos A, Haile G (2020) Indigenous ways of environmental protection in Gedeo community, Southern Ethiopia: a socio-ecological perspective. Cogent Food Agric 6(1):1766732. https://doi.org/10.1080/23311932.2020.1766732
Massiri SD, Malik A, Golar H, Nugroho B (2020) Institutional capacity of forest management unit in promoting sustainable community-based forest management. Case study of forest management unit in central sulawesi province, Indonesia. Jurnal Manajemen Hutan Tropika 26(2):169–177. https://doi.org/10.7226/JTFM.26.2.1695
Mattor KM, Cheng AS, Kittler B, McDonough M (2020) Assessing collaborative governance outcomes and indicators across spatial and temporal scales: stewardship contract implementation by the United States Forest Service. Soc Nat Resour 33(4):484–503. https://doi.org/10.1080/08941920.2019.1665762

Michel CL (2017) Collaborative governance : beyond mere participation. In: Third International Public Policy Conference, pp 1–34

Ming'ate FLM, Letema S, Obiero K (2019). Designing institutional arrangements for collaborative governance of forests in Kenya Using a Delphi Process. J Sci Res Rep, 25(4)1–11. https://doi.org/10.9734/jsrr/2019/v25i430191

MoEF (2023) Skema Hutan Adat. Ministry of Environment and Forestry2. https://gokups.menlhk.go.id/

Morley L, Cashell A (2017) Collaboration in health care. J Med Imaging Radiat Sci 48(2):207–216. https://doi.org/10.1016/j.jmir.2017.02.071

Nurrochmat DR, Pribadi R, Siregar H, Justianto A, Park MS (2021) Transformation of agro-forest management policy under the dynamic circumstances of a two-decade regional autonomy in Indonesia. Forests 12(4):419. https://doi.org/10.3390/f12040419

Nuswantoro (2023) Sulit Dinalar , dalam Semenit Bumi Kehilangan Hutan Hujan Seluas Ini. 2023. https://www.mongabay.co.id/2023/07/03/sulit-dinalar-dalam-semenit-bumi-kehilangan-hutan-hujan-seluas-ini/

Pane E, Yanis AM, Susanto IS (2021) Social forestry: the balance between welfare and ecological justice. Int J Criminol Sociol 10(January):71–78. https://doi.org/10.6000/1929-4409.2021.10.10

Peluso NL (1992) (Mis)Management and development of an extractive rainforest product. Conserv Biol 6(2):210–219

Premauer JM, Berkes F (2015) A pluralistic approach to protected area governance: Indigenous peoples and Makuira National Park , 4(June):1–16. https://doi.org/10.15451/ec2015-5-4.4-1-16

Quazi SA, Bushley BR, Miles WB (2008) Introduction: participation and the collaborative management of protected areas in Bangladesh. In: Connecting communities and conservation: Collaborative management of protected areas in Bangladesh. USA: United States Agency for International Development (USAID), pp 1–25. https://www.eastwestcenter.org/sites/default/files/fileadmin/resources/publications/PDF/connectingcommunities02introduction.pdf

Ramadhan B (2022) Miris, 60 Persen Hutan di Jambi Sudah Rusak _ Republika Online. Republika. https://www.republika.co.id/berita/rcqln2330/miris-60-persen-hutan-di-jambi-sudah-rusak

Ran B, Qi H (2018) The entangled twins: power and trust in collaborative governance. Admin Soc 51(4):607–636. https://doi.org/10.1177/0095399718801000

Reuters (2023) Deforestasi Hutan Tropis Dunia Melejit pada 2022. VOA Indonesia. https://www.voaindonesia.com/a/deforestasi-hutan-tropis-dunia-melejit-pada-2022-/7154428.html

Ritchie H, Roser M (2021) Forests and deforestation. OurWorldInData.Org.

Robitaille PA, Shahi C, Smith MA, Luckai N (2017) Growing together: a principle-based approach to building collaborative Indigenous partnerships in Canada's forest sector. For Chronicle 93(1):44–57. https://doi.org/10.5558/tfc2017-010

Roengtam S, Agustiyara A (2022) Collaborative governance for forest land use policy implementation and development. Cogent Social Sci. https://doi.org/10.1080/23311886.2022.2073670

Rude B, Niederhöfer B, Ferrara F (2021) Deforestation and migration. Cesifo Forum 22(1):49–57

Sahide MAK, Fisher MR, Erbaugh JT, Intarini D, Dharmiasih W, Makmur M, Faturachmat F, Verheijen B, Maryudi A (2020) The boom of social forestry policy and the bust of social forests in Indonesia: developing and applying an access-exclusion framework to assess policy outcomes. Forest Policy Econ. https://doi.org/10.1016/j.forpol.2020.102290

Santoso B (2018) Marak Perambahan Hutan TN Kerinci Seblat, Jambi Minta Bantuan Pusat. Liputan 6:1–4. https://www.liputan6.com/regional/read/3375690/marak-perambahan-hutan-tn-kerinci-seblat-jambi-minta-bantuan-pusa

Saputra D (2019) Enam desa di Jangkat sepakat selamatkan TNKS melalui pemanfaatan lahan kritis. AntaraJambi, 14–16. https://jambi.antaranews.com/berita/364919/enam-desa-di-jangkat-sepakat-selamatkan-tnks-melalui-pemanfaatan-lahan-kritis

Saputra R (2017) Ini Cerita Camat Terkait Maraknya Pendatang Liar Melakukan Perambahan, Meresahkan Hingga Konflik Sosial. Metrojambi.Com. https://metrojambi.com/read/2017/01/16/17593/ini-cerita-camat-terkait-maraknya-pendatang-liar-melakukan-perambahan-meresahkan-hingga-konflik-sosial-/%0AIni

Setra MS (2013) Indigenous peoples in Indonesia: The Struggle for "Legal" Recognition (Issue September). http://www.rightsandresources.org/wp-content/exported-pdf/minasusanasetrathestruggleforlegalrecognition.pdf

Silva AMda, Rodgers J (2018) Deforestation across the world: causes and alternatives for mitigating. Int J Environ Sci Dev 9(3):67–73

Sitienei J, Manderson L, Nangami M (2021) Community participation in the collaborative governance of primary health care facilities, Uasin Gishu County, Kenya. PLoS ONE 16:1–19. https://doi.org/10.1371/journal.pone.0248914

Situmorang ROP, Yen T-M (2022) Comparison of benefits, stakeholders, implementation, and locations of community forestry in Taiwan and Indonesia based on experts' views. J for Res. https://doi.org/10.1080/13416979.2022.2038764

Solihin (2017) Konflik Taman Nasional di Jangkat Berakhir , TimDu dan Seteru Sepakati Beberapa Poin Penyelesaian. Tribratanews.Polresmerangin.Com, December, 2–3. https://tribratanews.polresmerangin.com/konflik-taman-nasional-di-jangkat-berakhir-timdu-dan-seteru-sepakati-beberapa-poin-penyelesaian/

Solmi (2020) Luas Hutan di Jambi Makin Susut. Media Indonesia. https://mediaindonesia.com/nusantara/370068/luas-hutan-di-jambi-makin-susut

Suhrke A (1993) Pressure points: environmental degradation, migration and conflict. Peace and Conflict, June 1991, pp 1–43

Susanto AF, Rahayu MIF, Muliya LS (2020) Law community of "Tatar-Sunda": preservation of forests and climate change. Utopia y Praxis Latinoamericana 25(Extra7):165–170. https://doi.org/10.5281/zenodo.4009636

Tadaki M, Sinner J, Stahlmann-Brown P, Greenhalgh S (2020) Does collaborative governance increase public confidence in water management? Survey evidence from Aotearoa New Zealand. Water Altern 13(2):302–323

Timko J, Webbe JR (2020) Indigenous forest governance: challenges, enabling conditions and factors for success

Tonelli DF, Sant'Anna L, Barcelar Abbud E, Aparecida de Souza S (2018) Antecedents, process, and equity outcomes: a study about collaborative governance. Cogent Bus Manag 5(1):1–17. https://doi.org/10.1080/23311975.2018.1469381

Ulibarri N (2019) Collaborative governance: a tool to manage scientific, administrative, and strategic uncertainties in environmental management ? 24(2)

Ulya NA, Martin E, Rahmat M, Premono BT, Malau LRE, Waluyo EA, Imanullah A, Lukman AH, Asmaliyah Saputra D, Hadi EEW, Azwar F, Siahaan H, Purwanto DM, Nuralamin Anjani R, Cahyono SA, Utami S, Agustini S (2022) Enabling factors of NTFP business development for ecosystem restoration: the case of tamanu oil in Indonesian degraded Peatland. Sustainability (Switzerland), 14(17). https://doi.org/10.3390/su141710681

Vihma P, Toikka A (2021) The limits of collaborative governance: the role of inter-group learning and trust in the case of the Estonian "Forest War." Environ Policy Gov 31(5):403–416. https://doi.org/10.1002/eet.1952

Wanna J (2008) Collaborative government: meanings, dimensions, drivers and. In: O'Flynn J, Wanna J (eds) Collaborative governance: a new era of public policy in Australia? Australian National University Press, pp 3–12

Wardani AK, Cahyono A, Martono DB (2016) Analisis Metode Delineasi Bidang Tanah Pada Citra Resolusi Tinggi dalam Pembuatan Kadaster Lengkap. Jurnal Teknik ITS 5(2):380–384. https://doi.org/10.12962/j23373539.v5i2.17173

15 Charting the Past and Present of Pari Island: Reshaping the Future of a Socio-Economically and Ecologically Impacted Small Island

Agus Heri Purnomo, Zainal Arifin, Tenny Apriliani, Mujiyanto, and Desmiwati

Abstract

Pari Island once had a robust ecology consisting of mangroves, corals, and seagrasses, which effectively sustained the subsistence activities of its inhabitants. The balance between humans and nature, however, has been disrupted by tourism development, which has improved quality of life on the islands but at a socio-economic and ecological cost. During the COVID-19 pandemic which led to a cessation of tourism activities, there was a lack of alternative job possibilities as resources allocated to tourism-related businesses had been exhausted. The purpose of this study is to raise awareness among the global community and derive insights regarding the effective management of small islands from the example of Pari Island. The accuracy of our field observations supports the notion that tourism has indeed caused degradation to ecosystems, limiting the possibilities for their restoration.

Keywords

Coastal ecosystems · Ecological degradation · Marine tourism · Socio-economic disruption · Small island

15.1 Theoretical Background

In the past, Pari Island possessed a strong ecology consisting of mangroves, corals, and seagrasses, which effectively sustained the subsistence activities of its inhabitants. The equilibrium between humans and nature was disturbed with the implementation of local government initiatives aimed at enhancing community well-being, particularly through the facilitation of tourism development.

A. H. Purnomo (✉)
Research Centre for Society and Culture - National Research and Innovation Agency, Jakarta, Indonesia
e-mail: agus175@brin.go.id

Z. Arifin
Research Centre for Oceanography - National Research and Innovation Agency, Jakarta, Indonesia
e-mail: zain001@brin.go.id

T. Apriliani
Research Center for Behavioral and Circular Economics, National Research and Innovation Agency, Jakarta, Indonesia
e-mail: tenn001@brin.go.id

Mujiyanto
Research Center for Conservation of Marine Resources and Inland Waters, National Research and Innovation Agency, Jakarta, Indonesia
e-mail: muji019@brin.go.id

Desmiwati
Research Center for Society and Culture – National Research and Innovation Agency, Jakarta, Indonesia
e-mail: desm001@brin.go.id

H. Warsilah and A. H. Purnomo (eds.), *Approaching Natural Resources for Sustainable Development in Indonesia*, Sustainable Development Goals Series, https://doi.org/10.1007/978-981-96-6913-4_15

The implementation of these programs undeniably elevated the quality of life for residents of the islands, albeit at a cost in terms of socio-economic and ecological factors.

However, there was a rapid realization that an issue had arisen during the COVID-19 pandemic, which resulted in the cessation of tourism and a lack of other employment opportunities. This predicament was exacerbated by the prior depletion of resources that were allocated to the development of tourism-related enterprises.

Rahmawati et al. (2020) documented the detrimental effects of tourist activities, including the trampling of seagrass, on the ecological health of seagrass meadows on Pari Island's beaches. While the paper does not directly establish a causal relationship between the heightened environmental pressure and the deteriorating conditions of the peat on the island, it does suggest that inadequate management practices may lead to a decline in field conditions due to human-induced activities. In reference to alternative habitats present on the island, specifically coral reefs, Darmawan (2020) posits a comparable perspective, suggesting that the accelerated growth of tourism within the past five years on Pari Island has adversely impacted the ecological well-being of the mangrove habitat.

Another study conducted by Abrar and Wouthuyzen (2020) supported Dharmawan's assertion, documenting substantial degradation of coral reefs in the coastal regions surrounding Pari Island. Abrar and Wouthuyzen (2020) extensively documented multiple instances of damaged corals that, while perhaps resulting from natural factors, exhibit a notable anthropogenic influence. A decline in the extent of living coral coverage was observed, with reductions ranging from 30 to 50% on reef slopes and 5% to 20% in reef rivers. Moreover, the Indonesian Coral Reef Foundation's monitoring efforts conducted between 2004 and 2005 found that the state of coral reefs exhibited a downward trend. Specifically, the coral coverage in the southern region of Pari Island ranged from 29.13 to 38.13%, while in the eastern section of the island, it varied between 30.85 and 54.15% (p. 38). The coastal land is undergoing significant and intensive development, exerting a substantial strain on reef ecosystems. This pressure is mostly attributed to activities such as coral and sand mining, drilling operations, and land reclamation.

The impact of marine tourism on the communities and environment of Pari Island can be divided into three important aspects: (a) socio-economic, (b) cultural, and (c) environmental sustainability (biophysical). From a socio-economic perspective, the study found that marine tourism provided a positive impact on the community's economy. Employment in the island community shifted from a mix of fishing and seaweed farming to tourism-related businesses. The tourism industry has indirectly provided alternative livelihoods during the west monsoon season, when high waves prevent fishing.

From a cultural point of view, Pari Island communities, like other coastal communities, are considered an open society. The communities are made up of various ethnicities and races from several Indonesian regions. As in other regions of Indonesia, the collaborative culture of *gotong royong* and respect for outsiders are evident in practice. Tourism has slightly influenced the local community's culture, with local elders feeling there has been a shift in social norms. However, our field observations were unable to draw definitive conclusions as to the impact of tourism on local culture, hence in-depth and long-term observations will be an important part of future social studies.

Most studies find that marine tourism has a negative impact on the sustainability of environmental resources. Our field studies concluded that the sustainability of coastal ecosystem resources on Pari Island is threatened by marine tourism, due to irresponsible tourism activities, water pollution, and the increase in the amount of debris damaging seagrass and coral reef ecosystems. The seagrass in the conservation area was moderately dense (59–70% coverage) and rare-moderate (11–30% coverage) in the tourist destination areas. Similarly, the accumulation of marine debris in residential areas was highest compared to the conservation and tourist destination area. The majority of the debris was plastic bottles and plastic wrapping.

The accumulation of debris has been partially mitigated by a provincial government program of collecting and transporting all debris from Pari Island to mainland Java Island.

The development of marine tourism in Pari Island has had a positive impact on the economy of communities, but a negative impact on coastal ecosystems. So far, the focus of the development of marine tourism is the economic benefit, which has not been tested for its sustainability. The magnitude of the negative impact needs further studies, especially on the carrying capacity of the island for developing facilities for marine tourism activities. An appropriate marine tourism management model needs to be formulated through research and studies so that the objectives of social, economic, and environmental sustainability on Pari Island can be achieved.

The local government can play a better role in the conservation of coastal habitats and maintenance of the ecosystem services by implementing a 3R (reduce, reuse, and recycle) program and promoting composting programs. Composting programs will potentially benefit soil fertilization and improve soil conditions on the island, as well as complementing the provincial government's program of transporting all debris to mainland Java. We recommend, further, that the provincial government develops the concept of the tourist carrying capacity of Pari Island as part of the ecotourism industry. By imposing a maximum number of tourists, the pressure of the industry on ecosystem services will be eased, and environmental conditions will finally be improved. The role of tourist operators is crucial in sustainable tourism, therefore, we recommend that they should be obliged to provide basic education for tourists on the state of Pari Island, such as ecological conditions, sensitive areas of the island, and debris status. Ultimately, Pari Island will be a more attractive tourist destination as a result of partnership between tourist operators, local/provincial government, and academics.

Some research has observed harm to aquatic organisms, including crustacea (Pratiwi 2020). A study conducted by Widyastuti and Fitrian 2020) also examined the impact on the crab population. In the past, there was an abundance of crustaceans and other species in many environments, including coral reefs, seagrass beds, and mangroves. Species that were once readily captured in coastal areas are becoming significantly challenging to detect. In addition, estuary waters are experiencing environmental degradation as a result of population growth, which is further compounded by the development of resorts and tourism activities.

15.2 Tracing Pari Island to Its Past

Pari Island is an administrative area of Pari Island Village, which includes Pari Island and Lancang Island. With an area of 41.32 km^2, Pari Island is currently home to 406 households with 1,402 people, 713 men and 689 women (Kecamatan Kep. Seribu Selatan 2022). Like the people on the surrounding islands, the people of Pari Island originally made their living from fishing: capture fisheries and fish farming. Now they are generally entrepreneurs or work in tourism. Table 15.1 provides data on fisheries production on Pari and surrounding islands prior to the surge in the tourist industry. Table 15.2 summarizes the business characteristics of the two types of fisheries in the same period.

Table 15.1 Production of capture and aquaculture fisheries, 2014

	2011	2012	2013	Total
Capture fisheries	1.584	1.786	1.555	4.925
Aquaculture	798	774	822	2.394

Source (Y. F. Christian 2018)

Pari island is situated within the administrative jurisdiction of the Seribu Islands Regency, to the north of the coasts of Jakarta Province. The community used to be adequately sustained by the presence of three crucial ecosystems, mangroves, corals, and seagrasses, its inhabitants engaged in subsistence livelihoods. Historically, individuals relied on rudimentary practices such as simple aquaculture, coastal biota collection,

Table 15.2 Business characteristics of capture fishery and fish farming, 2014

No	Type	Species	Selling (IDR thousand/kg)	Direct buyer	Gear	Institutional
1	Fifish fishery	Rabbit fish	20	Local trader	Fishing rod	Fisher Business Group, @ 10 members
		Tuna	15–20		Gill net	
		Cutlassfish	11		Gill net	
		Spannish mackerel	40–60		Gill net	
		Mackerel	20–35		Gill net	
		Baracuda	12		Fishing rod	
2	Crab fishery	Swimming crab	40	Local trader	Pots	Bubu Fisher Business Group, @ lO members
3	Seaweed farming	Wet	7	Local trader	Longline	2 fish farmer groups, 20 members
		Dry	25		Longline	
4	Grouper farming	Sunu grouper, duck, coral	35	Local trader	24 floating cage dan 3 stepped cage	1 group, @ 27 members

Source (Y. F. Christian 2018)

artisanal fishery, and terrestrial crop cultivation for their livelihoods and to meet their essential requirements. External factors exerted a relatively minimal impact, allowing for a harmonious coexistence between nature and humans.

Like other inhabitants of remote areas in general, Pari Island consisted of communities with simple lifestyles in harmony with nature. The inhabitants are generally Banten and Sunda ethnic groups, different from those who live on the islands in the north which are Buginese, Mandarese, and Makassarese. From information transmitted verbally, it is known that fugitives from Banten became the early inhabitants of Pari Island during the period 1886–1900 (Christian and Desmiwati 2019). Authority over Jayakarta passed from the Banten Sultanate to the Dutch East India Company in 1619 (Ricklefs 2001). The conditions at the time prompted the Banten community to seek refuge on smaller islands to evade the compulsory labor imposed by the Dutch colonial authorities.

Subsequently, Mandar and Bugis sailors from Kalimantan and Sulawesi arrived in the area and established harmonious relations with the early inhabitants of Pari Island. The early inhabitants who initiated farming practices on the island were residents of the nearby islands, Panggang Island and Tidung Island. Around 1940, Dutch and Japanese military forces arrived and settled on the island (Christian and Desmiwati 2018). Christian et al. (2017) state that available reports show that the alien group that had taken ownership of land and established residence on the island of Pari comprised three distinct familial units. The presence of ancient family burial sites that remain intact on the island serves as evidence for this assertion. Owners of land have what is known as a letter-C testimony. The first recorded land registration on Paris Island by a community leader called Jaro occurred in 1960 (Christian and Desmiwati 2018). This landmark changed the formerly non-formal, communally based tenure system on the island, meaning that it could not be sold to foreigners and becomes private property (Christian et al. 2018). However, Pari Island experienced land issues in 1985, when the department staff engaged in a 'letter-C whitening process' under the pretense of obtaining enhanced legal status (Christian et al. 2018).

After independence, the livelihoods of the inhabitants of Pari Island experienced minor changes, but continued to primarily focus on subsistence activities. Seaweed farming, especially of *Euchema,* evolved as a prominent form of aquaculture, becoming a notable symbol for the island. The inhabitants effectively utilized a raft system to exploit the marine resources of the island's coastal seas. Post-harvest aquaculture treatment was characterized by its simplicity. Underwater activity was conducted in a freshwater environment and accompanied by exposure to sunlight. The outcome was a commodity referred to as 'white seaweed', primarily intended for use as a revitalizing drink. The main target market for this product were beverage manufacturers in mainland Jakarta, approximately two hours' journey from Pari Island.

By 1960–1970, the population of Pari Island had reached an estimated 40–50 families (Christian and Desmiwati 2018). During that period, the residents of Panggang Island and Tidung Island remained engaged with the population of Pari Island. The population continued to be reliant on conventional fishing and the cultivation of coconuts (Christian et al. 2017). Pari Island experienced an influx of new inhabitants from Bone (South Sulawesi), Pandeglang (Banten), and other communities from East Java who were granted authorization to construct residences and establish settlements on the island to revitalize the local economy and foster the development of villages (Christian 2018).

During the 1970s, there was a proposal to establish the island of Pari as a refuge for individuals suffering from morphine addiction. According to Sas-Kompas (1976), the proposal was declined by Dr. F.X. Subagyo, the director of Seribu Islands Health Public Center, noting potential negative consequences for the island's 805 inhabitants.

By the later 1990s, seaweed continued to hold promise as a valuable product for the residents of Pari Island, particularly in light of the difficult conditions surrounding the fishing industry. At that time the majority of the coastal waters were utilized as sites for cultivating seagrass. Despite the proliferation of seagrass production, prices experienced fluctuations ranging from Rp. 300 to Rp. 700 per kilogram (Ssd-Kompas 1998). This industry has been in decline since the early 2000s, mostly because of the deterioration of water quality caused by pollution, rapid population expansion, and extensive land development around the northern coast of Java, specifically in the Gulf of Jakarta and Seribu Islands (Abrar and Wouthuyzen 2020).

In 1999, there were reports of enduring favorable social and ecological conditions. In the realm of ecological indicators, substantial quantities of seagrass, coral, and mangrove were found distributed throughout the island's coasts (Kawaroe 2009). Moreover, a study conducted in 2019 showed that the seagrass coverage on Pari Island exhibited temporal oscillations, ranging from 312 to 793 hectares, with an average size of 312 hectares. Notable change is mainly observed in areas characterized by the predominance of seagrass species possessing small leaves, rendering them particularly vulnerable to environmental modifications. According to Kawaroe (2009), the observed value fell within the specified range of variation and was found to be 25% lower than the value recorded in the previous 15-year period.

The situation at the time depicted social and ecological issues as not suited to a significantly elevated quality of life. The island of Pari is one of the most economically disadvantaged regions within the province. However, its communal existence has remained active for a considerable time. Meanwhile minimal natural damage had occurred, which attracted visitors from beyond the island keen to appreciate its presence.

15.3 Methodology

This research uses a mixed-method approach, a combination of quantitative and qualitative. Our online survey of 180 participants and in-depth interviews showed that stakeholders demonstrated an optimistic perspective regarding the actions they should take. They expressed their belief that ecosystem restoration, diversification of livelihoods, and development of tourism based on local traditions can contribute to improving the future socio-economic and ecological conditions of their islands.

15.4 Socio-Environmental Crisis of Pari Island's Resources

15.4.1 Attempts at Welfare Improvement Through Marine Tourism Development

Following a period of little change, the local government realized how crucial it was to raise people's standards of living on Pari Island in order to move them from a state of mere survival to one of increased prosperity. The strategy subsequently employed involved the development of tourism focused primarily on leveraging the island's pristine beach landscapes. A decree from the Minister of Marine Affairs and Fisheries designates the Pari coastline regions as marine coastal tourist destinations, mangrove protection sites, and research and education centers, which supports the policy (Abrar and Wouthuyzen 2020).

The policy that addresses the development of tourism is regarded as a suitable choice, taking into account the diverse benefits associated with Pari Island, which include convenient transportation links to and from popular tourist destinations, as well as exceptional attractions such as sandy beaches, coral reefs, mangroves, and beautiful natural landscapes. This program succeeded in attracting people to travel to Pari Island. Tourism rose significantly during the period 2011–2013, then leveled off, and was corrected only during the Covid-19 pandemic (2020–2022) (Fig. 15.1).

15.4.2 The Emergence of Negative Tourism Impacts on the Socio-Economic and Ecological Situation in Pari Island

15.4.2.1 Impacts on Ecology

One of the consequences associated with tourism activities is the accumulation of waste. While the majority of literature does not explicitly address the association between collected waste and tourism, certain sources do acknowledge its significance. A notable study within the literature, conducted by Sahwan et al. (2004), documents the significant rise in the transportation of garbage from mainland Jakarta to other locations within the Seribu Islands. In the case of Pari Island (Yuliani et al. 2018), a range of 500–600 cubic meters of waste every month is specifically mentioned, which is equivalent to 40 tons per day (Nailufar 2018). Utami et al. (2018) also substantiate the notion that tourism has assumed a significant role in expediting the increase of waste accumulation. The popularity of tourism activities has led to a notable increase in plastic bottles used for beverage packaging. Our field study identified nine distinct categories of debris in Pari Island: (i) plastic bottles, (ii) plastic food wrap, (iii) plastic cups, (iv) non-food plastic wrap, (v) glass bottles, (vi) cans, (vii) sandals, (viii) straws, and (ix) fishing nets. During the survey conducted at Sampling Site #1, which is a designated conservation area, four distinct categories of trash were observed: plastic food wrap, plastic cups, sandals, and straws. Sampling Site #2, located in a residential neighborhood, exhibited a total of eight distinct categories of waste: plastic bottles, plastic food wrap, plastic cups, non-food plastic wrap, glass bottles, tins, straws, and fishing nets. In Sampling Site #3, located in a tourist area, four distinct types of waste were identified: plastic bottles, plastic food wrap, plastic cups, and non-food plastic wrap. These materials were found dispersed within the seagrass habitats located at our three designated sampling points. Several instances of debris were discovered partially immersed in the sand, visible due to their presence among seagrass roots. Several examples of debris were also observed drifting above the seagrass beds. Figure 15.2 illustrates the diverse array of marine debris encountered throughout the fieldwork, including sandals, liquid soap bags, snack bags, plastic bags, fishing nets, and soft drink cans.

According to Sahwan et al. (2004), a significant amount of the debris discovered in the Seribu Islands, particularly Pari Island, originates from mainland Jakarta rather than the nearby islands. With every year that passes, more trash is washing up on the shores of the islands. With

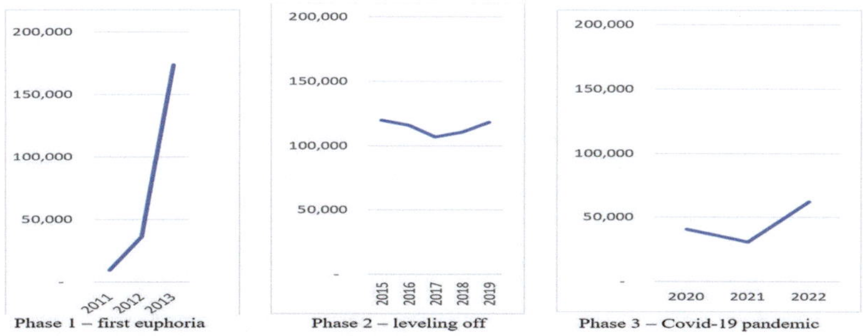

Fig. 15.1 Tourist visits to Pari Island Prior to Covid-19. (*Source* Regional Statistics of Seribu Island, 2012–2023)

Fig. 15.2 Types of marine debris on Pari Island. (*Photo taken by* Mujiyanto, June 2020)

the rise in the popularity of tourism-related activities, garbage accumulation has grown, and plastic beverage bottles that were previously uncommon have started to appear more frequently (Utami et al. 2018).

Further observations show that the prevailing category of trash was plastic food wrap, followed by plastic cups and non-food plastic wrap. Figure 15.3 provides a comparative analysis of the prevalence of different types of debris seen at various sampling stations. Meanwhile, Fig. 15.4 presents a weight-wise comparison, illustrating that residential areas exhibit the highest value, specifically 37 g/m^2.

The results gathered from our discussions with the local inhabitants indicate that the quantity of debris deposited on the islands' shorelines varies in accordance with seasonal patterns. During the western monsoon, a reduced amount of debris is observed due to its eastward displacement towards more distant locations. During the east monsoon

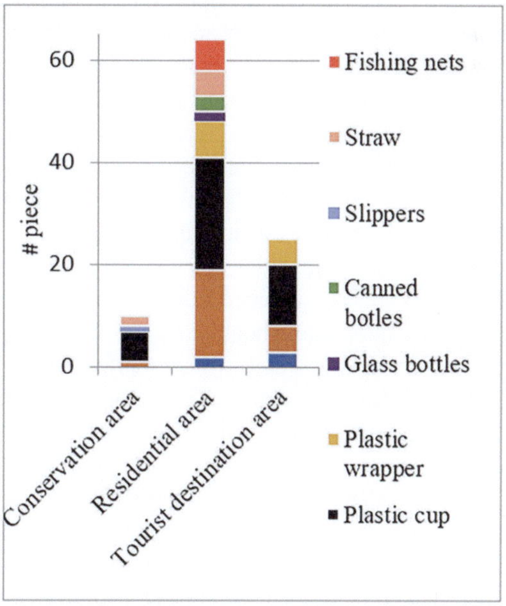

Fig. 15.3 Composition of types of debris at each areas

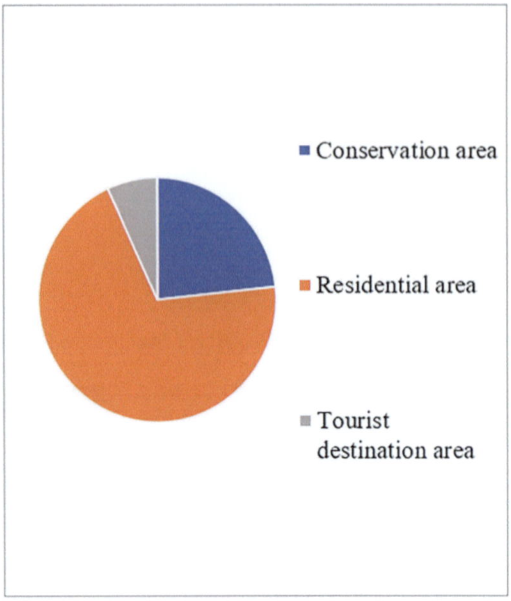

Fig. 15.4 Marine litter weight (g/m^2) at three sampling areas

season, however, the prevailing currents originating from the east exhibit a tendency to transport and deposit the accumulated waste materials originating from mainland Jakarta onto the coastal regions of the island. Currently, the management of debris is assisted by the provincial administration, which arranges collection of garbage from multiple collection stations across the island. Subsequently, the debris is transported to temporary shelters and ultimately conveyed by dump ship to landfills in mainland Jakarta. However, this system is susceptible to inclement weather conditions, which can result in the suspension of dump ship operations and waste left unattended.

An initial examination of the results (see Fig. 15.5) shows a counterintuitive trend in the weight of litter throughout the three locations, particularly in tourist areas where marine litter is found to be less prevalent. The implementation of a comprehensive beach clean-up initiative proved to be the solution. Indeed, spurred on by the fervor of local residents to promote marine tourism in their region, the primary focus has shifted towards the cleaning of beaches in popular tourist destinations. Efforts were made to allocate resources in the designated areas of the island.

A reduction in seagrass coverage was also observed. The recorded seagrass coverage on Pari Island in 2019 exhibited a range of 312 hectares, with seasonal variations spanning from 312 to 793 hectares. Significant variations are mainly observed where seagrass species with tiny leaves are prevalent, which are susceptible to alterations in their surrounding environment. The observed value within the range mentioned above was 25% lower than in the corresponding area in the preceding 15-year period (Kawaroe 2009).

The results of our field observation indicate that the distribution of seagrasses within the study region was patchy, with occurrences observed in the conservation area, residential area, and tourist attraction area (see Fig. 15.5). Seven different types of seagrass were found. They were *Thalassia hemprichii* (Th), *Halophila ovalis* (Ho), *Halodule uninervis* (Hu), *Enhalus acoroides* (Ea), *Cymodocea rotundata* (Cr), *Syringodium isoetifolium* (Si), and *Cymodocea serrulate* (Cs). Stands representing various species were discovered within the conservation area, perhaps because of the safeguarding from potential disturbances by both local populations and tourists. Conservation areas exhibit the highest level of seagrass species diversity, harboring a total of seven distinct species. In comparison, residential areas host five seagrass species, while tourist regions support a relatively lower diversity with only two

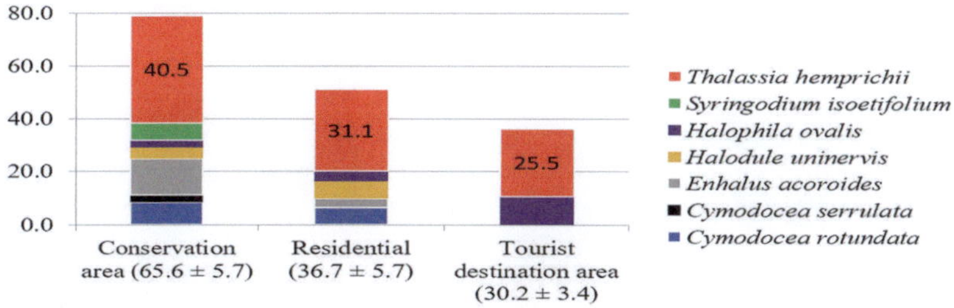

Fig. 15.5 Percentage of seagrass stands in the three areas

species. The main species observed across all sampling sites is *Thalassia hemprichii*, with the greatest proportion of stands occurring in conservation areas (40.5%). Residential areas have the second largest percentage of stands (31.1%), while tourism regions have the third highest percentage (25.5%).

Our observations of the current seagrass coverage serve as evidence for our assessment of the relationship between tourism-related activities and the condition of the seagrass ecosystem. Among the various categories, it was observed that the reserve region exhibited the largest proportion of seagrass coverage of 65.6 ± 5.7. The residential area followed, showing a moderate condition with a seagrass coverage of 36.7 ± 5.7. The tourist attraction area, on the other hand, exhibited a relatively lower seagrass coverage ranging from rare to moderate, with a value of 30.2 ± 3.4 (refer to Fig. 15.5). This finding demonstrates the magnitude of stress applied to the site. The level of vegetation coverage in conservation areas is significantly higher than in tourist districts, mostly due to the limited extent of human activities there. Conversely, in tourism zones, the vegetation coverage is considerably lower due to the enormous impact exerted by the presence of tourists.

The seagrass ecosystem serves a crucial function as a marine habitat for a diverse range of marine organisms. During the field study, a total of eight groups of marine megabenthos were identified, including crabs, fish, trochidae, shrimps, gastropods, sea cucumbers, and starfish. Figure 15.6 shows the notable prevalence of fish adolescents compared to other age groups. The study observed that young rabbit fish (*Siganus sp.*) were consuming periphyton, a type of algae that grows on the surface of seagrass leaves belonging to the *Enhalus acoroides* species. In addition, it was observed that young reef fish and fish belonging to the family *Chaetodontidae* sought refuge and established their habitat within branched coral colonies, specifically those of the *Acropora sp.* species. Numerous colonies of branched corals exhibit growth patterns inside the seagrass beds located in both conservation and residential areas. The megabenthic organisms commonly observed inside the sampling sites included sea urchins, sea cucumbers (*Holothuria leucospilota* and *Stichopus chloronatus*), and starfish (*Linckia laevigata*). The organisms were observed inhabiting the seagrass substrate that is commonly found in close proximity to branched coral reefs (*Acropora sp.*), as well as other types of corals such as massive, submassive, and fungus corals.

15.4.2.2 Impact of Tourism on Livelihoods

The inhabitants of Pari Island anticipate a promising enhancement of their quality of life through active participation in tourism-related activities. There is a diverse range of employment opportunities within the field of information technology that have the potential to yield a higher level of income than they currently enjoy. Some examples of occupations in the transportation sector include working as a provider of three-wheeled motorbikes, serving as a tour guide, and offering rental bicycles for visitors (Fig. 15.7).

A significant number of individuals have therefore moved from fisheries-related occupations to

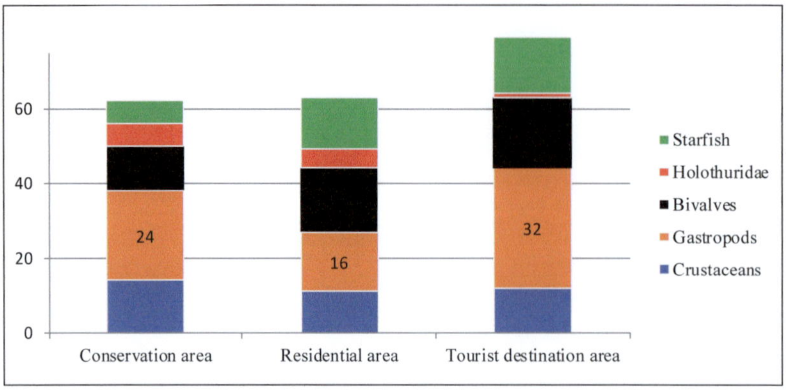

Fig. 15.6 Composition of megabenthic types

Fig. 15.7 New occupation in Pari Island during tourism development

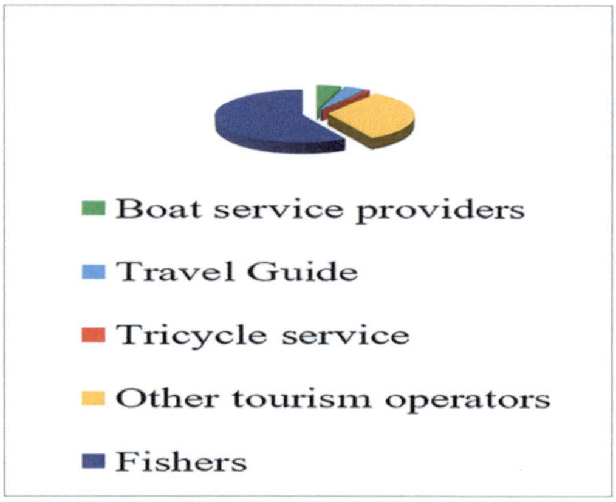

livelihoods associated with tourism. Fishing business assets are liquidated in order to generate immediate capital for the acquisition of tourist assets. Figure 15.8 illustrates the significant decrease in the population of individuals engaged in fishing activities from 2012–2018.

The COVID-19 pandemic, which peaked in 2020–2022, proves that the decision to convert all the important assets of fishing in Pari Island into a tourism business was a very bad idea. In fact, the tourism business was practically paralyzed as visitor numbers from outside dropped to zero. On the other hand, people could not immediately return to their primary livelihood of fishing because the ecology has been severely degraded. They no longer owned fishing boats, and their finances had been exhausted during the pandemic.

15.5 Recommendation and Conclusion

15.5.1 Conclusion

This study has analyzed the various impacts of marine tourism on the communities and environment of Pari Island, classifying them into three significant dimensions: (a) the socio-economic dimension, (b) the cultural dimension, and (c) the environmental sustainability dimension. The study revealed that marine tourism had an advantageous impact on the local economy, as viewed from a socio-economic perspective. The island community experienced a transition in its occupational structure from a combination of fishing and seaweed farming activities to a focus on companies related to tourism. The tourism sector

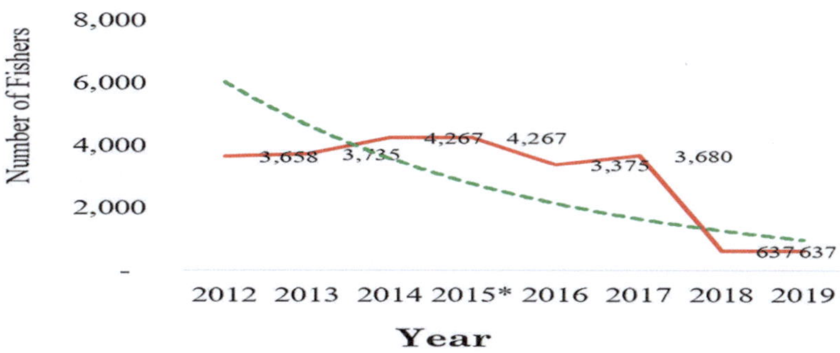

Fig. 15.8 Number of fishing activities

has indirectly facilitated the availability of alternate sources of income during the west monsoon season, a period when fishermen are unable to engage in fishing activities due to adverse wave conditions.

From a cultural perspective, the inhabitants of Pari Island, similar to other coastal villages, are seen as having an open societal structure. The communities consist of several ethnic and racial groups originating from various regions of Indonesia. Therefore, the presence of a collaborative culture known as *gotong royong* and a deep sense of respect towards individuals from other communities may be observed in various practices. According to the elders of the community, who have noticed a noticeable change in social norms, tourism has had a modest impact on the cultural fabric of Pari Island. Nevertheless, it is important to note that our field observation alone does not provide a comprehensive assessment of the influence of tourism on local culture. Consequently, conducting extensive and prolonged observations will be a crucial component of future social studies research.

In relation to the implications of maritime tourism on the sustainability of environmental resources, a majority of scholarly investigations have indicated a predominantly adverse impact. The results of our field research show that the impact of marine tourist activities is endangering the sustainability of coastal ecosystem resources on Pari Island. This threat arises from the presence of irresponsible tourism practices, water pollution, and the accumulation of debris, all of which have resulted in significant harm to the seagrass and coral reef ecosystems. The seagrass coverage within the conservation area was classified as moderate-dense, ranging from 59 to 70% coverage. In contrast, the coverage within the tourist destination area was categorized as rare-moderate, ranging from 11 to 30% coverage. In the same way, it was observed that the residential area exhibited the highest level of marine trash accumulation in comparison to both conservation and tourism attraction areas. Plastic bottles and plastic wrapping constituted the predominant portion of the waste. The issue of debris accumulation has been somewhat addressed with the implementation of a provincial government initiative whereby all debris on Pari Island is collected and sent to mainland Java Island.

15.5.2 Recommendations

The development of marine tourism in Pari Island has had beneficial implications for the economic well-being of local communities, although at the expense of negative impacts on coastal ecosystems. Currently, the primary emphasis in the development of marine tourism lies in its economic advantages, without adequate evaluation of its long-term viability. Additional study is required to determine the extent of the detrimental effects, particularly in relation to the island's carrying capacity, which will serve as a foundation for the development of tourism infrastructure. Thus, in order to fulfill the objectives of social, economic, and environmental sustainability, it is essential to develop a well-defined marine tourist management model through rigorous research and comprehensive studies.

The local government can enhance its contribution to the conservation of coastal habitats and maintenance of ecosystem services through the implementation of the 3R program as well as the promotion of a composting ram. The implementation of a decomposition program has the potential to yield benefits in terms of soil fertilization and the improvement of soil conditions. This will also serve as a complement to the provincial government program of collection and transportation of all waste to mainland Java. It is recommended that the provincial government pursue the development of a plan for visitor carrying capacity as an integral component of the ecotourism sector. By setting a maximum limit on visitor numbers, the industry's impact on ecosystem services will be mitigated, leading to an eventual improvement in environmental conditions. The involvement of tourist operators is pivotal here. We propose that it is imperative for them to furnish essential information on Pari Island to their customers. This information should encompass the ecological state, areas of sensitivity on the island, as well as the status of garbage present on the island. This knowledge constitutes an integral component of the educational experience for visiting travelers. Ultimately, through collaboration among tourist operators, local and provincial government entities, and scholars, Pari Island, situated in close proximity to the capital city of Indonesia, will be rendered more appealing to the general population as a tourist destination.

References

Abrar M, Wouthuyzen S (2020) Gugus Pulau Pari: Sebuah Pulau Sangat Kecil, Potensi, dan Ancaman. In Gugusan Pulau Pari, Kepulauan Seribu: Tinjauan Aspek Bio-Ekologi, Sosial-Ekonomi-Budaya, dan Pengelolaan Berkelanjutan. LIPI Press

Christian Y, Desmiwati D (2018) Menuju Urbanisasi Pulau Kecil: Produksi Ruang Abstrak dan Perampasan. J Regional Rural Dev Planning 2(1):45. https://doi.org/10.29244/jp2wd.2018.2.1.45-63

Christian Y, Desmiwati D (2019) From sea to exclusion: livelihood transformation of the Islanders. IOP Conf Ser: Earth Environ Sci 241(1):012007. https://doi.org/10.1088/1755-1315/241/1/012007

Christian YF (2018) EKONOMI POLITIK KONFLIK AGRARIA PULAU KECIL (Studi Kasus Pulau Pari, Kabupaten Administrasi Kepulauan Seribu, Provinsi DKI Jakarta). IPB University

Christian Y, Satria A, Sunito S (2017) Mode of dispossession of small Island's Agrarian resource. Int J Sci Basic Appl Res IJSBAR Int J Sci: Basic Appl Res 36(5):283–297

Christian Y, Satria A, Sunito S (2018) Konflik Agraria Pulau Kecil. Sodality: Jurnal Sosiologi Pedesaan 6(1):71–78

Darmawan IWE (2020) Kondisi Komunitas Mangrove Terkini Pada Gugusan Pulau Pari, Kepulauan Seribu. In: Wouthuysen S, Abrar M (eds) Gugusan Pulau Pari Kepulauan Seribu: Tijauan Aspek Bio Ekologi, Sosial Ekonomi-Budaya dan Pengelolaan Keberlanjutan. LIPI Press, pp. 11–24

Kawaroe M (2009) Perspektif Lamun Sebagai Blue Carbon Sink di Laut. Lokakarya Nasional I Pengelolaan Ekosistem Lamun

Kecamatan Kep. Seribu Selatan (2022) Laporan Bulan Januari 2022 Kelurahan Pulau Pari

Nailufar NN (2018) 10 Ton Sampah Diangkat dari Pulau Pari. Https://Megapolitan.Kompas.Com/Read/2018/11/28/08233761/10-Ton-Sampah-Diangkat-Dari-Pulau-Pari

Pratiwi R (2020) Kondisi Fauna Krustasea di Gugusan Pulau Pari, Kepulauan Seribu, Tahun 2010–2015

Rahmawati S, Hernawan UE, Hafizt M (2020) Kondisi padang lamun di Pulau Pari Kepulauan Seribu: Telaah Tiga Dekade. LIPI Press

Ricklefs MC (2001) Sejarah Indonesia Modern 1200–2004: Vol. Third Edition. Serambi

Sahwan FL, Di P, Pengkajian P, Penerapan D, Lingkungan T, Pengkajian B, Teknologi P (2004) Strategi pengelolaan sampah di kawasan kepulauan seribu. In: Strategi Pengelolaan Sampah. J Tek Ling P3TL-BPPT, Vol 5, Issue 1

Ssd-Kompas (1998) Rumput Laut Selamatkan Warga. Kompas

Utami H, Rahmat Darmawan Dampak Pariwisata terhadap Lingkungan di Pulau Tidung Kepulauan Seribu, dan, Darmawan R (2018) The impacts of tourism activities on the environment in Tidung Island, Kepulauan Seribu. Asia Pacific J Tourism Res 23(11):1062–1072

Widyastuti E, Fitrian T (2020) Kondisi Fauna Kepiting di Perairan Pulau Pari, Kepulauan Seribu. In: Wouthuyzen S, Abrar M (eds) Gugusan Pulau Pari, Kepulauan Seribu Tinjauan Aspek Bio-Ekologi, Sosial Ekonomi Budaya, dan Pengelolaan Berkelanjutan. LIPI Press

Yuliani S, Purwandari DA, Budiaman B (2018) Pemberdayaan Daur Ulang Sampah Laut dan Pesisir Pulau Pari, Kepulauan Seribu DKI Jakarta, pp 1112–1119

Socio-Ecological Risk Reduction for Communities in Coastal Areas: Comparative Study in Jakarta and Semarang

16

Henny Warsilah and Choerunisa Noor Syahid

Abstract

Jakarta and Semarang are densely populated cities, especially along the coasts of the two regions. As a result of climate change, these cities are facing a major socio-ecological crisis which may lead to disasters such as tidal floods, landslides, high waves, clean water crises, and plastic waste. However, infrastructure development in these two cities has made urban areas prone to disasters and has implications for social crises, such as marginalization, poverty, and social exclusion. This chapter aims to determine the impact of reducing the risk of tidal floods especially in the cities of Jakarta and Semarang, by using a community-centered paradigm that focuses on four variables: (1) local knowledge and wisdom about disaster risk; (2) stakeholder involvement in warning system services; (3) dissemination of the warning system to beneficiaries; and (4) the ability to act and react to disasters. Using qualitative methods with descriptive analysis based on in-depth interviews and FGDs with stakeholders, this chapter describes the local wisdom of the people in these two cities as a regional government approach to coastal city governance.

Keywords

Ecological social crisis · Community-based · Jakarta · Semarang

H. Warsilah (✉)
Research Centre for Society and Culture - National Research and Innovation Agency, Jakarta, Indonesia
e-mail: hennywarsilah@gmail.com

C. N. Syahid
Research Center For Area Studies - National Research and Innovation Agency, Jakarta, Indonesia
e-mail: c.runisa@gmail.com

16.1 Introduction

As a large archipelagic country located in Southeast Asia, Indonesia is very vulnerable to the impacts of climate change due to its geographical and environmental characteristics. Consisting of thousands of islands between the Indian and Pacific Oceans, geographically Indonesia is very diverse. While these geographical conditions contribute to the natural beauty of Indonesia, at the same time they also pose the challenge of natural disasters. Indonesia is more vulnerable to floods, landslides, weather changes and extreme waves, abrasion, and drought which can result in forest and land fires (BNPB 2020:13). Java Island is the most vulnerable to the impacts of climate change due to a combination of geographical area, environmental degradation, and socio-economic factors, especially flooding.

H. Warsilah and A. H. Purnomo (eds.), *Approaching Natural Resources for Sustainable Development in Indonesia*, Sustainable Development Goals Series, https://doi.org/10.1007/978-981-96-6913-4_16

Java Island has a long coastline that directly faces the Java Sea and the north coast of Java. It contains most Indonesia's largest cities, including Jakarta, the capital city and Semarang, the barometer of Central Java. The rapid urbanization and inadequate planning that have taken place make Java Island the most vulnerable area, since the densest population is located in coastal areas.

Between January 2012 and December 2019, frequent flood disasters occurred in Indonesia (Table 16.1). According to the BNPB Disaster Value Index, Central Java Province and DKI Jakarta Province are particularly at risk of flooding.

Table 16.1 Frequency of flood events in Indonesia

No	Year	Number of flood disaster event	Number of fatalities
1	2012	581	134
2	2013	710	232
3	2014	588	153
4	2015	523	48
5	2016	820	239
6	2017	973	162
7	2018	672	119
8	2019	984	298

Source Ministry of Public Works and Public Housing, 2019 (https://data.pu.go.id/infografis/banjir-di-indonesia-2012-2019)

DKI Jakarta, the capital city of Indonesia, and Semarang, the capital city of Central Java Province, both have dense populations and frequently experience tidal floods due to climate change and social problems. The coastal part of both cities has become vulnerable, since many migrants occupy coastal areas, mainly because land is more affordable and can be used in line with their skills. This finding is in line with global population patterns, whereby coastal areas are more urbanized, with 34% of the global population living in the near-coastal zone in 2010 (Kummu et al. 2016).

As the capital city of Indonesia, Jakarta is a target for an influx of migration from all over Indonesia. The population of Jakarta in 2023 was 11,248,839, an increase of 1.47% on the previous year. In general, the population of Jakarta has increased significantly each year (see Table 16.2). Given an area of 664.01 km^2, it can be calculated that the population density in Jakarta was around 17,013 people per square kilometer (km^2). The densest was Central Jakarta (23,249 people/km^2), followed by West Jakarta (20,717 people/km^2), East Jakarta (17,650 people/km^2), South Jakarta (16,449 people/km^2), North Jakarta (12,652 people/km^2), and the Thousand Islands (2770 people/km^2) (Databoks 2022).

Table 16.2 Total population of DKI Jakarta from 1950 to 2023

Year	Number of population	Growth rate (%)
2023	11,249,000	1.57
2022	11,075,000	1.47
2021	10,915,000	1.35
2020	10,770,000	1.23
2019	10,639,000	1.16
2018	10,517,000	1.12
2017	10,401,000	1.11
2016	10,287,000	1.12
2015	10,173,000	1.11
2014	10,061,000	1.11
2013	9,951,000	1.12
2010	9,626,000	14
2000	8,390,000	2
1990	8,175,000	27
1980	5,984,000	43
1970	3,915,000	56
1960	2,679,000	28
1950	1,452,000	0.00

Source https://www.macrotrends.net/cities/21454/jakarta/population

However, due to the impact of climate change, the coastal areas have become more vulnerable to the risk of disasters, especially landslides, tidal floods, clean water crises, slum areas, and various social problems. This situation is exacerbated by the existence of unplanned urban settlements around the delta river area. In the case of Jakarta, the impact not only affects the coastal areas but also paralyzes the city of Jakarta in general if it rains.

The local government has introduced several policies to mitigate the impact of climate change. Since the 1990s, the local government has been developing reclamation on the north coast with the goal of mitigating not only floods but also landslides, and replanning coastal areas in North Jakarta. However, the development process has caused conflict between the government and communities, which has made managing the impacts of climate change challenging, especially in the coastal areas of Jakarta (Fig. 16.1).

Like Jakarta, the city of Semarang, the center of industry and trade in Central Java, faces its own challenges and has become a city destination for urbanization for its surroundings. The unplanned growth of residential and industrial areas makes Semarang more vulnerable to disasters, especially floods, tidal floods, and subsidence. It has an area of about 373.70 km^2 and a population that reached 1,653,524 by 2020, making it one of the most populous cities with a density of 4424 people per km^2. As the third largest port in Indonesia, the Tanjung Mas area is a magnet for unskilled labor migration. Due to climate change, Semarang is currently at higher risk of disasters, especially sea level rise that frequently contributes to tidal floods. Semarang is only 2 m below sea level and most of its soil contour is swamp (BPBD Semarang 2020). Every year it experiences land subsidence reaching more than 10–11 cm, with sea level rising about 0.8–1 cm/year to 7.8 mm/year.

In 1992 and 1993 the Ministry of Public Works developed Jati Barang Reservoir for flood control, but flooding remains a problem in Semarang. In 2016, the area of tidal flood reached 2600 ha, with a maximum accumulation duration of 9 h, and an average height of 50 cm (Perda No. 6 of 2016, Semarang City Medium Development Plan/ RPJMD 2016–2021).

The local governments in both Jakarta and Semarang have taken several steps towards flood mitigation. However, the massive population growth and development have caused huge damage to the ecology as well as social problems. This chapter considers the socio-ecological impact of tidal floods in coastal areas and examines how disaster risk reduction has been taken up by the coastal community based on their own perspective.

16.2 Methodology

This study uses an interdisciplinary approach to analyze the phenomenon of tidal flood in the context of resilient cities, particularly in coastal urban areas. Jakarta and Semarang are the research locations chosen because of the physical and socio-economic impacts on their development caused by flooding, especially tidal floods. The descriptive analysis takes the variables involved into account. Both primary and secondary data were collected and analyzed.

Primary data were obtained through observation, semi-structured interviews, and focus group discussions with stakeholders encountered in the field. The data collection process is the result of accumulated field research from 2016 to 2018 conducted by a team from the Indonesian Institute of Sciences (LIPI). In Jakarta, data were collected through focus group discussions with scholars and NGO activists directly involved in government programs. This was an important perspective in view of these stakeholders' role in protecting local communities and ecological conservation in coastal areas of Jakarta. In Semarang, the data were collected in Kampong Kemijen, East Semarang, and Kampong Tambak Lorok, North Semarang, both of which are located near Tanjung Mas Port.

16.3 Theoretical Framework

16.3.1 Urban Coastal Communities and Disaster Risk Reduction

Ruth and Gasper (2008) state that the impact of climate change will be felt in urban areas because of demographic conditions, namely an increase in the number of urban residents. Population density in both Jakarta and Semarang triggers

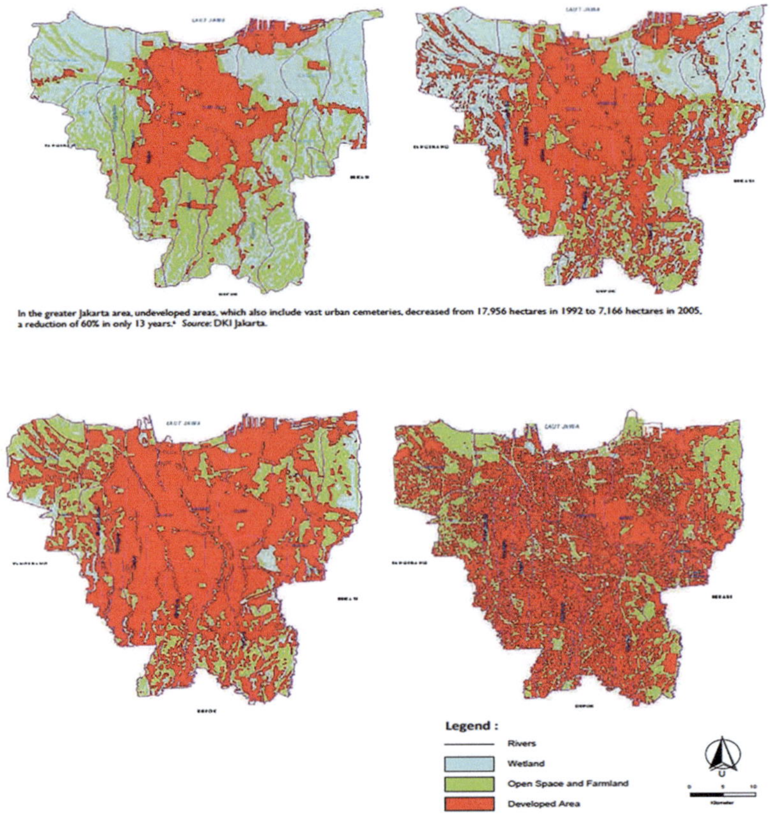

Fig. 16.1 Land-use development of Jakarta,1970s–2000. (*Source* World Bank Report, 2010)

flooding and several social impacts that are worsening in coastal areas. Kodoatie and Roestam (2010: 53–54) state that a disaster can be understood as an event, whether natural artificial, or progressive, that has a devastating impact on the community which then responds with ordinary external actions. Meanwhile, Cuny (1983) refers to disasters as a meeting between threats and vulnerabilities which become threats and events that harm humans. In Indonesia, the definition of disaster in Chapter 1 of Law Number 24 of 2007, is an event or series of events that threatens and disrupts people's lives and livelihoods, caused by natural or non-natural factors that result in losses. The impacts can take the form of human lives, environmental damage, property loss, and psychological trauma. In summary, a disaster is an event – natural, human, or a combination of both – that has an impact on people's lives and livelihoods.

To minimize the impact of disaster risk, it is necessary to implement disaster reduction. According to the United Nations Office for Disaster Risk Reduction (UNDRR), disaster risk is defined as the potential loss of life, injury, or destruction or damage to assets that can occur in a system, community, or communities within a certain period of time, which is determined probabilistically as a function of hazards, exposure, vulnerability, and capacity. In disaster risk terminology, disasters are no longer referred to as something ordinary. A complete understanding is needed, not only of the dangers that will occur now but also of the dangers ahead.

Disaster risk reduction (DRR) is a systematic approach to identify, recognize, assess, and reduce the risks arising from disaster events. The definition of disaster risk reduction is also explained by John Twigg (2009) as a systematic approach to identify, assess, and reduce disaster risk. DRR aims to reduce socio-economic vulnerability to disasters and address environmental and other hazards that trigger them (Twigg 2009: 10).

16.3.2 Social Resilience and Community-Based Disaster Risk Reduction

Social resilience is a key concept in social analysis that relates to the interaction of the natural environment with humans, commonly referred to as social-ecological systems (SES). Social resilience aims to explore the interaction between humans and climate factors, especially related to social issues, such as quality of life, identity, and socio-cultural values in relation to transformation towards sustainability. Obrist et al. (2010), more specifically, state that social resilience means simultaneously increasing physical ability to respond to adverse external conditions and developing collective actions aimed at changing the parts of the external social structure that limit the resilience of institutions. This means not only overcoming and adapting to bad conditions (in the form of reactive capacity) but also seeking and making choices in the form of proactive capacity to deal with disaster.

Speranza (2013) developed the concept of social resilience to protect people's livelihoods from negative impacts by dividing resilience into three interrelated parts (see Fig. 16.2):

1. *Buffer capacity.* Changing capacity and emerging opportunities for better livelihoods, such as poverty alleviation (Speranza 2013). Buffer capacity is also described as the amount to be absorbed by a system while maintaining the same structure, function, identity, and input function and structure (Carpenter et al. 2001; Resilience Alliance 2010). Capacities are described as livelihood modalities and community dynamics to reflect the formation of buffer capacities.
2. *Social self-organization.* It is concerned with autonomy, freedom of action, collective action, self-reliance, independence, power, and control over one's own actions, all of which can foster new identities and beliefs, and contribute to empowerment. Social self-organization includes:
   a. *Institutions:* social norms and rules (Ostrom 1990) as well as formal institutions such as groups, organizations, and government institutions.
   b. *Cooperation and networks:* interactions between actors in a socio-ecological system (SES). It creates rules, norms, and values (institutions), builds trust, and reduces dependence on external actors for information, innovation, and capital (Speranza 2010). Building trust and increasing social capital can support livelihoods when shocks or stress occur (Pretty and Smith 2004; Speranza 2010).
   c. *SES network structure*: can influence system dynamics and management outcomes through social media and collaboration (Cumming 2011).
   d. *Dependency (trust)*: the main resources that are not available locally and can integrate knowledge from outside (COMPAS 2004).
3. *Learning capacity.* This relates to adaptive management ability, which implies resilience as a learning system that combines past experiences into current actions and has memory. Learning capacity relates to knowledge or skills. Fiol and Lyles (1985:811) define learning as "the development of insight, knowledge, and relationships between past actions, the effectiveness of those actions, and future actions". The ability to learn at the level of individual livelihoods and systems is critical to building resilience through:
   a. *Knowledge of potential threats and opportunities*: the ability to analyze potential threats and opportunities to livelihoods.
   b. *Community collective vision for livelihood systems*: can contribute to resilience transformation. Collective vision has demonstrated the extent of existing institutions and conditions that encourage exchange of information and services, sharing of

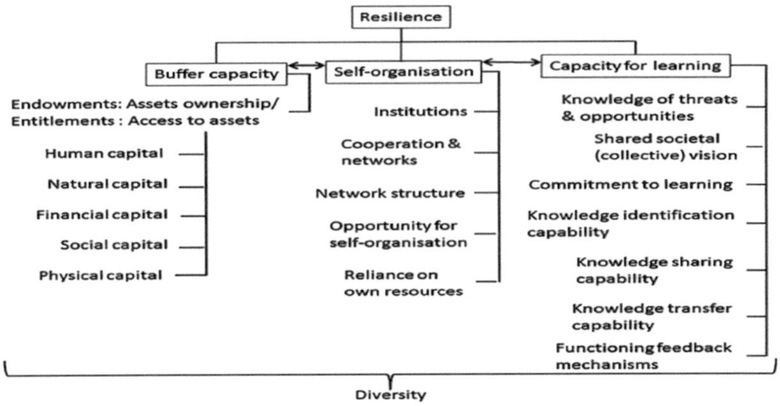

Fig. 16.2 Applicability of the Social Resilience Concept by Speranza. (*Source* Speranza (2013))

knowledge, perceptions, beliefs, and collective actions that empower individual actors (Jerez-Gómez et al. 2005).

c. *Commitment to learning and developing a culture*: encourages the acquisition, creation, and transfer of knowledge as a fundamental value (Jerez-Gómez et al. 2005). It involves government policies and regulations in supporting community livelihoods and involves actors in making decisions about their livelihoods.
d. *Knowledge identification capacity (KIC):* the ability to identify knowledge in the external environment, valuable for survival and development (Li et al. 2008). In this part, there is an opportunity to adapt livelihood strategies to dynamic conditions related to existing natural resources.
e. *Knowledge sharing*: the extent to which an actor might spread knowledge to others.
f. *Transferability of knowledge:* the degree of knowledge being internalized externally.
g. *Functional mechanisms*: feedback to disseminate knowledge and enhance social memory through interactions between actors.

In relation to community-based disaster risk reduction, these efforts encourage the community to manage disasters at the local level. It is an approach that requires public interpretation to evaluate all disaster risks, determine management priorities, plan disaster risk reduction activities, and evaluate their performance in disaster risk reduction efforts (Lassa et al. 2009: 8). This approach is expected to produce a community that is independent and has the ability to manage disaster risk. Preparedness is an important part of efforts to reduce disaster risk and demands the capacity and participation of people living in disaster-prone areas.

16.4 Findings and Discussion

16.4.1 The Jakarta Case: Flood Mitigation Strategy and Its Impact on the Coastal Community

As a coastal area, Jakarta cannot be separated from the threat of flooding that can strike at any time and cause environmental damage. Abidin et al. (2011), and Putra and Marfai (2012), have shown the significant social, economic, and environmental effects or damage experienced by residents of Jakarta. In fact, the floods in Jakarta have also weakened economic growth throughout Indonesia with, for example, delays in distribution of commodities, especially food, increasing prices during flood seasons, and rising inflation (Djunaedi 2014). Based on data from the Chamber of Commerce Indonesia (2014), flooding in Jakarta has caused industry losses of up to IDR1.5 billion per day, and repairs to infrastructure damage due to flooding cost more than hundreds of billions. Meanwhile, tidal floods have inundated some coastal parts in the north of

Jakarta, with subsidence reaching up to 20–25 cm per year.

Work by the local government includes the construction of the polder model system, normalizing the river flow, rearrangement of irrigation systems, and empowering people to make Jakarta free from flooding. However, with a population explosion occurring in Jakarta, these efforts have not been sufficient to solve the problem of floods. The worst flood years in Jakarta in recent times were 1996, 2002, 2007, and 2013 (Prihatin 2013) (see Fig. 16.3). On these occasions, the entire town was inundated and all activities paralyzed.

The worst flood in Jakarta, also recorded as the worst disaster in the twenty-first century, occurred in 2007. The National Development Planning Agency (BAPPENAS) reported that 590,000 residents had to be relocated, 79 people died, and the total damage and direct losses due to the flood reached IDR5.2 billion and IDR3.6 billion. Floods occurred again in 2013 and 2014. In 2013, 18,000 people had to be relocated, with 15 fatalities, and an impact on the economy of IDR7.5 billion (BNPB 2014). The following year 38,000 people were relocated and there were 23 fatalities reached 23 people, although the economic losses were smaller, at IDR5 billion (interview with BNPB 2014).

A strategic step taken by the Jakarta regional government is to initiate cooperation with the Rotterdam city government in a sister city scheme. One of the programs initiated by the Jakarta regional government is to continue the National Capital Integrated Coastal Development (NCICD) project, which is in three stages: (1) raising and strengthening 52 km of sea wall north of Jakarta and carrying out reclamation involving 17 islands; (2) construction of the outer ring of the embankment; and (3) the much talked-about Garuda Seawall project (Fig. 16.4).

Unfortunately, the reclamation project led to economic, social, and ecological degradation in Jakarta Bay. Traditional fishermen lost their jobs, leading to a decline in economic activity. Social disparity increased as these traditional fishermen and their communities moved out and their former homes became residences for the elite. Poverty rose in coastal areas and Jakarta's small islands in Jakarta, since the reclamation project only benefited the local government and the private sector. Rising inequality leads to structural poverty, rising conflict, increased crime, and loss of livelihoods, proof that reclamation creates unbalanced development and increases inequality. As far as ecological and environmental degradation are concerned, the existing ecosystem in Denpasar-Bali and North Jakarta will change due to the reclamation work. The abundance of natural metabolism in the protected mangrove area will be disturbed. The project also changed the course of the river further upstream, affecting the economic activities of traditional fishermen (Warsilah 2020).

16.4.2 The Semarang Case: Tidal Flood and Urban Kampong Communities Strategy

The city of Semarang in Central Java has experienced disasters and ecological crises. Tidal floods are a deep-rooted problem, mostly affecting settlements in the northern part of Semarang, especially in the Kemijen and Tambak Lorok areas. In these areas, tidal floods can happen twice a day due to the sea levels. Once it occurs, seawater gets into residential areas and paralyzes economic activities in the central business district, markets, schools, health centers, ports, and others. The flood may be followed by land subsidence of 1–2 cm a month, forcing Kemijen and Tambak Lorok residents to raise their houses at least every five years – or leave.

In Kemijen, a village in East Semarang regarded as a slum area, the total population of 13,413 people in 1245 households were affected by a tidal flood which was estimated to have affected up to 39 hectares of land. The area is bordered by a river and the main economic activity is fishing, since about 40% of the area is swamp used as fish ponds. Since the milkfish presto is a traditional Semarang food, many women's groups in Kemijen work as presto milkfish makers. Other Kemijen residents work in Tanjung Mas Harbor, the traditional herbal medicine (*jamu*) industry, and at PT Indonesia Power.

The population of Tambak Lorok was recorded in 2020 as 9024, most of whon work as fishermen

Fig. 16.3 Floods in Central Jakarta, 2007 and 2013. (*Source* travelerien.com)

Fig. 16.4 Garuda Sea Wall Project. (*Source*: https://www.abitare.it/en/habitat-en/urban-design-en/2016/01/24/the-great-garuda-the-masterplan-to-save-jakarta/)

so the people are very dependent on marine products. The population is increasing, causing overcrowding and slums in the area. Tidal floods are a frequent occurrence and Tambak Lorok has experienced land subsidence of about 9–10 cm per year.

The Semarang city government has taken several steps concerning physical and social infrastructure. In physical infrastructure, the city government has collaborated with international donors to build East and West Flood Canals, Kali Banger Polder in Kemijen, and also to dredge and widen the river. The local government initiated joining the 100 Resilient City Network (100 RC), connecting Semarang with other cities in this network. The local government also developed *Kampung Bahari* in Tambak Lorok to accelerate and engage the community with marine tourism as a mitigation strategy (Warsilah and Syahid 2023).

Kemijen and Tambak Lorok, both villages in the land of PT Kereta Api Indonesia (PT KAI), are real cases of urban influx migration to abandoned land and coastal areas. The development of urban infrastructure in these two kampongs leads to conflicts between the local government and PT KAI since the government wants to mitigate the tidal flood in this area by building water retention to accommodate the water overflow from the Banger Polder, especially during tidal floods or high-intensity rain in Semarang. The community felt marginalized in their hometown, even though they were living there illegally, and rejected the relocation plan as they considered the land value compensation offered inadequate.

No solution has yet been found. However, this vulnerable community facing tidal floods almost every day lacks social resilience. Kemijen and Tambak Lorok are regarded as slum areas since most of the community worked in the informal sector and have no savings. Meanwhile, it is impossible for them to move or even build a proper new home. They face a choice between meeting their basic needs or raising their houses every five years due to landslides.

The impact of flooding in these two areas on the living conditions of the community is both social and ecological. The environmental degradation suffered by the community in Kemijen and Tambak Lorok encompasses tidal floods, landslides, clean water crises, and uncontrolled plastic waste. Lack of access to clean water and proper waste disposal is leading to a low quality of public health, with tuberculosis and skin disease affecting adults and children.

The social issues caused by these environmental conditions includes significant problems in education, with a high number of children dropping out of school in Kemijen and Tambak Lorok as parents can only afford elementary and junior high school. The decision to raise their house rather than send their children to school has led to an increasing number of unemployed and unskilled people. Problems in Kemijen and Tambak Lorok include drug abuse, alcoholism, and prostitution. Although part of their survival, it endangers their social capital, especially on trust issues among the community.

16.4.3 Disaster Risk Reduction-Based Coastal Community in Jakarta and Semarang

Knowledge and understanding of disasters in both communities remains limited, so that people do not really know what to do when disaster strikes. However, they do have ways of dealing with the environment for their survival, even if these are unsuitable. Their coping mechanisms are derived from experience of flood events passed down through the generations through socialization processes based on the level of their understanding.

All elements of cultures and their influence on the environment result from human behavior which is controlled by certain decisions and choices. The main idea, referring to Speranza, is a community adaptation strategy for resilience. Bennett (1976: 270) describes adaptation as planned behavior to achieve goals and to meet needs, and having consequences for other people, society, and the environment. Adaptation is defined as the strategic behavior of humans in dealing with the natural environment and society, and how humans fulfill themselves and deal with their environment.

Bennett (1976: 3–5) states that the adaptation factor is an effort to actively involve humans in dealing with their natural environment. Adaptation is dynamic and continuously develops the way humans fulfill their lives, how they adapt and deal with their environment, and how they shape the environment to achieve their goals. There are three key ways for people in Jakarta and Semarang to adapt to environmental changes:

1. *Adaptive behavior*. A form of behavior adjustment to achieve goals, determine choices, and refuse to take action or involvement in adaptation.
2. *Strategic action*. Specific action by adjusting efforts to achieve the goal, focused on the utilization process. It includes rationalization, mechanization, progress orientation, and prioritizing human behavior.
3. *Adaptive strategy*. Actions and decision-making processes chosen by humans, since events can be predicted.

It is important to add the community capacity factor since it is the basis of community-centered disaster risk reduction. According to Brown in Rohdewald and Poppe (2005:11), capacity building is the process of increasing people's, organizations', or systems' ability to achieve the set goals. Increasing the capacity of the people in Jakarta and Semarang means increasing their own ability, organizations, and systems for the goals they want to achieve. Sessions (1993: 15) states that capacity building is usually understood as helping governments, communities, and individuals to develop the skills and expertise they need to achieve goals. The community is very adaptive in terms of transmitting disaster risk knowledge to the public by providing concise information about floods, tidal waves, landslides, and other disasters that they face.

16.5 Recommendation and Conclusion

16.5.1 Recommendation

The communities in both Jakarta and Semarang are very vulnerable and prone to disaster, especially of a socio-economic and environmental nature. Community-centered disaster risk reduction will reduce disasters more broadly through capacity and knowledge. In their disaster risk reduction (DRR) strategies, the local governments of Jakarta and Semarang should focus on the community through:

1. Starting to collect data on the local community related to disaster risk reduction (DRR) practiced by them.
2. Continuously updating disaster early warnings to the public, especially in coastal areas, such as the risk of subsidence, reducing the use of plastics, and preventing environmental damage.
3. Collaborating with the regional government in continuing to build sea embankments in abrasion and flood zones.
4. Involving the waterfront communities in North Jakarta and Semarang, to minimize marginalization and social exclusion.

16.5.2 Conclusion

The coastal areas of Jakarta and Semarang face social-ecological risks of tidal floods and land subsidence. The top-down approach still fails to recognize local needs, ignoring potential capacities and resources and increasing vulnerability. The coastal community has its own adaptive cultural values in response to disaster.

This chapter has described how the local government approaches the management of coastal cities. However, the local community knows better the level of damage or loss that is caused by disaster. Community-based risk reduction can reduce the impacts of disaster through four variables: (1) local knowledge about disaster risk; (2) stakeholder involvement in early warning systems; (3) socialization to beneficiaries; and (4) community capacity to respond to disasters.

References

Abidin HZ, Andreas H, Gumilar I, Fukuda Y, Pohan YE, Deguchi T (2011) Land subsidence of Jakarta (Indonesia) and its relation with urban development. Nat Hazards 59:1753–1771

Application of Disaster Risk Management Systems in Indonesia. Semarang City Regional Development Work Plan 2014. Semarang: Semarang City Government; http://www.rasikasemarang.com.

Badan Nasional Penanggulan Bencana (2020) Indeks Risiko Bencana Indonesia Tahun 2020. Available online on https://inarisk.bnpb.go.id/pdf/BUKU%20IRBI%202020%20KP.pdf

Barnett AJ, Anderies JM (2014) Weak feedbacks, governance mismatches, and the robustness of social-ecological systems: an analysis of the Southwest Nova Scotia lobster fishery with comparison to Maine. Ecol Soc 19(4):39. https://doi.org/10.5751/ES-06714-190439

Bennett CF (1976) Analyzing impacts of Extension programs. Washington, DC: U.S. Department of Agriculture Extension Service, No. ESC 575

BPBD Kota Semarang (2020) Tren Bencana 2012 - 2020. Retrieved from http://bpbd.semarangkota.go.id/po-content/uploads/trend_bencana_2012-2020.pdf Accessed on 8 Oct 2021

BNPB (2014) BNPB Regulation No. 11 of 2014 concerning Community Participation in Disaster Management. Jakarta: National Disaster Management Agency

Carpenter S, Brian W, Marty JA, Nick A (2001) From metaphor to measurement resilience of what to what?. Ecosystems. Springer-Verlag. Available online: https://www.researchgate.net/publication/226326551_From_Metaphor_to_Measurement_Resilience_of_What_To_What

Chinwe, Ifejika Speranza (2010) Drought Coping and Adaptation Strategies: Understanding Adaptations to Climate Change in Agro-pastoral Livestock Production in Makueni District Kenya The European Journal of Development Research 22(5) 623–642 https://doi.org/10.1057/ejdr.2010.39

Chinwe, Ifejika Speranza (2013) Buffer capacity: capturing a dimension of resilience to climate change in African smallholder agriculture Regional Environmental Change 13(3) 521–535 https://doi.org/10.1007/s10113-012-0391-5

Compas BE (2004) Processes of risk and resilience during adolescence: linking contexts and individuals. In: Lerner RM, Steinberg L (eds) Handbook of adolescent psychology. Wiley, pp 263–296

Cumming GS (2011) Spatial resilience: Integrating landscape ecology, resilience, and sustainability. https://www.researchgate.net/publication/226080293_Spatial_resilience_Integrating_landscape_ecology_resilience_and_sustainability

Cuny FC (1983) Disasters and development. Oxford University Press

Databoks. "Kepadatan Penduduk di Provinsi DKI Jakarta menurut Wilayah (2022)". Available online on https://databoks.katadata.co.id/datapublish/2022/10/09/jakarta-pusat-jadi-wilayah-terpadat-di-ibu-kota-per-juni-2022#:~:text=Secara%20keseluruhan%2C%20luas%20wilayah%20Provinsi,mencapai%2017.013%20jiwa%2Fkm%20persegi.

Disaster Risk (DRR) for Districts/Cities. 2019. Available at: www.docstoc.com. Accessed 23 Feb 2019

Djunaedi A (2014) Pengantar Perencanaan Wilayah dan Kota. Gadjah Mada University Press, Yogyakarta

Fiol CM, Lyles M (1985) Organizational learning. Acad Manag Rev 10:803–813

Guidelines for Preparation of Risk Maps. 2009. National Planning Agency

Ichwan M, Anwar K (2018) Model Kesiapsiagaan Masyarakat dalam Pengurangan Risiko Bencana Tanah Longsor di Desa Tugumukti , Kecamatan Cisarua Kabupaten Bandung Barat Abstrak Community Preparedness Model in Landslide Disaster Risk Reduction in Tugumukti Village , Cisarua Subdistric. Asian Social Work J 3(4):19–30. Available online on https://www.researchgate.net/publication/348167747_Model_Kesiapsiagaan_Masyarakat_dalam_Pengurangan_Risiko_Bencana_Tanah_Longsor_di_Desa_Tugumukti_Kecamatan_Cisarua_Kabupaten_Bandung_Barat/link/5ff1d85f92851c13fee3cc3e/download

Infografis Ministry of Public Works and Public Housing (2019) Banjir di Indonesia 2012–2019. Available online on https:/data.pu.go.id/infografis/banjir-di-indonesia-2012–2019

Jerez-Gómez P, Jose C-L, Ramon V (2005) Organizational learning capability: a proposal of measurement. J Bus Res 58(6):715–725. February 2005. Available online on https://www.researchgate.net/publication/222808435_Organizational_Learning_Capability_A_Proposal_of_Measurement

Kodoatie RJ, Roestam S (2010) Water layout, Publisher CV. Andi Offset, Yogyakarta

Kummu M, de Moel H, Salvucci G, Viviroli D, Ward PJ, Varis O (2016) Over the hills and further away from coast: global geospatial patterns of human and environment over the 20th–21st centuries. Environ Res Lett 11:034010

Lassa J, Puji P, Djuni P, Eko TP, Amin M, Hening P (2009) Kiat Tepat Mengurangi Risiko Bencana: Pengelolaan Risiko Bencana Berbasis Komunitas (PRBBK). Jakarta, Grasindo

Law No. 26 of 2007. Spatial Planning. Ministry of Home Affairs, 2007. -Law No. 27 of 2007. Management of Coastal Areas and Small Islands

Li S, Sheng G, Wenzhi L, Hongyin H, Quanfa Z (2008) Water quality in relation to land use and land cover in the upper Han River Basin, China. Catena 75:216–222. Elseiver. Available online on https://www.researchgate.net/profile/Wenzhi-Liu-3/publication/223218291_Water_quality_in_relation_to_land_use_and_land_cover_in_the_upper_Han_River_Basin_China/links/5aa77869aca27232682605cf/Water-quality-in-relation-to-land-use-and-land-cover-in-the-upper-Han-River-Basin-China.pdf

Macrotrends. Jakarta, Indonesia Metro Area Population 1950–2023. Available online on https://www.macrotrends.net/cities/21454/jakarta/population

Marfai MA (2011) The hazard of coastal erosion in Central Java Indonesia: an overview. Geogr Malaysia J Soc Space 7(3):19

Mills JP, Buckley SJ, Mitchell HL, Clarke PJ and Edward

Milestad R, Darnhofer I (2003) Building farm resilience: the prospects and challenges of organic farming. J Sustain Agric 22(3):81–97

Obrist B, Constanze P, Robert H (2010) Multi-layered social resilience: a new approach in mitigation research. Sage Publication. Available online on https://www.researchgate.net/publication/255889595_Multi-Layered_Social_Resilience_A_New_Approach_in_Mitigation_Research

Ostrom E (1990) Governing the commons: the evolution of institutions for collective action. Cambridge University Press

Prihatin RB (2013) Banjir Jakarta: Warisan Alam dan Upaya Pengendalian. InsistPress, Yogyakarta

Pretty J (2003) Social capital and the collective management of resources. Science 302(5652):1912–1914. https://doi.org/10.1126/science.1090847

Pretty J, Smith D (2004) Social Capital in Biodiversity Conservation and Management\Capital Social en la Conservación y Gestión de la Biodiversidad. Conserv Biol 18:631–638. https://doi.org/10.1111/j.1523-1739.2004.00126.x

Putra, Dian Rasmana dan Muh Aris Marfai (2012) Identifikasi Dampak Banjir Genangan (Rob) Terhadap Lingkungan Permukiman di Kecamatan Pademangan Jakarta Utara. Jurnal Bumi Indonesia Vol 1 Nomor 1

Resilience Alliance (2010) Assessing resilience in social-ecological systems: Workbook for practitioners. Version 2.0. http://www.resilience.org/3871.php.

Rohdewald R, Poppe M (2005) Guidelines on capacity building in the regions (Version 2.0) Module A: support for decentralization measures. GTZ-SFDM, Jakarta

Ruth M, Rebecca G (2008) Water in the urban environment: meeting a challenges of a changing climate. In:OECD Conference Proceedings "Competitive Cities and Climate Change" in Milan, Italy. Milan: ISPI - Istituto per gli Studi di Politica Internazionale

Sessions K (1993) Building the capacity for change: the world stands ill prepared to address problems that cut across sectors and boundaries. EPA J 19(2):15–19

Speranza CI et al (2014) An indicator framework for assessing livelihood security in the context of socio-ecological dynamics. Glob Environ Chang 28(1):109–119

Twigg J (2009) Characteristics of a disaster-resilient society: an introductory note. OXFAM
Williams K (2010) Sustainable cities: research and practice challenges. J Urban Sustain Dev 1(1–2):128–132
World Bank (2020) Jakarta Urban Challenges in a Changing Climate: Mayor's Task Force on Climate Change, Disaster Risk and The Urban Poor. World Bank Report
Warsilah H, Syahid NC (2023) Climate change and flooding: community vulnerability and resilience. In: Weather Development and Climate Change. ELSEVIER
Warsilah H (2020) The resilience of coastal city facing climate amendment: Benoa and North Jakarta Bay Reclamation Cases. Springer Nature

Photos

https://www.travelerien.com/2013/01/banjir-jakarta-januari-2013.html
https://www.abitare.it/en/habitat-en/urban-design-en/2016/01/24/the-great-garuda-the-masterplan-to-save-jakarta/

Territorial Claims on Floating Net Cage Sites: Early Symptoms of the Tragedy of the Commons in Lake Batur, Bali

17

Ary Wahyono, Agus Heri Purnomo, and Retno Widihastuti

Abstract

Fishing and the cultivation of inland waters tend to get out of control when the number of fish caught is no longer in balance with their recovery capacity. There are no social-traditional institutions that regulate or grant rights to use fishing gear and there are no local government regulations, which further encourages damage to the function of the lake. Claiming the location of individual floating nets is the beginning of the "tragedy of the commons", because there are no shared rules, whether by custom, village, or even the state. The purpose of this chapter is to describe the process of claiming the location of marine cages. This research approach uses descriptive statistical approaches, multidimensional scaling, and interpretative descriptive approaches to analyze the current trend in fisheries performance. The following recommendations are made as a result of this research: (1) claims on aquaculture locations need to be reorganized; (2) access rights should be provided to groups of aquaculture farmers; and (3) rights to fish resources should be recognized. In conclusion, the community is used to customary-based rights guarantees, and indeed customary-based rights guarantees can be relied on to create order.

Keywords

Lake Batur · Freshwater · The tragedy of the commons · Bali Island

17.1 Introduction

Common pool resources are resources whose natural condition or nature makes it very difficult to assert control rights over them, both in terms of the boundaries of the area of control and of localizing the resources they contain. The first difficulty with natural water resources is granting exclusive control, as well as obligations, only to those who own the control rights over the resources, thereby excluding other people who are not subjects of the rights. Second, the benefits that a person obtains from agrarian resources reduce the benefits that can be enjoyed by others (Barnes 2009); in other words, it prohibits or limits actors from exploiting resources.

A. Wahyono (✉) · A. H. Purnomo · R. Widihastuti
Research Centre for Society and Culture - National Research and Innovation Agency, Jakarta, Indonesia
e-mail: ariw001@brin.go.id

A. H. Purnomo
e-mail: agus175@brin.go.id

R. Widihastuti
e-mail: retn032@brin.go.id

H. Warsilah and A. H. Purnomo (eds.), *Approaching Natural Resources for Sustainable Development in Indonesia*, Sustainable Development Goals Series, https://doi.org/10.1007/978-981-96-6913-4_17

The purpose of this chapter is to analyze the issue of claims on lake waters used for fishing. The study was conducted in villages surrounding Lake Batur in Bali, which has exclusive floating net cage planting areas in the waters. It's important to understand the environmental impact caused by human activities. While many studies focus on the natural processes of lake environmental problems, few have addressed how human activities affect lake sustainability.

Currently, Lake Batur in Bangli district, Bali is facing the challenge of how to increase the economic function of the lake for food supply without damaging the environment. The lake, with an area of 6.28 ha and an undeveloped potential of 77.07 ha, covers up to 5% of the total fisheries water (Budiasa et al. 2018).

Aquaculture in Lake Batur aims to improve food security, as capture fisheries activities exceed environmental carrying capacity limits. However, uncontrolled aquaculture causes environmental problems and disrupts socio-economic functions. The main factors that contribute to the mass death of farmed fish in Lake Batur are high stocking density and nutrient accumulation from agricultural and anthropogenic activities (Wiradana et al. 2022).

The lack of spatial planning for fish farming in Lake Batur has given rise to environmental problems. Fish waste containing ammonia from fish farming, and organic waste from excretion, accumulates at the bottom of the water, leading to shallowing due to erosion at the rate of 1930 tons per year, causing a reduction in water capacity and pollution (Irdhawati et al. 2020). The natural phenomenon known as "sulfur plume", rising from the bottom of the lake to the surface of the water, also affects water quality, causing seasonal natural disasters. The lake water has been turning increasingly white since 2010, indicating a decline in water quality.

This seasonal natural disaster has intensified since 2010, leading to mass fish deaths that are detrimental to lake fisheries activities. The reduction in the population of caught fish in Lake Batur is also due to the emergence of redfish predators (Red Devils), which prey on small fish seeds in the lake waters. The introduction of aquaculture in Lake Batur has displaced capture fisheries, as fishing areas for capture fishermen have been narrowed down. Waters for fishing or netting must be kept away from fish farming locations. The existence of fish farming also causes natural fish to move away and concentrate under cage nets. This has reduced the sustainability of the environmental function of lakes and capture fisheries.

17.2 Methodology

The research area is located in four villages in Kitamani, Bangli Regency, Bali Province. Data were collected using surveys, expert consultations, and in-depth interviews. Descriptive statistical approaches (Montgomery and Runger 2003), multidimensional scaling (Borg and Groenen. 2005), and interpretative descriptive approaches, were used to analyze the three different categories of data. To analyze the current trend in fisheries performance at the research site, regression analysis (Bingham and Fry 2010) was used; multidimensional scaling was adopted to evaluate its sustainability prospects; and interpretive description was used to review the available management options.

17.3 Property Rights: Theoretical Perspective

Property rights are a set of rights determined by a set of rules. Schlager and Ostrom (1992) created three categories of resource ownership: property owned by the government, property owned by no one, and property owned and defended by a community of resource users. Meanwhile, Fenny et al. (1990) created four categories of resource ownership for common property resources: open access, private property, communal property, and state property.

There are no significant differences between these two frameworks of common property resource ownership. For example, marine resources are the same as the term 'open access', meaning that there are no regulations that specifically regulate the ownership of marine resources. Everyone is free to exploit natural resources everywhere. The two parallel concepts of property owned by the government and state property define resource ownership rights that are determined exclusively by the state through decisions that regulate the level of exploitation so that there are equal use rights for every citizen.

Communal resource ownership means resources controlled by certain resource user community groups, and such ownership is part of their identity. Ownership rights are exclusive or non-transferable, and often the rights between members are the same when it comes to accessing or utilizing existing resources. Meanwhile, open access is ownership rights that are not clearly defined. Access to resources is free and open to anyone and there are no regulations governing it.

In private property, the rights to natural resources only apply to the owner of the rights and these rights can be transferred to other people. Individual resource ownership, however, is recognized and regulated by the government. In private ownership, regulations regulate the owner's rights in resource utilization, resource management is exclusive, and ownership rights can be transferred.

The management of marine and fisheries resources can take the form of open access or controlled access regulation. Open access allows fishermen to catch fish and exploit other biological resources anytime, anywhere, in any quantity, and with any equipment. Empirically, this regulation has negative impacts, including the tragedy of the commons, both in the form of damage to marine and fisheries resources and in conflicts between fishermen. Controlled access regulation can take the form of input restrictions or output restrictions. One formulation of the input barrier is territorial use rights which emphasize the use of fishing rights (the right to utilize fisheries resources) in a certain area within a clear jurisdiction.

This system, which leads to the formation of marine plots, places a premium on protecting the interests of small fishermen who operate in coastal areas. It is important to preserve the function of resources as the focus of attention.

The government has implemented regulations controlling access to maritime resources, rather than an open access system. UU no. 32 of 2014 sets out the rules for provincial maritime jurisdiction up to 12 miles and regencies/cities up to 4 miles, and the concept of marine plots is still in use as a tool to control access, even though Law no. 23 of 2014 grants maritime jurisdiction only to provinces. This principle falls under territorial use rights, which limit access to resources.

In contrast to Hardin's concept of private ownership of resources, where the state is responsible for managing resources, the government's centralized resource management has some weaknesses, such as the need for well-defined property rights, competitive markets, and cost-effective enforcement of property rights. However, in reality, these conditions are rarely met, and it remains unclear if private ownership would increase efficiency.

The government's centralized resource management has resulted in a lack of monitoring tools, infringements by users, and disregard for environmental concerns. Nevertheless, the customary rights of Indigenous peoples, which involve community property rights, have shown promise in effectively managing marine resources. These practices demonstrate how Indigenous communities can collaborate to prevent excessive resource exploitation and could provide a viable alternative to the government's centralized approach.

17.4 Result and Discussion

17.4.1 Floating Net Cages

Lake Batur has recently seen the emergence of aquaculture, particularly the cultivation of fish in floating net cages. Residents who previously relied solely on agriculture have been blessed with a new economic opportunity. The cultivation of fish in floating net cages offers more attractive and promising returns compared to onion farming. For instance, a farmer who invests in nine plots of nets with tilapia seeds and fish food worth a combined 15 million rupiahs can earn up to 35 million rupiahs in a maintenance period of six to seven months, doubling their income.

The respondents in our survey all claimed to have lived around Lake Batur for more than 10 years. This means that the owners of floating net cage(s) are native residents and not outsiders. The majority of respondents (72%) who own floating net cages live less than 1 km from the lake, and only around 13.6% of floating net cage(s) owners live more than 2 km away. These native residents have occupied the land around the lake for a long time, having inherited it from their parents or other family members.

The land under control appears to be close to the floating net cages, which are located on the edge of the lake waters or close to settlements or agricultural land (Fig. 17.1). Interviews with the owners suggest that the placement of these floating net cages is related to land control. Those cages on the edge of the lake are only intended for residents who control land near the edge of the lake (closed access). This means that not everyone can place floating net cages in the waters at the edge of the lake; anyone who wants to do so must obtain permission from the person who controls the land close to the lake. However, the information obtained from interviews shows that only those who control the land or people related to them are planting floating net cages there, so the waters around the lake are closed to people outside the village. This limitation on use only applies to floating net cages, and not to the use of gillnets and fishing rods. People are free to fish and net fish without obtaining permission from the people who control the land. Likewise, the waters that jut out into the middle of the lake are free for other people to use, including for planting floating net cages.

The land in the lake water area belongs to the state and is recognized as such by residents around the lake. The land around the lake, however, is land controlled by community members and is identified as such by a certificate provided by the traditional village, especially by whoever first took control of clearing the forest or their descendants. However, this land can be transferred to other people, so the role of custom is limited to providing social recognition to someone who controls the land. All members of the community recognize that a certificate from the traditional village conveys land ownership. No resident dares to violate this customary provision.

Claims to the lakeside water area as floating net cages fishing grounds can be explained as follows. First, claims are related to the proximity of the land under control. The boundary of the water is determined by the furthest distance of that a stone is thrown by a person standing on the edge of the lake. This distance is considered the length of the claimed water area from the edge of the lake. Meanwhile, the width of the claimed water area is the boundary of the controlled land. So, the water area of the lake which is claimed is determined from the right–left boundary of the controlled land drawn to a point as far as the length of a stone's throw. According to information from several floating net cage(s) owners, these claims originate from unwritten provisions that have developed in the community.

However, some interviewees said that this claim was not an *awig-awig* from the traditional village, nor was it an agreement between the traditional villages around the lake. In fact floating net cages in Lake Batur are estimated to have started to appear as recently as early 2014. One informant said that floating net cages were developed because a fish food company agent from Paiton, Probolinggo, East Java invited residents to develop floating net cages and at the same time a Minapolitan program was developed by the Ministry of Maritime Affairs and Fisheries.

This subjective claim appears to be a narrative produced by floating net cages owners to justify their existence in this marine area. The waters of the lake belong to the state and the community believes that they can be used by everyone. According to the alternative argument, however, the peripheral waters area is an exclusive floating net cage location because it was previously controlled land.

Lake Batur is a closed water system that does not have an outlet channel, so material input greatly influences the quality of its waters. Lake Batur is part of the Mount Batur caldera. The area at the edge of the lake, previously agricultural land and residential areas, was flooded due to rising lake water levels. This area cannot be separated from the land it controls. One discourse that is

Fig. 17.1 Lake Batur, Bali and location of floating net cages. (*Source* Batur Kintamani Geopark Museum)

developing in the community is that the expansion of the lake area has occurred due to the swallowing of the lake bed because of mud originating, among other things, from feces and leftover fish feed from floating net cages. Another discourse that has emerged in the community is that the weight of the floating net cages and anchors has caused the lake water level to rise. According to this argument, the disruption of the lake's environmental function and the rise in lake water levels, which are detrimental to agricultural landowners, are caused by floating net cages. The fact that floating net cages are increasingly seen as a source of environmental problems shows that not all community members agree with their location in Lake Batur.

Figures 17.2 and 17.3 show the number and distribution of floating net cages in Lake Batur. The number of floating net cages grew significantly between 2006 and 2019, driven by the potential for aquaculture profits to reach as high as 700%. In 2022 there will be around 10,902 net plots spread across nine villages around Lake Batur, with the highest concentration of floating net cages in the villages of Truyan, Abang Datu Wall, Buahan and Btur Tengah. There are around 502 floating net cage farmers, each of whom controls around 21 nets over a cage net plot. The ratio of control over floating net cages is above average in the villages of North Batur and Central Batur, with each farmer controlling on average 33 and 35 net cages. Floating net cage owners typically reinvest their profits into expanding their net boxes and adding new cages each year. This has eliminated the need for external investment and prevented the emergence of patron–client relationships or monopolistic marketing practices in the Lake Batur aquaculture industry.

Women are typically the primary fish sellers who purchase directly from floating net cage owners and sell their products to hotels, restaurants, and other local businesses. While field surveys indicate that the number of floating net cages is relatively small compared to the overall population, only residents with capital, such as onion farmers, can take advantage of these opportunities. The dynamics of the Lake Batur aquaculture industry can be unpredictable, with both ups and downs in recent years. Since 2020, there has been no significant increase in the number of floating net cages due to challenges in obtaining fish seeds and the high cost of feed. As a result, profits for floating net cage owners have been relatively small, and some have chosen to pass their businesses on to family members.

17.4.2 Exclusive Aquaculture Location

Location claims for aquaculture are usually related. First, the area (fishing ground) that is the location of the aquaculture, which is not only limited to area boundaries, but also territorial exclusivity, applicable not only to certain types of marine resources and types of fisheries, but also to the technology operated, the level of exploitation, and temporal limitations (Fig 17.4). Second, the social unit that owns the rights (rights

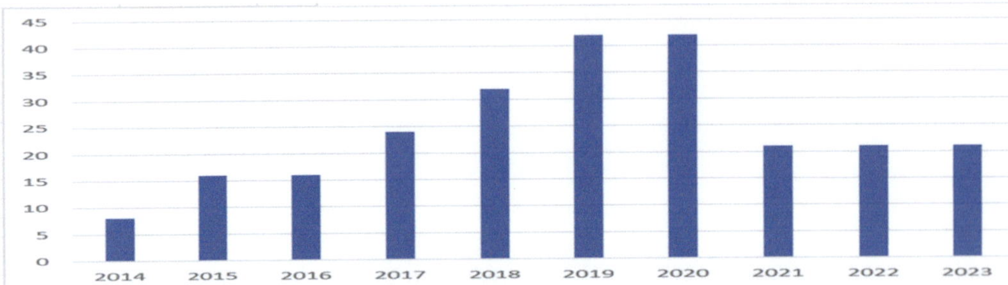

Fig. 17.2 Number of floating net cages in the waters of Lake Batur, Bali, 2014–2023. (*Source* Fisheries cultivation group on Lake Batur)

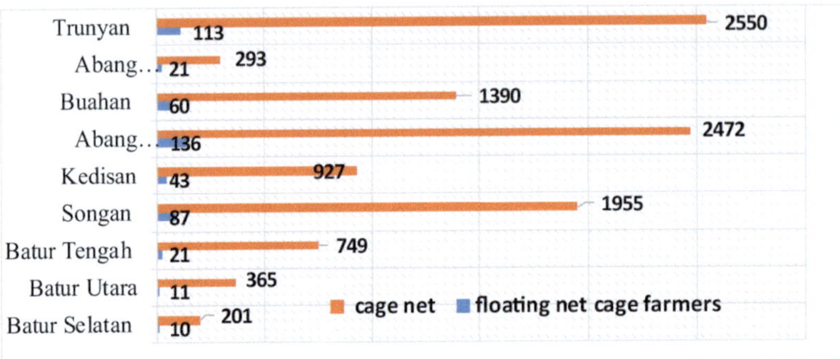

Fig. 17.3 Distribution of the floating net cages in Lake Batur, Bali. (*Source* Department of Agriculture, Food Security and Fisheries, Bangli Regency, 2022)

holding unit), whether an individual or a kinship group, is the holder of the rights. An important element in the social unit of rights holders is transferability, with institutions that regulate how exploitation rights are transferred from one party to another, and there is an equity aspect, or division of rights into a unit of rights holders (see Sudo 1983; Akamichi 1991; Pollnac 1984; Polunin 1983). Third, the existence of social unit holding rights can also be seen from a legal perspective. The legal basis underlying the implementation of the local regulatory system in aquatic communities is sometimes in written form, but extra-legal practices are also common, with legal rules that are based on the habits of the community, such as belief systems or myths regarding certain resources.

According to land control regulations, the ideal location for aquaculture is at the edge of a lake, where floating net cages can be used to raise fish. However, only those who control the land around the lake can put cages in position, since this location is considered part of the land. Before installing a cage, permission must be obtained from the land controller.

Land that is flooded with water is considered an asset for agricultural land and settlement owners. Aquaculture waters are perceived as a property that has investment value, labor, and active ownership of the location, creating a sense of entitlement amongst cultivators. Onion farming is the main source of income for those who live around the lake, and it is regarded as more stable than other farming types.

Even though the land around the lake is the subject of a certificate from the traditional village, some residents feel they control it, having inherited it from their parents. However, the transfer of fish cultivation rights does not occur, and those who manage cultivation are usually closely tied to the owner of the land. There are almost no outsiders who manage the cultivation, except for close relatives of the cultivation landowner.

Fig. 17.4 Fish farming locations claimed as exclusive waters. (*Source* Primary Data, 2023)

The specific narrative about the fish farming location is not clearly defined as per the customary rule or community agreement mentioned in the *awig-awig*. This leads to complications, particularly with regard to preserving the lake's environmental function. The residents believe that the lake's waters are a resource controlled by the state, and the state has the right to regulate its use. The lake is a common property resource and its fisheries belong to everyone. However, around 2006, some people started claiming ownership of water areas for fish cultivation sites without proper regulations or state control (Fig. 17.5).

As a result, the growth of aquaculture is unregulated, leading to a situation where not everyone can place floating net cages in the lake waters, and those who wish to do so must first seek permission from the person who controls the land close to the lake. Although this provision is an unwritten rule, it is not followed by the residents and not explicitly stated in the customary village *awig-awig* rule.

The control of water land for the placement of exclusive floating net cages has been made possible by the expansion of the lake area, which has flooded both agricultural and residential land. The location of fish cultivation areas, which are lakeside waters, is intertwined with the agricultural land and settlements that surround them, resulting in territorial claims in the lake water area where fish farming takes place. However, the emergence of property rights in lake use has not been accompanied by any regulations that regulate the owner's rights in resource use. Resource management is exclusive, and ownership rights can be transferred without any guarantees from the government.

Therefore, it is crucial that the government provides guarantees for the implementation of the rights attached to a resource. The state should own the rights of access and exclusive use of these resources, and the government should determine the implementation of these rights to resources. The government should also make decisions regarding access, level, and nature of exploitation or use of these resources.

The absence of the state highlights the importance of the practices of Indigenous communities in managing natural resources. The role of custom has so far been more about providing social recognition to community members who control land. However, traditional villages need to make *awig-a wig* regarding the management of fisheries cultivation in Lake Batur. The practice of the *awig-awig* in Balinese traditional communities has proved that Indigenous communities can work together and refrain from excessive exploitation of marine resources to avoid the social-ecological crisis of a collective tragedy of the commons in lake resources.

17.4.3 Created Limited Property

The issue of control of water land for the placement of exclusive floating net cages has occurred due to the expansion of the lake area which has inundated agricultural land and residential areas. The location of lakeside waters – fish cultivation

Fig. 17.5 Flooding in agricultural and residential areas caused by rise in lake water levels. (*Source* Primary data, 2023)

areas – cannot be separated from land – agricultural land and settlements. This is the basis for territorial claims in the lake water area which is the location for fish farming. Individual resource ownership can be seen as one of the solutions to the problem of open-access lakewater fisheries (Berkes 1985). However, the emergence of property rights in the use of the lake has not been accompanied by clear regulations regarding ownership rights of floating net cages.

However, in the context of Lake Batur, it is this situation that is the cause of environmental problems, not the excesses of open access. Rights of access and exclusive use of these resources should be determined by the government. The government considers decisions regarding access, level and nature of exploitation, or use of resources (Ostrom 1990). The absence of the state in the case of aquaculture in the lake has created a vacuum in regulations, leading to uncontrolled growth in the number of floating net cages. It is hoped that government regulations or customary *awig-awig* rules can regulate aquaculture in lakes in the form of "created limited property", or a closed resource ownership arrangement (Berkes 1985).

In private ownership, regulations clearly define the owner's rights to resource utilization, resource management is exclusive, and ownership rights can be transferred. In Lake Batur, fisheries cultivation business rights occur within the family, with almost no transfer of rights occurring. The implementation of the rights attached to a resource is guaranteed by the government.

There are two forms of regulation in the management of marine and fisheries resources, namely open access and controlled access regulation. Open access is a regulation that allows fishermen to catch fish and exploit other biological resources anytime, anywhere, in any quantity, and with any equipment. This regulation is similar to the "law of the jungle" and "free market".

Empirically, open access regulations have negative impacts, including what is known as the tragedy of the commons, both in the form of damage to marine and fisheries resources and conflicts between fishermen. On the other hand, controlled access regulation is a controlled regulation (which) that can take the form of (1) input restrictions that limit the number of actors, number of types of vessels and types of fishing gear, and (2) output restrictions that limit the number of catches for each actor based on a quota. One formulation of the input barrier is territorial use rights, which emphasize the use of fishing rights (the right to utilize fishery resources) in a certain area within a clear jurisdiction.

This fishing rights system gives rights holders the right to carry out fishing activities in an area, while those who do not have fishing rights are not permitted to operate in that area. It also regulates when and with what equipment fishing activities are carried out. This system, which leads to the form of marine plots, focuses attention on the protection of the interests of small fishermen and the preservation of the function of resources. Created limited property is basically rights-based fisheries management.

The system allocates guaranteed rights to take a certain volume of catch, to use a certain amount of fishing effort, or to access certain fishing areas, generally to address excess capacity and/or overfishing in a fishery.

In addition to allowing a certain amount of catch from a particular fish stock, the system also gives rights holders the ability to sell, trade, or lease those rights to other parties. A rights-based systems approach, one form of which is individual transferable quotas, has been quite successful in reducing capacity and achieving other management objectives, although scientifically established catch limits and more efficient and effective regulations and law enforcement underlie this success. However, the application of the rights system in lake fisheries cultivation needs to be done carefully, bearing in mind the possibility of negative excesses, especially for small-scale fishermen who are still operating.

17.5 Concluding Remarks

17.5.1 Conclusion

The absence of regulation of control rights for lake fisheries cultivation can have an impact on the uncertainty of guaranteeing community rights to natural resources. Based on the tenure system, customary-based rights guarantees are necessary for two reasons. First, the community is used to customary-based rights guarantees and indeed customary-based rights guarantees can be relied on to create order. Second, the lack of statutory regulation of fish cultivation rights can create irregularities in a community environment that is used to and relies on customary-based rights guarantees. When formal guarantees of rights over natural resources in their territory are not effective, there is no other option for customary law communities and individual rights (as part of customary law communities).

Conceptually, the right to control over natural resources is a legal relationship between the subject of the rights and the natural resources, giving rise to both rights and obligations. This concept of control rights becomes very important as it provides guaranteed certainty of rights for each party over the natural resources they own or control. Ideally, tenure rights are accommodated in the preparation of government regulations.

17.5.2 Recommendation

The claim to fishing resource rights in the case of aquaculture in Lake Batur can be seen as an alternative fisheries management system that can reduce the risk of the tragedy of the commons. The authors have three recommendations. First, claims on aquaculture locations need to be reorganized because they are useful for controlling fishing areas in waters, especially fishing grounds, that overlap with other activities in lake utilization. Second, providing access rights to groups of aquaculture farmers will bring an end to the problem of maximizing profits from aquaculture and at the same time protect fish stocks and habitats. Third, recognition of rights to fish resources in this case can be used to control the growth in the number of floating net cages.

References

Akimichi T (1991) Territorial Regulation in the Small-Scale Fisheries of Ittoman, Okinawa. Dalam Maritime Institution in Western Pacific, Osaka: National Museum of Ethnology

Berkes F (1985) The common property resources problem and the creation of limited property rights. Human Ecol 13:187

Barnes R (2009) Property rights and natural resources. Hart Publishing, London

Bingham NH, Fry J (2010) Regression: Linear Models in Statistics. London: Springer

Borg I, Groenen P (2005) Modern Multidimensional Scaling: Theory and Applications (2nd ed.). New York: Springer-Verlag. pp. 207–212

Budiasa IW, Santosa IG, Agung Ayu Ambarawati IG, Suada IK, Sunarta IN, Shchegolkova N (2018) Feasibility study and carrying capacity of Lake Batur ecosystem to preserve tilapia fish farming in Bali, Indonesia. *Biodiversitas* ISSN: 1412-033X 19(2):613–620. https://doi.org/10.13057/biodiv/d190232

Feeny D, Berkes F, Bonnie JM, James MA (1990) The tragedyof the commons: twenty-two years later. Human Ecol 18(1):1–19

Irdhawati Manuntun Manuruang M, Reichelt-Brushett A (2020) Trace metals and nutrients in lake sediments in the Province of Bali, Indonesia: a baseline assessment linking potential sources. Mar Freshw Res 72(1):66–75. https://doi.org/10.1071/MF20003

Machoherson CB (1989) Pemikiran dasar tentang hak milik. Yayasan Bantuan Hukum Indonesia

Mathew S (1990) Fishing legislation and gear conflict in Asian countries, Samudra monograph. Brussels, Samudra Publications

Montgomery DC, Runger GC (2003) Applied Statistics and Probability for Engineers (3rd Edition). Hoboken: John Wiley & Som, Inc.

Ostrom E (1990) Governing of commons: the evolution of institutions for collective action. Cambridge University Press, Cambridge

Pollnac RB (1984) Investigating Territorial Use Right Among Fishermen. In: Maritime Institutions in Western Pacific, National Meseum of Ethnology, Osaka

Pollunin NVC (1983) Do traditional marine "reserves" conserve? A view of Indonesian and New Guinean Evidence. In Buku R, Kenneth, Johannes RE (eds) Traditional marine resources management in the Pacific Basin, an Anthology, Jakarta, UNESCO/ROSTSEA

Satria A, Matsuda Y, Sano M (2005) Contractual solution to the tragedy of property rights in coastal fisheries. Mar Policy 30:226–236

Schlager, Edella and Ostrom, Elinor (1992) Property-Rights Regimes and Natural Resources: A Conceptual Analysis Land Economics 68(3) 249–262. https://doi.org/10.2307/3146375

Sudo K-I (1983) Social organization and type of sea tenure in Micronesia. In Buku R, Kenneth, Johannes RE (eds) Traditional marine resources management in the Pacific Basin, an Anthology, Jakarta, UNESCO/ROSTSEA

Wiradana PA, Yudha IKW, Mukti AT (2022) Mass tilapia (*Oreochromis mossambicus*) mortality in floating net cages at Batur Lake, Bangli Regency, Bali Province: a case report. IOP Conf Ser: Earth Environ Sci 1036:012068

Zerner C (1995) Common property and biological diversity in the marine and Coastal Zones of Southeast Asia and the South Pacific. Draft Manuscript. World Resources Institute

Part V

Cases of Socio-Ecological Crisis in Southeast Asia as a Comparison for Indonesia

18 Sustaining Recreational Activities Depending on Marine Resources in a Changing Environment: Adaptation Measures at Clam-Picking Parks in Osaka, Japan

Kazumi Wakita

Abstract

Recreational clam picking, called *shiohigari* in Japanese, is a traditional marine recreational activity which is still enjoyed by people today in Japan. For several decades, fisheries cooperatives have managed clam-picking parks appropriately to make sure that visitors to the parks can take enough clams back home and that clams in the parks are safe to eat, in other words, that shellfish toxin detected in the clams are within regulation limits. However, due to the decrease in the wild clam population in coastal areas in Japan and the more frequent detection of shellfish toxins in excess of the regulation limits, sustaining recreational clam picking in Japan has been hanging in the balance. Through interviews with managers of the clam-picking parks in Osaka and textual analysis of newspaper articles on *shiohigari* and shellfish toxin, the author found that the system of exchanging clams taken from the parks for those with shellfish toxin test safety certificates prepared by managers is socially accepted in Osaka. An online survey of 814 residents in Japan showed that appreciation of marine recreational activities such as *shiohigari* was the most influential among various marine ecosystem services in driving attitudes towards marine conservation. This suggests that efforts by clam-picking park managers, mainly fisheries cooperatives, not only provide valuable recreational clam-picking opportunities for people but also promote positive attitudes to marine conservation.

Keywords

Marine ecosystems services · Clam picking · Marine conservation · Shellfish toxin

18.1 Introduction

Clam picking, or *shiohigari,* has been enjoyed by people for hundreds of years in Japan. This traditional cultural marine recreation can be found depicted in Ukiyo-e drawings from the Edo era in the 1850s. Even today, *shiohigari* is enjoyed by people, especially by families. Clam picking is a good introduction to the sea for small children, for swimming is not required and it is considered safer and easier than other marine recreational activities.

In Japan, as well as natural coastal areas where people can enjoy recreational clam picking freely, there are recreational clam-picking parks

K. Wakita (✉)
School of Marine Science and Technology, Tokai University, Shibuya, Japan
e-mail: wakita@tokai.ac.jp

H. Warsilah and A. H. Purnomo (eds.), *Approaching Natural Resources for Sustainable Development in Indonesia*, Sustainable Development Goals Series, https://doi.org/10.1007/978-981-96-6913-4_18

in coastal areas that are usually managed by fisheries cooperatives based on fisheries rights. The general system of recreational clam-picking parks is shown in Fig. 18.1. Visitors pay a picking fee at the park entrance to enjoy clam picking and take the clams back home to eat. At night, the park operator, usually a fisheries cooperative, scatters clams in the park so that there will be enough for visitors to pick the next day.

However, *shiohigari* is at risk in Japan these days. This study focuses on major two reasons: the reduction in the clam catch by fishermen and the increase in the detection of shellfish toxins in excess of the regulatory limit (Fisheries Agency of Japan 2022; The Mainichi Newspaper, 2018). Although clam-picking parks are widely distributed in coastal prefectures of Japan, the example presented in this chapter is an initiative taken by operators of clam-picking parks in Osaka prefecture (Fig. 18.2). The operators of clam-picking parks in Osaka prefecture continue to open the parks by taking food safety measures even when shellfish toxins in excess of the regulation limit are detected.

This chapter first reviews long-term clam catch trends in Japan and in Osaka. Next, details of the initiative taken at clam-picking parks in Osaka and its social acceptance are explained. Lastly, the implications of this initiative are discussed in the context of other research conducted by the author and colleagues on the relationship between the value attached by people to marine ecosystem services and their attitudes towards marine conservation.

18.2 Methodology

This research used a quantitative approach. Interviews with shellfish-picking park managers in Osaka and analysis of newspaper articles about *shiohigari* and shellfish toxins show that the system of exchanging shellfish in Osaka picked at the park for those with shellfish toxin test safety certificates prepared by managers is socially acceptable. An online survey of 814 residents in Japan showed that appreciation of marine recreational activities such as *shiohigari* was the most influential among various marine ecosystem services in driving attitudes towards marine conservation.

18.3 Basis of Analysis

18.3.1 Long-Term Commercial Clam Catch Trends Over 100 Years in Japan and in Osaka

Clams picked by people today in *shiohigari* are mainly Japanese littleneck, i.e., *Ruditapes philippinarum*. It is a common clam eaten by people, for example as an ingredient of miso soup. In spite of its familiarity, the commercial catch of *Ruditapes philippinarum* is on the decline. Figure 18.3 shows long-term trends of the *Ruditapes philippinarum* commercial catch over 100 years in Japan and Osaka (Wakita and Maruyama 2020). Peaking at 160,000 tons in 1983, it has decreased rapidly since and the commercial catch of *Ruditapes philippinarum* in Japan is now only around 10,000 tons. Among the reasons for this decline discussed by researchers are a reduction of habitat due to coastal reclamation and poor management of the commercial catch by fishermen (Toba 2017).

In Osaka prefecture, a rapid decrease in the commercial catch of *Ruditapes philippinarum* has happened twice, in the mid-1930s and the 1950s. The first is thought to have occurred because of habitat loss due to the construction of storm surge barriers and embankment after Typhoon Muroto hit Osaka in 1943 (Wakita and Maruyama 2020). The second is considered to be due to loss of tidal flats due to the development of coastal industrial areas and the deterioration of water quality due to the discharge of untreated water from industries and the increase in water discharge from households (Wakita and Maruyama 2020).

The reduced *Ruditapes philippinarum* catch in Osaka prefecture suggests that wild clams in recreational clam-picking parks in Osaka would not have been sufficient for the number of visitors. According to operators of clam-picking parks in Osaka prefecture, they had already started to purchase clams to provide enough clams for

Fig. 18.1 General system of recreational clam-picking parks in Japan

visitors by 1970 (Wakita and Maruyama 2020). Today, due to the scarcity of wild clams, the purchase of clams by operators of clam-picking parks occurs almost everywhere in Japan except Hokkaido (Okoshi 2012).

18.3.2 Food Safety Initiatives Taken by Operators of Clam-Picking Parks in Osaka Prefecture to Address Shellfish Toxin

Another challenge of sustaining *shiohigari* is the increase in the detection of shellfish toxins in excess of the regulation limit. Toxic algal blooms are a natural phenomenon and cannot be stopped. The shellfish toxin is caused by many different toxic plankton(s). In Osaka Bay, the large-scale proliferation of toxic dinoflagellate *Alexandrium tamarense* in spring has become problematic in recent years (Yamamoto 2019). In Osaka prefecture, the first shellfish toxin to exceed the regulation limit was detected in 2002 and this has been detected almost every year since (Fisheries Division of Osaka Prefecture 2023). Generally, operators of clam-picking parks voluntarily refrain from opening the parks to prevent any cases of food poisoning whenever shellfish toxins surpassing the limit are detected. The more frequently such toxin levels are detected, the more frequently clam-picking parks are voluntarily closed, leading to a reduction in opportunities for people to enjoy clam-picking.

Clam-picking parks in Osaka prefecture are an exception to these practices. Now, all three operators of the clam-picking parks – Nishikinohama, Tannowa, and Hakotsukuri – keep the parks open even when shellfish toxin in excess of the regulatory limit is detected. With the agreement of the Osaka prefectural government, operators of the clam-picking parks have taken food safety measures resulting in no cases of food poisoning by shellfish toxins among visitors to the clam-picking parks in Osaka prefecture (Wakita 2021; Wakita and Fukuyo 2020).

Figure 18.4 shows the framework of collaboration between the prefectural government of Osaka, municipal governments, and operators of clam-picking parks in Osaka prefecture to prevent food poisoning cases by shellfish toxin. The Research Institute of Environment, Agriculture and Fisheries of Osaka Prefecture monitors planktons in the coastal seawater to find toxic algal blooms. The results of the monitoring are passed to the Fisheries Division of Osaka Prefecture, and shellfish toxin tests are conducted as appropriate. The test results are reported to the operators of clam-picking parks in Osaka prefecture, whether shellfish toxin surpasses the regulatory limit or not. At the same time, the Healthcare Division of Osaka prefecture asks the local municipal governments to provide guidance to operators of the clam-picking parks ensuring that careful food safety measures are taken to prevent any

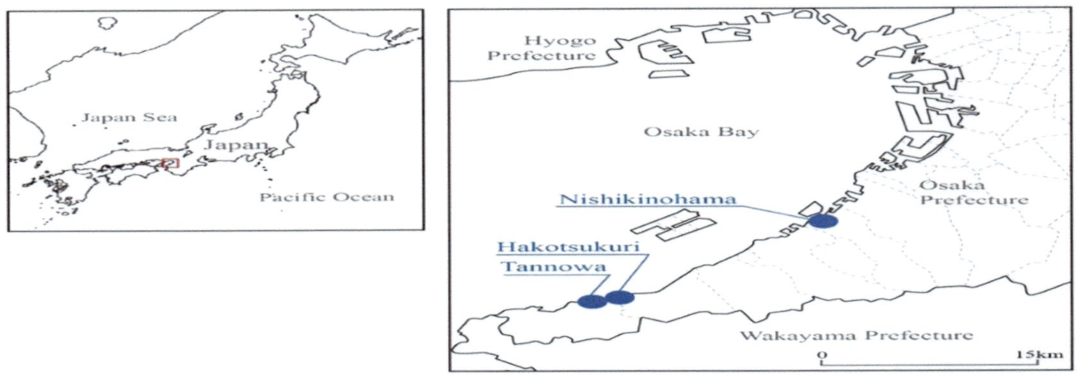

Fig. 18.2 Clam-picking parks in Osaka prefecture. (modified from a digital map of the Geospatial Information Authority of Japan)

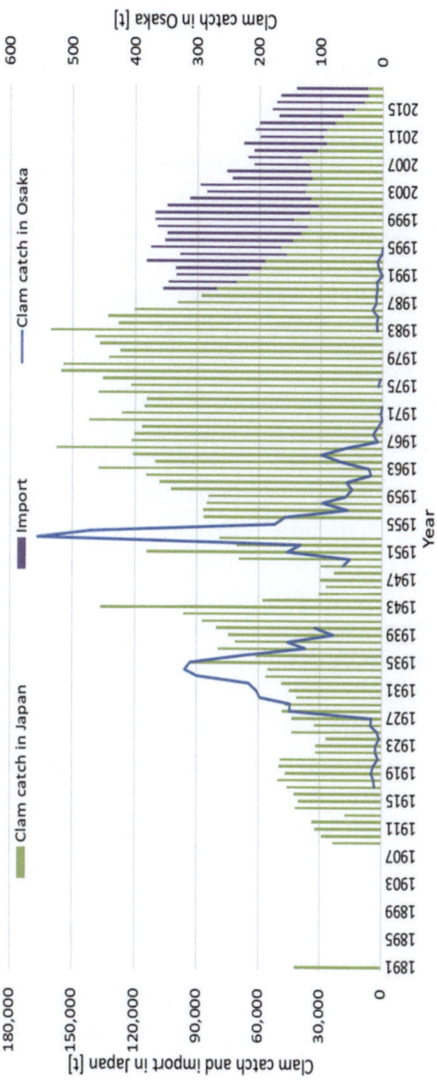

Fig. 18.3 Long-term trends of commercial clam catch over 100 years in Japan and in Osaka. (*Source* Wakita and Maruyama (2020))

cases of food poisoning by shellfish toxin. The close relationships between the Fisheries Division of Osaka prefecture and the operators of clam-picking parks in Osaka are worth noting. By the time an operator is exploring the possibility of opening a clam-picking park once excessive shellfish toxin has been detected, they have already discussed the pros and cons, especially the risks of food poisoning cases, with the Fisheries Division of Osaka prefecture. This close communication between the Fisheries Division of Osaka prefecture and the operator of the clam-picking park is considered to be one of the keys to making the initiative possible. Another important reason why the operators of clam-picking parks have been able to keep their parks open even when shellfish toxin exceeding the regulatory limit is detected is that, legally, the closing of the parks is not mandatory but voluntary (Wakita and Fukuyo 2020).

Nishikinohama is the only one of the three clam-picking parks that is located on a natural sandy beach. It is operated by the Nishikinohama Tourism Association which is composed of descendants of fishermen there. They introduced a clam exchange system (see Fig. 18.5) in response to requests from citizens who wanted to visit the clam-picking park even when shellfish toxin levels exceeding the regulatory limit were detected (Wakita and Fukuyo 2020). At the Nishikinohama, clams are exchanged only when the level of shellfish toxins exceeds the regulatory limit. At the Tannowa and Hakotsukuri clam-picking parks, however, clams are exchanged whether or not shellfish poisoning in excess of the limit is detected, because these two parks are located at artificial beaches where it is difficult for clams to grow (Wakita 2021; Wakita and Maruyama 2020).

18.3.3 Social Acceptability of Clam Exchange System

From interviews with operators of Nishikinohama and Tannowa clam-picking parks, it appears that visitors have become used to the clam exchange system and have a good understanding of the food safety measures taken at the parks to prevent shellfish poisoning cases (Wakita and Fukuyo 2020). At the same time, textual analysis of newspapers reporting on *shiohigari* and the opening of the clam-picking parks even when excessive shellfish toxins are detected reveals that food safety measures under the clam exchange system keep on being reported, showing that the clam exchange system is generally accepted (Wakita and Fukuyo 2020). One article quoted a visitor of the Nishikinohama clam-picking park, "I do not care about the exchange of clams as long as my children can enjoy clam-picking" (The Asahi Newspaper 2017).

Social acceptance of the clam exchange system is further proved by the number of visitors at the clam-picking parks in Osaka prefecture (Fig. 18.6). Between 1983 and 1998, 50,000–80,000 people visited the clam-picking parks. After shellfish toxins surpassing the regulatory limit were first detected in 2002, the number of visitors fell dramatically to less than 27,000 in 2003. Then, after the introduction of the clam exchange system to all the clam-picking parks in Osaka prefecture in 2007, the number of visitors recovered and since 2012 it has remained stable between 40,000 and 80,000 in spite of the detection of shellfish toxins surpassing regulatory limits almost every year.

18.3.3.1 The Potential of Clam Picking to Enhance Marine Conservation Behaviors

Without healthy marine ecosystems, traditional recreational clam-picking or *shiohigari* cannot be enjoyed by people in Japan, for it relies on clams, a natural marine resource. Under the Millennium Ecosystem Assessment (2003), the benefits people obtain from ecosystems are called "ecosystem services". In general, there are four categories of ecosystem services: "provisioning services" such as food, "regulating services" such as climate regulation, "cultural services" such as spiritual and recreational benefits, and "supporting services" such as nutrient cycling (Millennium Ecosystem Assessment 2003).

The same four categories are also applied to marine ecosystem services. For example, fish and

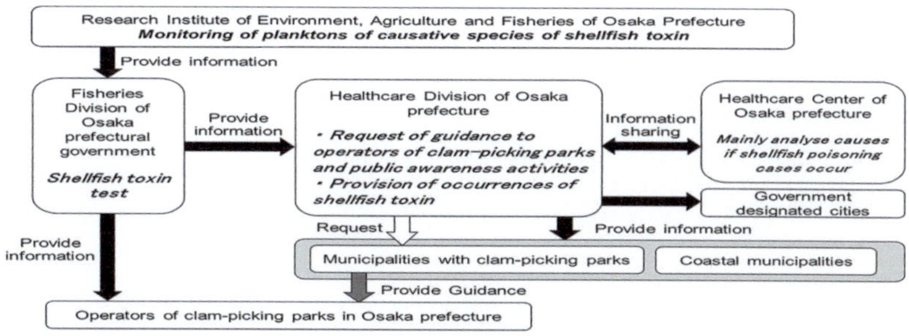

Fig. 18.4 Long-term trends of commercial clam catch over 100 years in Japan and in Osaka. (*Source* Modified from Wakita and Fukuyo (2020))

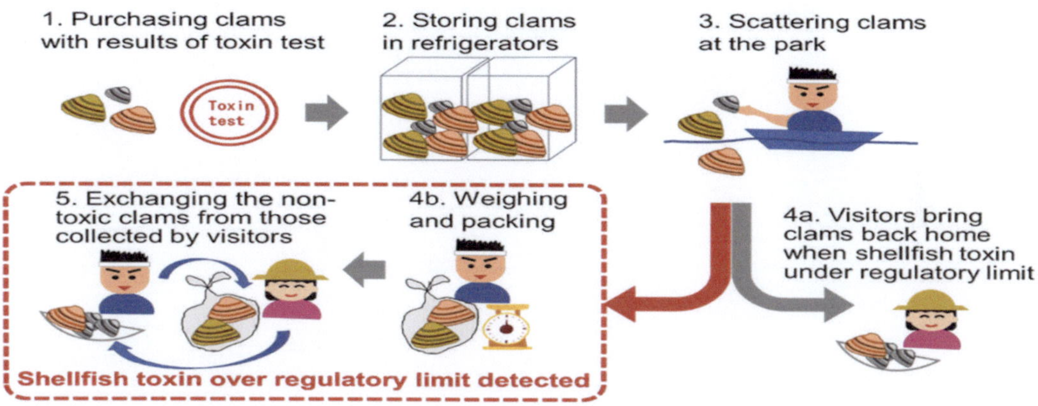

Fig. 18.5 Clam exchange system Nishikinohama clam-picking park

shellfish serve as food for people, hence they are categorized as provisioning services. Sandy beaches and mangroves weaken the power of waves which provides safety to people, hence they are categorized as regulating services. Marine recreation such as snorkeling and recreational fishing makes people happy, hence they are categorized as cultural services. The oceans are the very places where nutrient cycling occurs, which is the basis for the lives of people, and the nutrient cycling by the ocean is categorized as supporting services. These are some examples of the various benefits people obtain from the marine ecosystems. According to this categorization, *shiohigari* comes under cultural services.

The world's ecosystem services can also be valued in monetary terms. According to Costanza et al. (2014), the value of total global ecosystem services was estimated at US$125 trillion/year in 2007, showing that ecosystem services have enormous value. However, as pointed out by Kumar and Kumar (2008), the general public's perception of ecosystems is quite different from what is conceptualized by conventional economists. Gregory et al. (2006) argue that people's perceptions of and the value they attribute to ecosystem services should be at the heart of environmental management.

The author and colleagues conducted an online questionnaire with 814 residents in Japan to establish the relationships between people's perceptions of marine ecosystem services and the influences of these services on people's attitudes and behaviors towards marine conservation (Wakita et al. 2014). Respondents' perceptions of marine ecosystem services can be divided into three categories (Fig. 18.7). The first is "indirect benefits", which consist of regulating services such as reduction of the height of waves and some provisioning services such as provisioning of minerals which are not close to daily lives of the respondents. The second is "essential benefits", consisting of

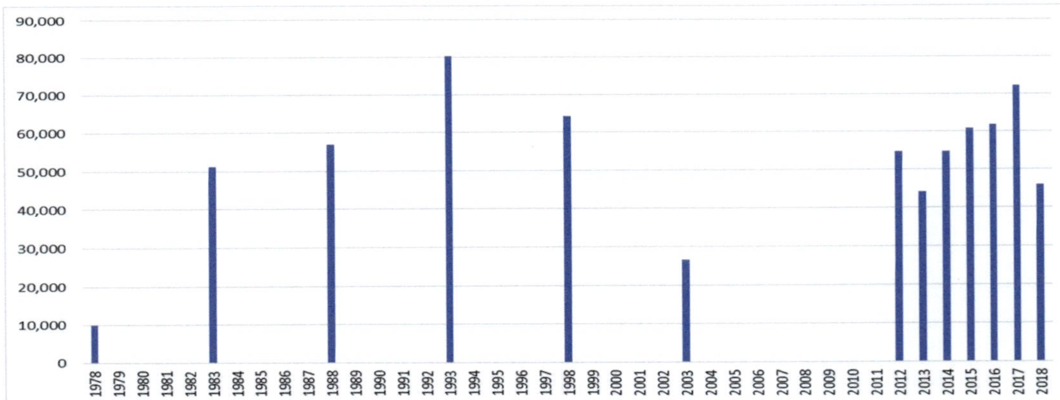

Fig. 18.6 Number of visitors to clam-picking parks in Osaka prefecture. (Red squares mean the years when shellfish toxin surpassing regulatory limit is detected. Years without bars means no data. Data courtesy of Port and Harbor Bureau of Osaka prefecture.)

supporting services such as nutrient cycling, food in provisioning services, and scenery in cultural services. The third category is "cultural benefits", which are cultural services such as recreational value, religious value, health values, etc. (Wakita et al. 2014).

Shiohigari is a marine recreation, and hence comes under "cultural benefits". As we can see in Fig. 18.7, people's appreciation of the *shiohigari* would have an influence on promoting behavioral intentions for marine conservation. This implies that the decision by the operators of the clam-picking parks in Osaka prefecture to keep the parks open by exchanging clams to secure food safety even when shellfish toxins surpassing the regulatory limit are detected could enhance people's appreciation of cultural benefits, and consequently promote positive behaviors and attitudes towards marine conservation. In other words, the food safety measures taken by the operators of the clam-picking parks in Osaka prefecture have the effect of providing more clam-picking opportunities to people, through which their appreciation of "cultural benefits" would be enhanced, which could eventually lead to promoting positive behavior towards marine conservation.

18.4 Recommendation and Conclusion: The Future of Recreational Clam Picking in Japan as a Lesson for Indonesia

This study focuses on an initiative by the operators of clam-picking parks in Osaka prefecture to keep the parks open through food safety measures even when excessive shellfish toxin levels have been detected. The study aslo examines social acceptance of the initiative and discusses the possible effect on people's behavior and attitudes towards marine conservation. There has been considerable recovery work regarding the decrease in *Ruditapes philippinarum* resources in Osaka and in Japan. Commercial clam catches were already banned between May and August as early as 1911 through revision of the Fisheries Regulatory Rules of Osaka Prefecture (Adachi 1997). The Osaka prefectural government started a project to transplant baby clams to increase the *Ruditapes philippinarum* resource some time after World War II, an initiative that lasted until 1967 (Motoki 1997). Efforts at national level include the releasing and nurturing of baby clams (NAAPS 2023; Wakita and Maruyama 2020). Despite all these efforts, recovery of *Ruditapes philippinarum* has been difficult.

Considering the tough situation for recovery of the clams and shellfish toxin detection trends, the future of *shiohigari* in Japan is very much at stake. The most desirable situation would be that people could enjoy clam-picking with plenty of natural clam resources and no detection of shellfish toxin. However, recovery of *Ruditapes philippinarum* would take many years, and the production of shellfish toxin itself cannot be stopped artificially because it is a natural phenomenon. Under these circumstances, scattering purchased clams at clam-picking parks and introducing an exchange system of clams to avoid any occurrences of shellfish poisoning cases in order to keep the clam-picking parks open is regarded as unavoidable. These efforts by the operators of clam-picking parks, mainly fisheries cooperatives, are important not only to keep providing opportunities for clam-picking to citizens regardless of shellfish toxin detection but also to bring traditional and cultural *shiohigari* to future generations.

Indonesia can learn lessons about cultivating edible shellfish by following Japan's example, first by tightening regulations on cultivating edible shellfish, second, by massive cultivation of young shellfish, and third, by providing reliable opportunities to pick shellfish, which is good for children and families in terms of education about the marine environment. Initiatives by shellfish-picking park managers to continue opening parks will increase public appreciation of marine recreation and the marine environment sustainably. It is hoped that it will also contribute to encouraging public attitudes and behaviors towards marine conservation and avoiding a socio-ecological crisis of marine resources.

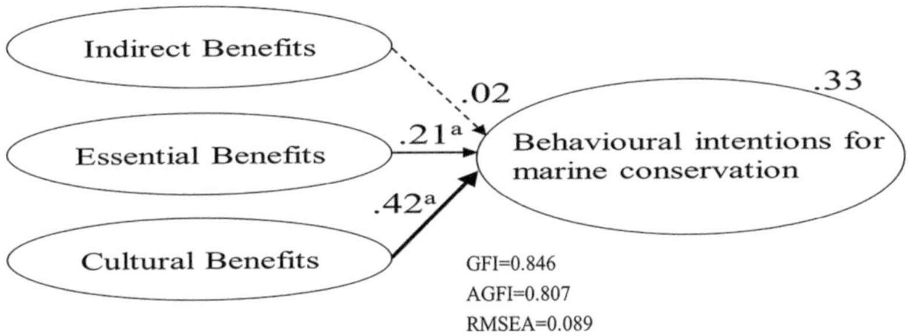

Fig. 18.7 Standardized estimated hypothetical model showing relationship between perceptions of marine ecosystem services and behavioral intentions towards marine conservation. (a)indicates significance at the 0.001 level. Dashed line indicates path that is not significant at 0.05 or better. (Simplified Fig. 1 in Wakita et al. 2014)

References

Adachi Y (1997) Transition of fisheries policies. In: Council of History of fisheries in Osaka prefecture (ed) History of fisheries in Osaka prefecture, pp 254–257. Dai-ichi Houki Shuppan.

Costanza R, de Groot R, Sutton P, van der Ploeg S, Anderson SH, Kubiszewski I, Farber S, Turner RK (2014) Changes in the global value of ecosystem services. Glob Environ Chang 26:152–158

Fisheries Agency of Japan (2022) Situation surrounding the clam. (Asari wo meguru jousei.) https://www.jfa.maff.go.jp/j/saibai/other/attach/pdf/asari-1.pdf. In Japanese.

Fisheries Division of Osaka Prefecture, https://www.pref.osaka.lg.jp/suisan/kai/index.html, Accessed on 18 Sept 2023. (**in Japanese**)

Gregory R, Failing L, Ohlson D, McDaniels TL (2006) Some pitfalls of an overemphasis on science in environmental risk management decisions. J Risk Res 9(7):717–735

Kumar M, Kumar P (2008) Valuation of the ecosystem services: a psycho-cultural perspective. Ecol Econ 64:808–819

Millennium Ecosystem Assessment (2003) Ecosystems and human well-being: a framework for assessment. Island Press, Washington DC

Motoki H (1997) Programs of stock enhancement in shallow sea water. In: Council of History of fisheries in Osaka prefecture (ed) History of fisheries in Osaka prefecture, pp 254–257. Dai-ichi Houki Shuppan

NAAPS (Koueki-shadan-houjin Zenkoku Yutakana Umidukuri Suishin Kyougikai) (2023) Stock enhancement. http://www.yutakanaumi.jp/saibai/. (**in Japanese**)

Okoshi K (2012) The impact of alien snail Euspira fortunei to the Manila clam culture in Japan. Nippon Suisan Gakkaishi 78(5):979–982

The Asahi Newspaper (2017) The SHIOHIGARI: Be careful about shellfish toxin, bringing clams back home is prohibited in Osaka, exchanging clams from ones picked by visitors to the ones from Kyusyu. (SHIOHIGARI, Kaidoku Goyoujin, Oosaka ha Mochikaeri Kinshi, Kyusyu-san to koukan) May 2, 2017. (**in Japanese**)

The Mainichi Newspaper (2018) Shellfish Toxin: Rage throughout Japan. (Kaidoku: Zenkoku de Moui.) May 3, 2018. In Japanese with an abstract in English

Toba M (2017) Revisiting recent decades of conflicting discussions on the decrease of Asari clam Ruditapes philippinarum in Japan: A review. Nippon Suisan Gakkaishi 83(6):914–941

Wakita K (2021) SHIOHIGARI and paralytic shellfish toxin in Japan: Initiatives by operators of clamming parks save traditional recreational clamming. UNESCO IOC IPHAB Harmful Algae Newsl 67:1–3

Wakita K, Fukuyo Y (2020) Food safety measures to prevent paralytic shellfish poisoning: an analysis of newspaper articles and stakeholders' effort at clamming parks in Osaka. J Coastal Zone Stud 33(3):59–69 in Japanese

Wakita K, Maruyama M (2020) Transition of commercial catch of Ruditapes philippinarum and operation of clamming parks in Osaka: State discovered through statistics and interviews. J School Mar Sci Technol Tokai Univ 18:11–20 in Japanese

Wakita K, Shen Z, Oishi T, Yagi N, Kurokura H, Furuya K (2014) Human utility of marine ecosystem services and behavioural intention for marine conservation in Japan. Mar Policy 46:53–60

Yamamoto K (2019) Long-term fluctuations in phytoplankton and marked bloom of the toxic dinoflagellate Alexandrium tamarense in Osaka Bay, Eastern Seto Inland Sea, Japan. Bull Coastal Oceanogr 56(2):63–72

The Ecological Crisis and the Sustainability of Migratory Shorebirds

19

Ahmad Ismail, Nurul Izza Ab. Ghani, and Faid Rahman

Abstract

The East Asian–Australian Flyway (EAAF) is among the important flyways for bird migration, especially migratory shorebirds. Shorebirds fly from the northern to the southern hemisphere and back to their breeding ground, escaping from cold winter to hotter areas for their food, survival, and other ecological functions. Along their migration routes, they stop over at important areas, especially coastal areas that provide them with enough food and shelter. Coastal areas are facing rapid development and habitat destruction, which means ecological risks to migratory shorebirds such as loss or declination of feeding areas, less abundance of food, and diet contamination from heavy metals, pesticides, plastic, and other new emerging chemicals produced by industrial and agricultural activities. Human economic activities in the coastal areas including ecotourism may cause an ecological crisis for the sustainability of shorebirds' population.

Keywords

EAAF · Ecology · Coastal · Migration · Pollution · Shorebirds

19.1 Introduction

Migratory shorebirds are under threat due to the disturbance of their habitat, and there has recently been increased discussion of this population due to habitat loss, chemical pollution, hunting, and climate change. Some species will become extinct if not well protected.

Ecological issues such as population size, diversity, distribution of organisms, and some behavioral responses are affected by environmental problems. The ecological issues are caused by environmental factors such as pollution, deforestation, global warming, and other factors that change or modify their habitat, creating competition for limited food and space or contaminated habitat.

A. Ismail (✉) · N. I. Ab. Ghani
Academy Science Malaysia, University Putra Malaysia, Serdang, Malaysia
e-mail: ismailmt56@yahoo.com.my

N. I. Ab. Ghani
e-mail: Nurul_izza@upm.edu.my

F. Rahman
Institute of Sustainability and Development (LESTARI), Universiti Kebangsaan Malaysia, Bandar Baru Bangi, Malaysia
e-mail: faidrahman@ukm.edu.my

H. Warsilah and A. H. Purnomo (eds.), *Approaching Natural Resources for Sustainable Development in Indonesia*, Sustainable Development Goals Series, https://doi.org/10.1007/978-981-96-6913-4_19

The Philippines, Indonesia, and Malaysia are known for their enormous biological diversity, with globally significant endemism. Like other countries along the East Asian–Australasian Flyway (EAAF), such as China, Vietnam, Thailand, and Indonesia, Malaysia also faces several environmental issues that affect migratory shorebirds, their habitat, feeding, and roosting sites.

Rapid developments in coastal areas due to high demand for economic growth are creating more pressure on waterbirds, including migratory species. Shorebirds in particular depend on intertidal mudflats for their feeding sites and mangroves for their roosting areas. Since they are long-distance migrants, they need a lot of energy to fly and therefore availability of feeding sites, high-quality diets, and suitable roosting sites are very important.

Coastal development involving the clearance of mangrove forests and land reclamation causes the destruction of productive coastal ecosystems. These mangrove forests and intertidal mudflats serve as crucial habitats for migratory shorebirds. Any changes in land use or forest conversion can pose a threat to the sustainability of migrating species.

Developments in the coastal environment and along the river system are also contributing to chemical pollution. Chemical pollution may be from atmospheric deposition, from inland, and from the sea. Industrial, agricultural, and urbanization activities involving heavy metals, organochlorine, and others are also contributing to chemical pollution. This pollution poses a serious threat to migratory shorebirds, causing significant harm to their growth and development. The toxic chemicals they encounter can disrupt embryonic development, and cause premature hatching and other deformities in young birds (De Luca-Abbot et al. 2004; Kertész et al. 2006; Ayas 2007; Kim and Koo 2007). Although the rate of decline for the different species and populations varies greatly (Studds et al. 2017), their reducing numbers are still alarming and many species are facing the risk of extinction. However, in terms of ecological crisis, there are no clear reasons for the population decline of shore birds, especially in specific stopping sites.

Globally, there are nine flyways that support migratory species: the Black Sea/Mediterranean Flyway, the East Atlantic Flyway, the West Asian–East African Flyway, the Central Asian Flyway, the West Pacific Flyway, the Mississippi Americas Flyway, the Pacific Americas Flyway, the Atlantic American Flyway, and the East Asian–Australasian Flyway. However, most research on migratory birds is focused on the American and European flyways. Importantly, research contributes knowledge on the ecological functions of bird migrations and the need for their protection and conservation. The EAAF is the most species rich and threatened flyway, supporting millions of migratory shorebirds annually, but studies on migration routes in this flyway are still lacking (Kirby et al. 2008; Young et al. 2021). A more holistic approach involving multidisciplinary studies is needed to answer unsolved and emergent issues still affecting key sites along the EAAF.

There are several ways in which the migratory behavior of animals can impact the ecosystem. As they move, eat, excrete, and sometimes die, their overall activity has a substantial impact on the trophic structure, function, and communities of the various ecosystems they visit (Bauer and Hoye 2014). Migrants also transport various materials, including energy, nutrients, seeds, macrobenthos, parasites, and pathogens between locations in their routes or flyways, contributing to the stability and sustainability of the community structure in general.

Bird migration may contribute to the survival of a population or species through many evolutionary factors such as mutations, genetic drift, and natural selection (Ismail et al. 2022). Migration is among the factors that shape genetic diversity of populations (Alcala et al. 2013), which in itself is essential for adapting to changes in the environment and ensuring the survival of those populations.

Indonesia and Malaysia are among the important stopovers along the flyway. Stopover sites such as Kuala Gula, Perak in the northern part of the Malay Peninsula (MNS-BCC 2018), and Bagan Percut, Sumatera (Putra et al. 2015) are all facing similar threats. Conservation efforts in these areas are limited by lack of studies and information on shore habitat, and the survival of migratory shorebirds is further threatened. Forest conversion and land-use change are among the important factors that threaten bird populations (Putra et al. 2017). Some species along the flyway are also under pressure as reported in West Papua, Indonesia by Yudha et al. (2024). According to BirdLife International, Indonesia receives 270 species of migratory birds, some of which are endangered or critically endangered.

The findings of this research can be used not only to assist or guide future conservation efforts but also to gain support from local communities. Seasonal migration of shorebirds attracts birders and promotes ecotourism. Although this can support conservation, guidelines are needed to avoid misinterpretation or manipulation allowing over-development that can disturb shorebirds' habitats.

This chapter will take into account the status and the issues faced by the East Asian–Australian Flyway network in its discussion of issues that are potentially contributing to the ecological and environmental crisis of migratory shorebirds, particularly in Peninsular Malaysia and Sumatera, Indonesia.

19.1.1 Mangrove Clearance

Mangroves provide good habitat for migratory shorebirds. Mangroves can be good places for birds to roost and good habitats for macrobenthos. Tide movement makes the mangrove ecosystem more productive. At low tide, tidal areas are exposed, and some trapped small fishes and shrimps can be diets for migratory shorebirds. Exposed tidal flats will give opportunities for shorebirds to feed on worms, crabs, and molluscs.

In Peninsular Malaysia, mangrove forests are distributed along the coastlines of Kedah, Perak, Selangor, and Johor facing the Straits of Malacca. On the east coast, mangrove forests can be found in Terengganu and Pahang. Perak has recorded the largest mangrove forest coverage (Forest Research Institute Malaysia 2012). Indonesia is also very rich in mangrove forest with 4.25 million ha. and quite a high rate of clearance (Onrizal and Kusmana 2008). The destruction of mangrove forests in North Sumatera threatens the livelihoods of traditional fishermen and the quality of the coastal environment, which may affect migratory shorebirds.

Mangrove clearance can create a very significant ecological crisis affecting not just humans but the whole mangrove ecosystem from sedimentation and pollution to changes in the population of benthic organisms, affecting the behavior and population of shorebirds and others in the food web system. Land-use changes and forest conversions are known to significantly threaten the crucial stopover points for migrating species. Rapid changes in the coastal landscape require urgent integrated research.

A lot of mangrove forest were destroyed for development either for human uses such as infrastructures for fishing activities, firewood, charcoal, housing, urbanization, or for ecotourism activities. Large parts of mangrove areas are also cleared for aquaculture activities like shrimp ponds, fish ponds, or other agriculture-related activities. Despite its direct impact on the shorebirds' habitats, there is no detailed information to explain the relationship between mangrove clearing and migratory shorebirds' behavior or movement patterns as a whole. The lack of recent scientific data, particularly on the EAAF, not only hinders conservation efforts but also poses a significant threat to the survival of species.

19.1.2 Chemical Pollution

Environmental pollution, especially from chemicals, has become a major threat to bird populations (Ma et al. 2022). Even though feeding and roosting areas are well protected, they still can be contaminated by chemicals through atmospheric deposition, surface leaching, direct input of urban and domestic waste, agricultural and industrial effluent, and oil pollution.

Chemical pollution can directly or indirectly poison birds. Chemicals such as heavy metals and organochlorines are accumulating in benthic organisms and sediments. Benthic organisms are important diets for migratory shorebirds. Eating a diet contaminated with toxic chemicals may have many effects on the physiological systems of the birds.

The toxicological effects of chemicals on birds will end up reducing their population in the long run (Ma et al. 2022). Apart from causing an ecological crisis, chemical pollution has already been shown to affect the behavior of birds, thinning of eggshells, and other impairments in their physiology that ultimately reduce their survival. Moreover, the accumulation of toxic chemicals in the tissues and food chain also affects the distribution and biodiversity of macro-invertebrates (Azrina et al. 2006). Macro-invertebrates are major food for migratory shorebirds. Thus, any ecological changes in their population will also affect shorebirds and other animals that feed on them.

In view of the facts highlighted above, chemical pollution monitoring in the coastal environment is very important. The now continuous inputs of chemical pollution to coastal ecosystems need to be monitored to give warnings of any elevated levels of contaminant. For effective monitoring, a specific indicator is probably needed, and surface sediment, surface water, and macrobenthic organisms are usually used. Ismail (2006) suggested that intertidal molluscs can be used for long-term programs monitoring contaminants such as heavy metals, pesticides, and organotin compounds in the coastal environment.

One of the successful projects in chemical contamination monitoring in the coastal areas of East Asia is the mussel watch program headed by Professor Shinsuke Tanabe of Ehime University, Japan (Tanabe 2000). This program used green-lipped mussels (*Perna viridis*) as a biological indicator for chemical pollution. Many articles have been published over the years in popular journals on using *Perna viridis* to monitor the contamination of persistent organochlorines, tributyltin, heavy metals, and others. After 25 years, a similar method can be used to compare data along the EAAF including Malaysia and Indonesia.

Other successful monitoring of chemical pollution, especially of persistent organic pollutants (POPs) such as PCBs and organochlorine pesticides, comes under the International Pellets Watch. This program focuses on POPs that accumulate in resin pellets (plastic raw material) from seawater (Takada 2006). Microplastics can be an important indicator that can carry toxic chemicals into the marine ecosystem (Campanale et al. 2020).

19.1.3 Plastic

The whole world is talking about plastic pollution in the ocean. Malaysia happens to be one of the main contributors to plastic pollution in this region (Ng et al. 2023). According to Jambeck et al. (2015), Southeast Asian countries generate more than 1.5 million metric tons of mismanaged plastics annually. Malaysia, Thailand, Vietnam, the Philippines, and Indonesia are among the top ten countries significantly contributing to plastic pollution.

Plastic pollution is among the recent things that scientists worldwide have investigated, especially the abundance and effect it has on migratory shore birds. Lamin et al. (2021) reviewed works on plastic and summarized the effects of macroplastics, microplastics, plastic-derived additives, and plastic-absorbed chemicals on birds. The accumulation of plastic waste and debris has caused significant environmental pollution globally over the past few decades, and it has affected numerous terrestrial and aquatic avian species. Ninety per cent of seabirds are eating plastic. In general, macroplastics are particles > 5 mm, mesoplastics for particles from 5 to > 1 mm, microplastics for particles from 1 mm to > 0.1 μm, and particles less than 0.1 μm are nanoplastics. Microplastics are widely found in the ecosystem as standard-size particles from the manufacturer or as a result of the breakdown of larger pieces of plastics into small particles through photodegradation, physical impacts, and other processes (Tanaka and Takada 2016).

19.1.4 Pesticides

In modern agriculture, the use of pesticides to control diseases, insects, and weeds has increased tremendously. Pesticides affect human health, wildlife including birds, other non-target organisms, and environmental quality in general. Some pesticides are hazardous and cause long-term effects on the environment. Pesticides accumulate in the organisms living in sediment and can be transferred to higher trophic levels through the food chain in the food web system. Pesticides cause behavioral changes, loss of safe habitat, local extinction, and population decline in several birds.

Some countries have already banned chemicals that are persistent in the environment. For example, dichlorodiphenyltrichloroethane (DDT) is known to cause serious effects for bird populations due to eggshell thinning (Mitra et al. 2011). Studies by Monirith et al. (2003) showed that DDT and its metabolites were found in mussels collected along the coastal areas of Asian countries. Even though DDT is banned it remains in the environment. Lee et al. (2001) found that CHLs and DDTs were strongly correlated with total organic carbon content in sediments while HCH residue levels showed differently. Organochlorine pesticides (OCPs) such as HCHs, CHLs, and DDTs were also known to be bio-accumulative due to their high lipophilicity and persistency (Kannan et al. 1992).

Due to their presence, and their persistent and accumulative levels in the environment, continuous monitoring is needed to evaluate contamination status, distribution, and possible sources of pollution, and to assess the risks to aquatic organisms, shorebirds, and humans. The availability of pesticides in different types of macrobenthos and their behavior are contributing to the ecological crisis of the complex ecosystems of intertidal areas.

19.1.5 Heavy Metals

Heavy metals are highly toxic and cause various physiological and biological abnormalities in contaminated organisms. They have been widely studied across the world in evaluations of ecosystem quality due to ease of their assessment in various types of samples.

Yong et al. (2015), in a review on conservation of shorebirds along EAAF, showed that there have been more documented studies on the Charadriidae and Scolopacidae families and toxic heavy metals such as cadmium, copper, lead, mercury, and zinc. Rahman and Ismail (2018) and Pandiyan (2022) state that waterbirds in general are known as a good indicator of harmful metal levels in aquatic environments. Non-invasive chemical pollution biomonitoring and habitat assessment can be carried out on the organs of birds such as kidneys, liver, feathers, and dead carcasses.

Heavy metals contaminate prey species such as crabs, shrimps, molluscs, and fish which are the main diets for water birds. They also constitute an ecological risk to birds' habitat. Once toxic chemicals contaminate the habitat, some benthic organisms can die, and some accumulate in their tissues at tolerable levels. Toxic metals accumulating in benthic organisms will transfer to their predators. Both biotic and abiotic components in certain ecosystems can be assessed to determine their contamination levels at lower trophic levels, enabling the ecological risk to be assessed.

Heavy metals have already been shown to accumulate in birds. Honda et al. (1986) found high concentrations of metals including Mn, Cu, Ni, Pb, Cd, and Hg in egrets' livers, kidneys, feathers, bones, and skin but relatively low values of those metals in the muscles and brain. About 50% of the Hg was reported in the feathers. Metals are excreted via feathers by molting. Agusa et al. (2007) reported elevated levels of V, Cr, Zn, Pb, Sb, Hg, and Bi in fish in Southeast Asian coastal waters. Buhari et al. (2022) also assessed heavy metals Cu, Zn, Pb, Cd, and Ni in surface sediments and giant mudskipper (*Periophthalmodon schlosseri*), which showed that accumulations of metals have occurred in the potential prey of migratory shorebirds. They further suggested that *P. schlosseri* be used as a biomonitoring agent for heavy metal pollution, especially in coastal areas. Toxic heavy metals can transfer through the food web system of intertidal mudflats. Stable isotope studies by Zulkifli et al. (2012) to determine mudskippers' (*P. schlosseri*) food preference showed fiddler crabs (*Uca* sp.), medaka fish (*Oryzias* sp.), and other juvenile fish species in their stomach contents. Ismail et al. (1991) described heavy metals studies in those benthic organisms that are potential diets for shorebirds.

In general, heavy metals are ubiquitous and persistent in the environment, accumulating in organisms and biomagnified in the food chain. Ismail et al. (1993) reported on the heavy metals background in surface sediment collected from the west coast of Peninsular Malaysia. In general, the background levels of heavy metals in sediment 30 years ago were within the normal levels found elsewhere. However, the levels are expected to increase because of the increases in industrial, urbanization, modern agriculture, and domestic effluent discharges. Naji et al. (2010), Udechukwu et al. (2015), and Abubakar et al. (2018) have shown some elevated levels of metals in the sediments receiving effluent from urban and industrial areas. Therefore, regular monitoring of toxic heavy metals is needed.

19.1.6 Butyltin

The use of organotin compounds on ships longer than 25 m has been prohibited since 1 January 2008. However, before tributyltin (TBT) compounds were banned in the shipping industry, much research on TBT was conducted and published, contributing to the long process of monitoring and research that has sought to understand both the short- and long-term impacts of TBT compounds in the environment.

In an Asia–Pacific mussel watch program, Sudaryanto et al. (2002) monitored butyltin (BT) contamination using green mussels (*Perna viridis*) as an indicator in coastal waters of developing Asian countries. Guruge et al. (1997) reported that BT compounds were also detected in most kidney and liver samples of seabirds. Senthilkumar et al. (1999) found that BT concentrations in feathers were much greater than in soft tissue body homogenates. Their study revealed that BT is a compound that is widespread in lower and higher trophic aquatic animals in coastal areas and open-ocean birds. Among BT derivatives, TBT was the predominant compound, indicating its ongoing usage and recent exposures in Asian coastal waters. Mohamat-Yusuff et al. (2014) studied imposex in intertidal molluscs and confirmed that gonad malformation in imposex snails is generated specifically by TBT. The results are correlated with levels of TBT in some sampling sites studied by Sudaryanto et al. (2002). Among the BT compounds examined, monobutyltin (MBT) occupied a major proportion, followed by dibutyltin (DBT) and TBT in most body tissues and feathers of resident and migratory birds (Senthilkumar et al. 1999). Kannan et al. (1992) further highlighted that TBT and its breakdown products, mono- and di-butyltin, were determined in the livers of water birds.

Bioaccumulation and biomagnification of TBT compounds were also seen in many non-target organisms, invertebrates and vertebrates, and affected their population. Many reports also showed that TBT induced significant morphological and histological malformations of gonadal tissue from female embryos which showed a virilization of the ovaries (Scheider et al. 2018). The reduction of the invertebrate population, especially, in specific ecosystems creates a functional ecological imbalance. The results of many studies that highlighted the toxicity of butyl tin compounds have led to TBT compounds being banned.

19.2 Important EAAF Stopover Points on the West Coast of Peninsular Malaysia

BirdLife International Partnership (2020) reported that 277 important bird and biodiversity areas (IBAs) are in danger in 48 countries in the world. Most of the issues are the same worldwide: habitat loss, land reclamation, pollution, and sea level rise causing the decline of the shorebird populations.

All coastal areas along the Straits of Malacca are important sites for migratory birds along the

EAAF route. It can be a strategic stopover location for migratory shorebirds during their migration southwards and northwards. Wide sloping intertidal mudflats on the west coast of Peninsular Malaysia provide good feeding areas for migratory shorebirds. A few spots are identified as popular sites for bird watching during the migration period. The Malaysian Nature Society (MNS), together with international organizations such as BirdLife International, conducts a waterbirds census in January every year. This activity is part of the citizen science program in the Asian Waterbird Census (AWC) to support the global waterbird monitoring program, the International Waterbird Census (IWC) (MNS-BCC 2021).

Among the important areas on the west coast of Peninsular Malaysia are Teluk Air Tawar in Penang, Kuala Gula in Perak, North Central of Selangor Coast (NCSC) and a few sites in the southern part towards Tanjong Piai, the southern tip of Asian land. During the waterbirds census, besides updating the checklist of bird species and the number of birds observed, their habitat quality also was assessed in terms of pollution, size, and other potential threats. Over many years of observation by the MNS Bird Conservation Council, coastal areas of Peninsular Malaysia have been facing the threat of development for human requirements and activities that cause habitat loss and a decline in habitat quality.

Deforestation creates a chain of effects from loss of feeding and roosting areas to contamination of habitats and changing habitat landscape. It modifies wetlands and drainage which can cause deterioration of coastal mudflats, important feeding areas for migratory shorebirds. Heavy metals, butyl tin compounds and pesticides have been reported to contaminate their main diet macrobenthos.

Teluk AirTawar in Penang and Kapar Power Station in Selangor are two important sites that can be good models of the migratory shorebird monitoring program and habitat assessment. These areas already fulfill some criteria as important sites for bird population assessments that should be protected and conserved. If Kapar Power Station Ash Ponds is accepted as an important indicator site for assessing migratory shorebirds' population, a report by MNS-BCC (2016) can be a good reference. Observations showed a 23% decline in the shorebird population between 1986 and 2006. Kapar Power Station supported 23,566 birds in November 2015 and 25,500 in January 2016. The aggregation of birds in Kapar Power Station Ash Ponds is probably because of the reduction in availability of suitable roost areas along the NCSC, and the area is protected. This recommendation needs to be considered, and more places are probably needed for birds roosting during high tide along the Selangor coast.

The NCSC (designated IBA MY011) is a traditional stopover along the flyway where migratory birds refuel before continuing their journey southwards or northwards. This NCSC covers the coastal mudflats and sandflats stretching from the estuary of Sungai Klang (including several islands) to the river mouth of Sungai Bernam, Selangor in the north. Chin et al. (2018) reported that the peak number of priority species can be seen in the area in October and November. The species are little ringed plover (*Charadrius dubuis*), Kentish plover (*Charadrius alexandrines*), Pacific golden plover (*Pluvialis fulva*), grey plover (*Pluvialis squatarola*), lesser sand plover (*Charadrius mongolus*), greater sand plover (*Charadrius leschenaultii*), bar-tailed godwit (*Limosa lapponica*), black-tailed godwit (*Limosa limosa*), Eurasian curlew (*Numenius arquata*), common redshank (*Tringa totanus*), great knot (*Calidris tenuirostris*) and curlew sandpiper (*Calidris ferruginea*). The critical and globally threatened species observed are the spoon-billed sandpiper (*Calidris pygmea*), Far Eastern curlew (*Numenius madagascariensis*), Norman greenshank (*Tringa guttifer*), and great knot (*Calidris tenuirostris*). The spoon-billed sandpiper (*Calidris pygmea*) is the world's rarest bird, with the least known breeding population, and may become extinct – only about 400 mature individuals are reported globally.

The use by shorebirds of artificial (human-made) wetlands, such as ash ponds, salt production sites, power production sites, aquaculture, and other man-made wetlands in the EAAF, are not well reviewed (Jackson et al. 2020). Over 80% of wetlands on the EAAF are threatened. There are 176 artificial sites in the EAAF. However, there have been no detailed studies on the ecology of shorebirds such as species habitat, species abundance, species richness, and species diversity in the artificial habitat. Artificial habitat is a new challenge in shorebird conservation strategies and it needs to be suitable for shorebirds and well managed.

Teluk Air Tawar, Penang consists of a small area 2 km wide and 10 km long and is another important stopover point for migratory shorebirds along the EAAF. This is a natural habitat close to the important estuary of Sungai Muda, a busy port and industrial and urban areas. Hence, Teluk Air Tawar is another challenging area, not only facing rapid development but also receiving freshwater input from Sungai Muda, and experiencing changing landscape, pollution, and other climate change issues. The area needs more attention, not just bird population monitoring and public awareness programs but also ecological studies on the changes of the ecosystem due to natural and anthropogenic activities.

The Malaysian Nature Society has identified key points making Teluk Air Tawar-Ulu Muda an important bird and biodiversity area (IBA, MY003). Among them are (i) it supports over 200 species of birds that depend on the coastal mudflats, mangrove, and inland area, (ii) thousands of migratory shorebirds use the site each year, (iii) it is the second-most important migratory water wintering site in Peninsular Malaysia after North Central Selangor Coast (IBA MY011), (iv) it supports six globally threatened waterbird species, and (v) the critically endangered spoon-billed sandpiper has been recorded at Teluk Air Tawar coast.

MNS-BCC (2018) reported 12 main species depending on Teluk Air Tawar IBA stopover site. The species are spoon-billed sandpiper (*Calidris pymaea*), Norman's greenshank (*Tringa guttifer*), Far Eastern curlew (*Numenius madagascariensis*), great knot (*Calidris tenuirostris*), Chinese egret (*Egretta eulophotes*), and lesser adjutant (*Leptoptilos javanicus*). Many of the following are threatened species: Eurasean curlew (*Numenius arquata*), black-tailed godwit (*Limosa limosa*), bar-tailed godwit (*Limosa lapponica*), curlew sandpiper (*Calidris ferruginea*), whimbrel (*Numenius phaeopus*), and little egret (*Egretta garzetta*).

Similar monitoring activities are conducted to assess the migratory shorebird population passing through long coastal areas in Sumatera. Studies by the EKSAI Foundation (2021) at important sites from North to South Sumatera showed that migratory waterbirds are facing the threat of habitat loss due to reclamation, port development, industrial use, agriculture, and urbanization, as well as hunting, pollution, and human disturbance.

19.3 Conclusion and Recommendations

Continuous monitoring by researchers, NGOs and authorities in Malaysia provide good records of migratory shorebirds on the west coast of Peninsular Malaysia and on the North–South Sumatera coast to West Papua in Indonesia. There have been many publications reporting habitat disruptions, chemical pollution, and land reclamation in the area.

As discussed above, habitat destruction such as mangrove clearance, land reclamation, and chemical pollution will cause ecosystem imbalance and lead eventually to ecological crisis. With coastal habitats under continuous threat, more scientific studies are urgently needed to assess the problems of the declining migratory shorebird population. Scientific evidence will support the CEPA (communication, education and public awareness) program to protect and conserve habitat.

In the case of chemical pollution, specific indicators need to be used to measure chemical pollution and ecological changes. Since intertidal areas are very complex ecosystems, close observation of the potential ecological crisis affecting the sustainability of the migratory shorebirds is needed.

Collaboration and information sharing between the countries along the EAAF is important, since all countries are seeking to protect and conserve the same group of migratory shorebirds.

References

Abubakar US, Zulkifli SZ, Ismail A (2018) Heavy metals bioavailability and pollution indices evaluation in the mangrove surface sediment of Sungai Puloh, Malaysia. Environmental Earth Sciences, 77 (225): https://doi.org/10.1007/s12665-018-7399-1

Agusa T, KunitoT SA, Monirith I, Kan-Atireklap S, Iwata H, Ismail A, Sanguansin J, Muchtar M, Tana TS, Tanabe S (2007) Exposure assessment for trace elements from consumption of marine fish in Southeast Asia. Environ Pollut 145(3):766–777

Alcala N, Streit D, Goudet J, Vuilleumier S (2013) Peak and Persistent Excess of Genetic Diversity Following an Abrupt Migration Increase. Genetics 193(3):953–971. https://doi.org/10.1534/genetics.112.147785

Ayas Z (2007) Trace element residues in eggshells of grey heron (Ardea cinerea) and black-crowned night heron (Nycticorax nycticorax) from Nallihan Bird Paradise. Ankara-Turkey. Earth and Environmental Science 16(4):347–352

Azrina MZ, Yap CK, Ismail AR, Ismail A, Tan SG (2006) Anthropogenic impacts on the distribution and biodiversity of benthic macroinvertebrates and water quality of the Langat River. Peninsular Malaysia. Ecotoxicology and Environmental Safety 64(3):337–347

Bauer S, Hoye BJ (2014) Migratory animals couple biodiversity and ecosystem functioning worldwide. Science 344 (6179) https://doi.org/10.1126/science.124255

Buhari TR, Ismail A (2022) Determining the correlations between Geo-chemical Speciation of Heavy Metals in Surface Sediments and Their Concentrations in Giant Mudskipper. Current Advances in Geography, Environment and Earth Sciences 3:1–11

Campanale C, Massarelli C, Savino I, Locaputo V, Uricchio VF (2020) A Detailed Review Study on Potential Effects of Microplastics and Additives of Concern on Human Health. Int J Environ Res Public Health 7(4):1212. https://doi.org/10.3390/ijerph17041212

Chin CL, Khoo SS (2018) Kapar Waterbirds Survey 2015/16 and Kapar Waterbirds survey 2016/17 (a supplementary). Kuala Lumpur: MNS Conservation Publication No. 19. 48pp.

De Luca-Abbot SB, Wong BSF, Peakell DB, Lam PKS, Young L, Lam MHW, Richardson BJ (2004) Review of effects of water pollution on the breeding success of waterbirds. Earth Environ Sci 10(6):327–349

EKSAI Foundation (2021) Far Eastern Curlew Potential Site Survey at Sumatra Island, Indonesia (2020). EAAFP Secretariat, Hwaseong City.

Forest Research Institute Malaysia (2012) Status of Mangroves in Peninsular Malaysia. (O. Hamdan, Ed.) Malaysia: Gemilang Press Sdn Bhd.

Guruge KS, Iwata H, Tanaka H, Tanabe S (1997) Butyltin accumulation in the liver and kidney of seabirds. Mar Environ Res 44(2):191–199

Honda K, Min BY, Tatsukawa R (1986) Distribution of heavy metals and their age-related changes in the eastern great white egret, Egretta alba modesta, in Korea. Arch Environ Contam Toxicol 15:185–197

Ismail A, Badri MA, Ramlan MN (1991) Heavy metal contamination in fiddler crabs (Uca annulipes) and hermit crabs (Clibanarius sp.) in a coastal area of northern peninsular Malaysia. Environmental Technology, 12(10): 923–926.

Ismail A, Badri MA, Ramlan MN (1993) The background levels of heavy metal concentration in sediments of the west coast of Peninsular Malaysia. Sci Total Environ 134(1):315–323

Ismail A (2006) The use of intertidal molluscs in the monitoring of heavy metals and organotin compounds in the west coast of Peninsular Malaysia. Coastal Marine Science 30(1):401–406

Ismail A, Ab Ghani NI, Arifin W (2022). Conservation Genetics for Managing Biodiversity in Protected Area Management. Recent Advances 15. IntechOpen

Jackson MV, Choi C-Y, Amano T, Estrella SM, Lei W, Moores N, Mundkur T, Rogers DI, Fuller RA (2020) Navigating coasts of concrete: Pervasive use of artificial habitats by shorebirds in the Asia-Pacific. Biol Cons 247:108591

Jambeck JR, Geyer R, Wilcox C, Siegler TR, Perryman M, Andrady A, Narayan R, Law KL (2015) Plastic waste inputs from land into the ocean. Science 347(6223):768–771. https://doi.org/10.1126/science.126035

Kim J, Koo TH (2007) Heavy metal concentrations in diet and livers of Black-crowned Night Heron Nycticorax nycticorax and Grey Heron Ardea cinerea chicks from Pyeongtaek. Korea. Earth and Environmental Science 16(5):411–416

Kirby JS, Stattersfield AJ, Butchart SHM, Evans MI, Grimmett RFA, Jones VR (2008) Key conservation issues for migratory land- and waterbird species on the world's major flyways. Bird Conservation International 18:S49–S73. https://doi.org/10.1017/S0959270908000439

Kannan K, Tanabe S, Ramesh A, Subramanian A (1992) Tatsukawa R (1992) Persistent organochlorine residues in foodstuffs from India and their implications on human dietary exposure. J Agric Food Chem 40(3):518–524

Kertész V, Bakonyi G, Farkas B (2006) Water pollution by Cu and Pb can adversely affect mallard embryonic development. Ecotoxicol Environ Saf 65(2006):67–73

Lee KT, Tanabe S, Koh CH (2001) Distribution of organochlorine pesticides in sediments from Kyeonggi Bay and nearby areas. Korea. Environmental Pollution 114(2):207–213

Lamin W, Nabi G, Yin L, Wang Y, Li S, Hao Z, Li D (2021) Birds and plastic pollution: recent advances. Avian Research 12 (1): https://doi.org/10.1186/s40657-021-00293-2

Ma Y, Choi CY, Thomas A, Gibson L (2022) Review of contaminant levels and effects in shorebirds: Knowledge gaps and conservation priorities. Ecotoxicol Environ Saf 242:113868

Malaysian Nature Society-Bird Conservation Council (MNS-BCC) (2016) Charting the future of Malaysia's most important migratory waterbird roost site: the Kapar Power Station Ash Ponds. Report from the Asian Waterbird Census (AWC) Seminar 25 June 2016. Kuala Lumpur Malaysia Nature Society (MNS Conservation Publication No. 12), 31pp.

Malaysian Nature Society-Bird Conservation Council (MNS-BCC) (2018) Waterbirds, Mangroves and Mee Udang, Teluk Air Tawar-Kuala Muda Coast: Important Bird and Biodiversity Area. Malaysian Nature Society, Kuala Lumpur, p 54

Malaysian Nature Society-Bird Conservation Council (MNS-BCC) (2021) A Checklist of the Birds of Malaysia 2020 edition Kuala Lumpur; Malaysian Nature Society (MNS Conservation Publication No. 22). 123pp.

Mitra M, Chatterjee C, Mandal FB (2011) Synthetic Chemical Pesticides and Their Effects on Bird. Research Journal of Environmental Toxicology 5(2) https://doi.org/10.3923/rjet.2011

Mohamat-Yusuff FSZ, Zulkifli T, Otake H, Harino A, Ismail, (2014) Study on a new mechanism of sterilisation in imposex affected females of tropical marine neogastropod. Thais Sp. J. Environ Biol. 35(5):995–1003

Monirith I, Ueno D, Takahashi S, Nakata H, Sudaryanto A, Subramanian A, Karuppiah S, Ismail A, Muchtar M, Zheng J, Richardson BJ, Prudente M, Hue ND, Tana TS, Tkalin AV, Tanabe S (2003) Asia-Pacific mussel watch: monitoring contamination of persistent organochlorine compounds in coastal waters of Asian countries. Mar Pollut Bull 46(3):218–300

Naji A, Ismail A, Ismail AR (2010) Chemical speciation and contamination assessment of Zn and Cd by sequential extraction in surface sediment of Klang River. Malaysia. Microchemical Journal 95(2):285–292

Ng CH, Mistoh MA, Teo SH, Glassi A, Ibrahim A, Sipaut CA, Foo J, Seay J, Taufiq-Yap YH, Janaun J (2023) Plastic waste and microplastic issues in Southeast Asia. Front. Environ. Sci. Sec. Toxicology, Pollution and the Environment, 11. https://doi.org/10.3389/fenvs.2023.1142071

Onrizal KC (2008) Ecological study on mangrove forest in East Coast of North Sumatra. Biodiversitas 9(1):25–29

Pandiyan J, Poiyamazhi A, Mahboob S, Al-Ghamim K, Al-Misned F, Ahmed Z, Manzoor I, Govindarajan M (2022) Assessment of the toxic effects of heavy metals on waterbirds and their prey species in freshwater habitats. Toxics 10(11):641. https://doi.org/10.3390/toxics10110641

Putra CA, Hikmatullah D, Prawiradilaga DM, Harris JBC (2015) Surveys at Bagan Percut, Sumatra, reveal its international importance to migratory shorebirds and breeding herons. Kukila 18(2):46–59

Putra CA, Peritasari-Farajallah D, Mulyani YA (2017) Habitat Use of Migratory Shorebirds on the Coastline of Deli Serdang Regency, North Sumatra Province. HAYATI Journal of Biosciences 24(1):16–21

Rahman F, Ismail A (2018) Waterbirds: An important bioindicator of ecosystem. Pertanika Journal of Scholarly Research Reviews 4(1):81–90

Senthilkumar K, Tanabe S, Kannan K, Subramanian A (1999) Butyltin residues in migratory and resident birds collected from South India. Toxicol Environ Chem 68(1–2):91–104

Scheider J, Afonso-Grunz F, Jessl L, Hoffmeier K, Winter P, Oehlmann J (2018) Morphological and transcriptomic effects of endocrine modulators on the gonadal differentiation of chicken embryos: The case of tributyltin (TBT). Toxicol Lett 284:143–151

Sudaryanto A, Takahashi S, Monirith I, Ismail A, Muchtar M, Zheng J, Richardson BJ, Subramanian A, Prudente M, Hue ND, Tanabe S (2002) Asia-Pacific mussel watch: Monitoring of butyltin contamination in coastal waters of Asian developing countries. Environmental Toxicology and Chemistry: an International Journal 21(10):2119–2130

Takada H (2006) Call for pellets! International Pellet Watch Global Monitoring of POPs using beached plastic resin pellets. Mar Pollut Bull 52(12):1547–1548. https://doi.org/10.1016/j.marpolbul.2006.10.010

Tanabe S (2000) Mussel watch: Marine Pollution Monitoring in Asian Waters. Centre for Marine Environmental Studies (CMES), Ehime University.

Tanaka K, Takada H (2016) Microplastic fragments and microbeads in digestive tracts of planktivorous fish from urban coastal waters. Sci Rep 6:34351

Udechukwu BE, Ismail A, Zulkifli SZ (2015) Distribution, mobility, and pollution assessment of Cd, Cu, Ni, Pb, Zn, and Fe in intertidal surface sediments of Sg. Puloh mangrove estuary. Malaysia. Environmental Science and Pollution Research 22:4242–4255

Yong DL, Heim W, Chowdhury SU, Choi CY, Ktitorov P, Kulikova O (2021) The state of migratory Land birds in east Asian flyway: distributions, threats, and conservation need. Front Ecol Evol 9:100. https://doi.org/10.3389/fevo.2021.613172

Yong DL, Liu Y, Low BW, Espanola CP, Choi C-Y, Kawakami K (2015) Migratory songbirds in the East Asian-Australasian Flyway: A review from a conservation perspective. Bird. Conserv. Int. 25(2015):1–37. https://doi.org/10.1017/S0959270914000276

Yudha RP, Husein AS, Putra CA (2024) The Importance of Bintuni Bay in West Papua. Indonesia for Migratory Shorebirds. Stilt 77(2024):23–29

Zulkifli SZ, Mohamat-Yusuff F, Ismail A, Miyazaki N (2012) Food preference of the giant mudskipper Periophthalmodon schlosseri (Teleostei: Gobiidae). Knowl Managt Aquatic Ecosyst 405:07

Civil Society and Social Capital in Community-Based Forest Management in Southeast Asia

20

Dundin Zaenuddin, Dwi Purwoko, and Dicky Sofjan

Abstract

Southeast Asian countries have fairly extensive tropical forests. Unfortunately, many countries in the region such as the Philippines, Thailand, and Malaysia, not to mention Indonesia, have experienced severe deforestation since the second half of the twentieth century, leading to a natural resource crisis. The most severe deforestation occurred between the 1960s and the 1990s and persists today. This was largely caused by government policies that pushed for timber export. As a result, forest cover in this region decreased dramatically from 60% in the 1960s to just 30% in the 2000s. Such negative developments have encouraged governments to issue logging bans to prevent further deforestation and shift production from natural to planted forests. Community-based forest management (CBFM), which incorporates social capital such as myths and local knowledge, is a program implemented by the government. This chapter is based on field research conducted in the Cordillera and Palawan regions of the Philippines and Chiang Mai in Thailand, two Southeast Asian countries where forests are well managed by different ethnic groups who have lived in harmony with the forest for centuries. The aim of the chapter is two-fold: to describe and to analyze how social capital and civil society are incorporated into CBFM. Its novelty lies in its demonstration of how the unique socio-cultural combination of social capital and indigenous knowledge incorporated in CBFM programs has enabled these countries to manage forests sustainably.

Keywords

Civil society · Social capital · Community-based forest management (CBFM) · Natural resource management (NRM) · Ethnic or Indigenous communities

20.1 Introduction

Within the Southeast Asian tropical forest belt, the Philippines and Thailand have their fair share of tropical forests. These forests, however, suffered tremendously during the era of devel-

D. Zaenuddin (✉) · D. Purwoko · D. Sofjan
Research Centre for Society and Culture - National Research and Innovation Agency, Jakarta, Indonesia
e-mail: dund001@brin.go.id

D. Purwoko
e-mail: d_purwoko2003@yahoo.com

D. Sofjan
e-mail: dickysofjan@ugm.ac.id

D. Sofjan
Indonesian Consortium for Religious Studies (ICRS), Universitas Gadjah Mada (UGM) Graduate School, Yogyakarta, Indonesia

H. Warsilah and A. H. Purnomo (eds.), *Approaching Natural Resources for Sustainable Development in Indonesia*, Sustainable Development Goals Series, https://doi.org/10.1007/978-981-96-6913-4_20

opment between 1961 and 1991. The Cordillera Ifugao and Palawan, for instance, experienced the worst deforestation in the region. In the late nineteenth century, 70% of the area was covered by dense forest. In the 2000s, the remaining forest merely covers 20%. The deforestation rate has been estimated at 100,000 hectares per annum (Lasco 2006: 2). In the 1960s, the forest area accounted for 60% of Doi Inthanon, but in the 1970s, forest cover was reduced to 23,096,354.17 acres or 45% of the total area (Yonariza 2009: 10). In 2006, forest area was 158,652 km^2 or about 30% (Forestry Statistics of Thailand 2007) (Figs. 20.1, 20.2).

Some of the main causes of deforestation in these countries are illegal and commercial logging, shifting cultivation, construction of roads and dams, and conversion of forests into shrimp ponds, salt, and agricultural land. This policy is the starting point for forest conservation management in Thailand. In 1992, the government revised the 1985 Thai National Forestry Policy (TNFP) which stipulated an allocation of 15% for forest production and 15–20% for conservation (Laungaramsri 2001: 52).

Previously, forests were usually considered uninhabited places, with boundaries separating humans from the animals, trees, and plants that required arrangement and management (Laungaramsri 2001: 54). However, the concept of forest changed at the end of the nineteenth century when British colonial expansion reached the Mon Burmese border with northern Thailand. The forest was no longer considered an isolated and mysterious place but became a prestigious asset. It was at this point that the exploitation of forest resources in Thailand began, to be precise in 1896 with the promulgation of several laws regulating the logging of teak wood (Kaiyoorawong 2005: 71). As a result, there has been a very large increase in deforestation in Thailand. In 1988, the total forest area covered 143,803km^2 or 28.03% of the country's area and about 47% of the northern part. According to Nyunt (2005), the rate of deforestation reached 38% in the period 1975–1988.

The Land Reform Law is also responsible for increasing deforestation because it gives the government permission to grant large concessions to private companies to exploit forests. However, in 1988–1998 the deforestation rate fell significantly, by 9.8%. A 9.8% reduction in deforestation in ten years is not very significant compared to the 28.03% deforestation of the 13 years between 1975 and 1988. Nevertheless, the progress needs to be acknowledged. This reduction is a direct result of the government's efforts to preserve forest areas. In 1985, the government began implementing the TNFP which sought to achieve a target of 15% of the total land area for conservation forests, 40% as protected forests, and 'only' 25% for commercial purposes (Nyunt 2005: 116).

20.2 Methodology

This research uses qualitative research methods through literature reviews, in-depth interviews, and direct observations in the Cordillera Ifugao and Palawan in the Philippines and Doi Inthanon in Chiang Mai, Thailand. Both are located in Southeast Asian countries where forests have been well managed for centuries by different ethnic groups in harmony with the forests.

20.3 Paradigmatic and Theoretical Framework

20.3.1 Theoretical Framework

Management of natural resources, which includes land, water, forests, minerals, and various ecosystems, primarily focuses on preserving both the environment and human life. This natural resource management (NRM) is in line with sustainable development. Sustainable development requires developmental processes that optimize the utilization of available natural resources for economic development and that consider social justice and ecological sustainability. Essential to this development is the capacity to satisfy the current generation's needs while safeguarding the health and well-being of future generations, rather than only meeting immediate requirements. Thus, "economic progress must harness natural resources

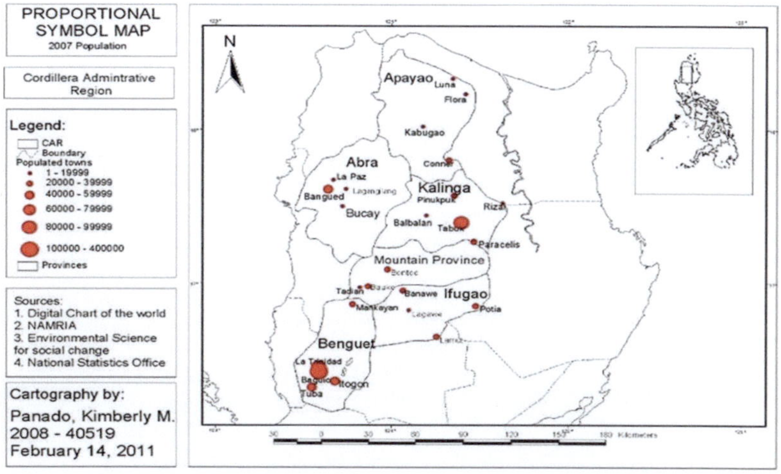

Fig. 20.1 Map of Coldirella and Terras System of Ifaguo. (*Source* www.pintrest.com)

while enabling subsequent generations to perpetuate ongoing development" (SMI 2022).

Local communities play an important role in NRM to achieve the goals of nature conservation and sustainable use of natural elements in a particular environment. The role of local communities in NRM stems from increasing awareness of the need for local community involvement in the management process. With increasing levels of environmental awareness, NRM is no longer considered the responsibility of governments or the private sector alone. In contrast, good NRM requires the active involvement of local communities.

Forests are an important source of human life, both economically and ecologically. Economically, forests provide biodiversity, timber, mining, oil, and other resources that support the lives of living species, such as plants, trees, animals, and humans. Ecologically, forests not only function as the main source of soil fertility and carbon storage, but also provide climate mitigation and genetic resources. Forests provide great benefits for human welfare, including recreation, protection and regulation of water, and prevention of erosion. The decline in forest function can and does contribute to the environmental crisis.

Managing the environmental crisis requires deliberate and integrated efforts to preserve environmental functions and ecosystem services. This includes planning, utilization, development, maintenance, restoration, monitoring, and environmental control policies. This process needs to ensure that all efforts integrate the environment, including resources, into the development process to ensure the capabilities, welfare, and quality of life of current and future generations.

Studies show that forests ideally serve as a medium for "reciprocal relations between humans and non-humans" (Reksohadiprojo 2000). The quality of life of forest communities also depends on local knowledge and that communities can live in harmony and coexist peacefully with nature (Zulueta and Liwag 2001: 43).

The term "local wisdom" began to emerge in the post-industrialization era as a result of large-scale exploitation of nature. The implementation of a market economy has also resulted in changes in the way we think about forest resources. Previously, forests were seen as a necessity for life. Since the advent of the market economy, forests have become a major asset, causing further environmental degradation. Local wisdom carries at least three important values related to environmental management. First, local wisdom contains traditional positive values which are associated with beliefs and behavior in NRM. Second, local wisdom holds many secrets and solutions to ensure the regeneration of the natural environment. Third, local

Fig. 20.2 Map of Thailand and Picture of Doi Inthanon. (*Source* www.pinterest.de)

wisdom is the last bulwark facing the crisis of modern industrial society which depends on the economy and technology.

While local knowledge and myths are almost interchangeable (Milton 1996: 222), not all local knowledge contains myths, and local knowledge cannot resolve all the problems of environmental degradation. Local knowledge is relevant to the notion of "mutual interdependence" between humans and nature, a system that integrates nature and culture through the institutions, knowledge, and practices of NRM.

Environmental degradation and natural disasters are not always ecological; they can also be economic and political. Paterson (2000) argues that the "political environment" is an approach that combines political and economic factors with ecology. Meanwhile, other scholars define environmental politics as a frame for understanding the complexity of the interconnectedness of the local, national, and global political economy and ecosystems (Blaikie and Brookfield 1987). To resolve problems of integrated forest management, it is therefore imperative to consider the complexity of interconnecting ecosystems with the actors involved as well as the participatory management.

Participatory management, according to Feyerabend and Taronwski (2005), means a partnership of social factors that are legitimized by interests, abilities, and commitments related to state ownership of natural resources. Collaboration here is vital, as it details the management functions and ensures the right measure of costs and benefits as well as rights and responsibilities. Furthermore, participatory management perfectly ties in with what is called community-based management. When this principle is adopted in the forest context it is known as community-based forest management (CBFM).

To achieve long-term forest sustainability in Southeast Asia, involving the local community and empowering it to manage natural forests, along with selective combinations of traditional and modern scientific management practices, might be the golden way to move forward for CBFM as part of sustainable development (Zaenuddin 2002). In this context, local knowledge, including myths, is considered a component of "social capital" (Uphoff 2000) that needs to be incorporated into the CBFM.

20.3.2 Myths as Social Capital

Myths provide interesting insights into philosophical and academic discussions. The importance of myths in human history cannot be underestimated. Myth is seen as a source of personal and political self-renewal through contact with the sacred dimension of existence, and this transcendental belief is reflected in the activities of everyday life. Through myths, humans can connect, accept, and understand what happens in the real world. This is in line with Habermas's (1987) idea of the "lifeworld". For him, it is the way people see, interpret, and understand everyday life.

Myths can also provide clues about what to do and what to avoid. In the context of this research, because of this rich multidimensional meaning, myths can determine and to a certain extent dictate people's attitudes and behavior towards nature, especially as regards how they cultivate the land and manage forests sustainably according to their myths or local knowledge. The conceptualization of myth as a shared norm has been included in the CBFM of the Ifugao subregion of the Cordillera of Northwestern Luzon.

20.4 Findings and Discussions

Local people believe that nature and forests contain spirits whose power cannot be ignored, especially considering the need for a harmonious relationship between humans, God, and nature. In the Ifugao tradition, for example, *payo* is a terraced rice field that symbolizes steps towards Heaven. *Payo* refers to any type of irrigable agricultural land that is hydraulically leveled and artificially demarcated. Therefore, the way they manage the forest is always related to spiritual beings.

This tradition is related to a land-use zoning system called *muyong* in the Philippines. *Muyong* comes from the local dialect and means forest or woodland. It is believed that the souls of people's ancestors will 'occupy' their *muyong* and remain part of their PHBM system and *muyong* practices are done to honor those ancestors' spirits. The *muyong* system has been recognized internationally, including by the United Nations Development Program (UNDP), as an ideal CBFM strategy, deeply rooted in Ifugao traditions and culture. The *muyong* system can be seen from various points of view, either as a forest conservation strategy, a watershed rehabilitation technique, an agricultural system, or what is called an assisted natural regeneration (ANR) strategy. Although these systems can be viewed from different perspectives, the role of culture in the development and sustainability of the system has deep roots. Although not obvious to the 'professional' observer, there is a complex network of relationships between human and non-human resources in the system, which transfer to higher spheres or 'verticalism' in the spirit world (Ngidlo 1998). The *muyong* myth, the traditional system of protecting forests, is considered social capital. In the *muyong* system, several areas are included in a group.

Dacawi (1982) reports that *muyong* generally ranges from several hundred square meters to about five hectares. Meanwhile, Klock and Tindungan (1995) found that on Mount Amuyao (the second-highest peak in Ifugao) the standard *muyong* land area ranged from 0.6 to 2.4 hectares. *Muyong* is a storehouse of flora and fauna (Ngidlo 1998). Research by Rondolo (2001) found that *muyong* contains 264 species, mostly native species, belonging to 71 plant families. Euphorbiacea is the most dominant family followed by Moracea, Meliacea, Leguminosae, Poacea, Anacardiacea, and Rubiacea. The number of species per woodlot ranges from 13 to 47, most of which are endemic to the region. Of the 264 species, 234 are considered useful and the remainder (mostly grasses) are reported to be of unknown use.

Muyong is a clear testament to the Ifugao people's knowledge in the fields of silviculture, agroforestry, horticulture, and soil and water conservation. The Ifugao successfully practiced ANR before it was known in the forestry sector as a forest regeneration strategy. They assign value to forests based on their cultural ways and practices. Current forestry development recognizes the importance of integrating customary forest management systems into alternative ANR frameworks and strategies.

The Ifugao are considered traditional practitioners of agroforestry. The Ifugao adopted agroforestry on forest land and double cropping on cultivated fields as economic insurance in case of crop failure on terraces. The integration of value-added trees and herbaceous plants with natural *muyong* vegetation and field cultivation has proved very suitable. The preferred species to be integrated into natural vegetation are rattan, coffee, santol, and orange, as well as banana, taro, and *cadios* (Cajanus cajan).

A study conducted by Rondolo (2001) found that almost all forest land contained commercial coffee crops (88%), bananas (66%), and oranges (49%). Edible rattan (Calamus manillensis, littuko) is also found in almost all woodlots. Activities carried out include thinning, cleaning, pruning, and cutting excess. This activity increases the growth and development of natural stands. The Ifugao have their own plant classification system. Plants are classified according to taxo-morphological characteristics and according to their uses. The Ifaguos' knowledge of rattan classification is more detailed and accurate than that of most formally trained botanists (Rondolo 2001).

In the past, the Ifugao migrated to areas where there were forests and from there began to convert the landscape to other productive uses. However, what is most convincing is the wealth of traditional knowledge systems used by *muyong* owners in managing forest resources. The Ifugao only harvest their trees when the leaves are ripe and not when the leaves have just changed. They believe that cutting down mature trees that have new or young leaves makes the wood vulnerable to insect attack. They also avoid cutting down trees during the full moon because they believe the wood is easily damaged by wood-boring insects.

In Palawan, there are no special concepts or myths such as *muyong* or *payoh*. Palawan is a province of the Philippines consisting of small islands bounded by the South China Sea to the northwest and the Sulu Sea to the east. Understanding myths cannot be applied in the context of Palawan. However, for several ethnic minority groups such as Tagbanua, Kuyunin, Batak, Honggo, and Tagalog, the substance of the myth is still maintained and practiced in the context of forest management, especially for those who live in nature reserve areas such as Campo Uno, Impapay, and Taga-ud.

Another interesting location is Doi Inthanon in northern Thailand. Thailand is one of the most homogeneous countries in Southeast Asia because around 95% of the population is Buddhist, more than 90% of the population speaks Thai, and more than 80% of the population are ethnic Thai (Buergin Reiner 2000). Minority ethnic groups in northern Thailand such as Karen, Hmong, Akha, Yao, and others have applied local knowledge consisting of myths, values, and norms contained in the traditional CBFM system, such as dividing the forest into three, namely: (1) Pa Duta forest, intended for the house of the spirit; (2) Pa Tikuaki, or a forest that has a spring with a water composition surrounded by forest trees; and (3) Pa Suah, or forest to be exploited, for example for wood to build houses and so on (Dheeraprasart 2005). For the Karen people, forest conservation relies on the interdependence between nature and humans. Local knowledge gained through everyday experience can differentiate between edible and poisonous foods. Local conservation depends on the way of life and "related knowledge of those living inside and outside the forest" (Ayuthaya 1997: 125).

20.4.1 Social Capital and CBFM

New thinking and practices are needed, especially to develop forms of social organization that are structurally appropriate for NRM and environmental protection at the local level. This usually means more than just reviving old institutions and traditions. More generally, new forms of association for collective action are required. There is now increasing recognition of the effectiveness of local groups and associations in achieving sustainable environmental and economic outcomes. These local associations are largely based on "social capital" (cf. Bourdieu 1986; Coleman 1988, 1990; Putnam 1993, 1995; Carney 1998; Flora 1998; Grootaert 1998; Ostrom 1998; Pretty 1998; Scoones 1998; Uphoff 2000).

Social ties and social norms are an important part of the basis of sustainable livelihoods. The value of social capital was identified by Jacobs (1961) and Bourdieu (1986), then given a clear theoretical framework by Coleman (1988, 1990), and received widespread attention by Putnam (1993, 1995). Coleman describes it as "a structure of inter-actor and intra-actor relations" that drives productive activity. These aspects of social structure and organization act as resources that individuals can use to realize their interests. Local institutions are effective because "they allow us to go about our daily lives with little repetition and costly negotiation" (Bromley 1993). Social capital can lower the costs of cooperation, and facilitate cooperation. Although there are many different descriptions of social capital, four main aspects can be identified: relationships of trust, reciprocity, and exchange; common rules; shared norms and sanctions and connectedness and networks.

Social capital provides the necessary impetus for sustainable and equitable NRM solutions consisting of a combination of existing natural resources and resources facilitated by external parties. External agencies or individuals may act or work with individuals to improve their knowledge and skills, leadership capacity, and motivation to act. They can act or collaborate with communities to create conditions for the emergence of new local associations with appropriate rules and norms for resource management. If this leads to the desired increase in natural capital, then this will have a positive impact on social capital.

Despite an emerging consensus that social and human capital realized in groups is beneficial (cf. Narayan and Pritchett 1996; Rowley 1999), there is surprisingly little research capable of comparing group and individual approaches in the same context. The investment of resources needed to help develop social capital is through a participatory approach that will produce benefits that sufficiently exceed the costs incurred (Grootaert 1998; Dasgupta and Serageldin 2000; Ostrom 1998).

In the Thai context, relations between the state and forestry communities are experiencing ups and downs. A community forest is defined as an area where part of the community agrees to plant and protect trees, and use them wisely and collectively. There are several reasons why local community involvement is very important in forest management. According to Hirsch (1998: 76), "Forest areas have clear boundaries, and this is recognized by all communities living in the surrounding areas. The main goal of this kind of forest management is to respond fairly to the survival needs of the members of each community. Therefore, resources are expected to be managed efficiently and sustainably."

In the Thai context, logging concessions have been awarded to outside companies. During 1969 and 1985, the area was logged for its timber for railway sleepers and electricity poles as well as to supply pulp raw materials. Meanwhile, officials are labeling local communities as forest destroyers. A second pressure began when a campaign to eliminate shifting cultivation was implemented by the government. This alienated villagers' lands, as the local community was assumed to be the major actor in severe deforestation. The Royal Forest Department divided land into that allocated for state forestry and village land. Every village had an area for permanent cultivation. The regulation stated that the right bank of the Mae Wang River would be solely allocated for forestry, and the left for cultivation. Not all villagers accepted this division because it meant that people on the right bank only had a little permanent rice land. This was one of the reasons for their preference for shifting cultivation.

The third period of pressure came when parts of the watershed areas were inaugurated as the Khun Khan and Doi Inthanon National Parks. Since 1992, the Thai government has implemented a policy of conservation forest, which covers 25% of the total land area of Thailand. People were thus forbidden to live in National Parks, resulting in the resettlement of local communities to new localities. These three periods signaled a downward trend in the relations between the state and community forestry.

The local community in northern Thailand practices *muangfai* (irrigated farming system) that is supplemented by upland short cultivation-short fallow swidden and permanent tree crop gardens such as *miang* tea gardens on hillslopes at middle elevations. Karen people subsist on irrigated farming systems and short cultivation-long fallow swidden. CBFM is inherent in the culture of the ethnic groups of Karen and Hmong. They still maintain King Mangrai's regulation that forbids lowland northern Thailand from violating Pa Sua Ban, areas of the forest dedicated to the guardian spirits of each community (Shalardchai et al. 2003). Karen and Hmong as well as other ethnic groups are committed to forest protection (Prasert 1991; Prasit 1991).

20.4.2 IPRA as Structural Social Capital

The Philippine 1997 Indigenous Peoples' Rights Act (IPRA), signed by then president Fidel V. Ramos, signaled a positive change in that the tenurial rights of ancestral domains for local communities were now formally recognized. It provided a legal framework for upholding indigenous land rights, especially over communal land. This legal development was not only important for security of tenure but also to further achieve sustainable development in the context of the state and the region. This is important since large-scale deforestation has been overwhelmingly occurring in this country. In 1996, for instance, the remaining forest was at 5.4 million hectares (18%), compared to the year 1521, when 90% of the country was covered with lush tropical rainforest (about 27 million hectares out of 30 million total land area). This represents about 200,000 hectares of average forest loss every year (Sajise 1999: 225).

Several factors have contributed to the massive destruction of the country's forest resources: indiscriminate and illegal logging practices, shifting cultivation, an unfavorable policy environment, the slow pace of rehabilitation, and weak law enforcement. Thus, IPRA was a good indication that the country was heading in the right direction. IPRA is the result of various consultations, consolidated bills related to ancestral domains and lands, and international agreements on the recognition of land or domain rights of Indigenous peoples. IPRA is based on the provisions of the UN Draft Declaration on Indigenous Peoples' Rights.

Another aspect of structural social capital is the civil society organizations (CSOs) that can be found in Palawan. Several CSOs work for the common good on this beautiful island. Chief among them is the Palawan NGO Network Incorporation (PNNI), which was formed by the mainstream CSOs and People's Organization in Palawan in November 1991. They intended to present a broader CSO consensus that would participate more in the policy-making process in various local government units in the province.

In developing an agenda for CSOs, PNNI worked toward sustainable development, which essentially balances the use, preservation, and protection of Palawan's natural resources while recognizing that local communities play a major role in resource management. Any development measure that did not promote sustainable development will be accepted.

The PNNI organizational structure is composed of 21 Palawan CSOs or People's Organizations. The General Assembly, its highest policy-making body, meets every year to map out the network's thrust for the year. Its officers are elected on two-year terms from among the member organizations. To actively monitor the Council decisions, particularly those concerning the strict implementation of the Environmental Critical Areas Network (ECAN), PNNI itself actively engages in the arena to directly benefit the communities as well as to influence the development direction of the province towards continuous protection of Palawan's environment. Currently, PNNI is the CSO representative to the PCSD.

20.4.3 Incorporation of Social Capital and CSOs in CBFMs

The CBFM's main formulation can be summed up as "people first and sustainable forestry will follow". The Philippines government believes that by addressing the needs of local communities, they will join hands to protect and manage the very source of their livelihood. Community-based resource management has evolved as a clear policy in NRM. CBFM's goals include promoting sustainable management of forest resources, social justice and improved well-being of local communities, and strong partnership between local communities and the Department of Environment and Natural Resources. CBFM applies to all areas classified as forest lands, including allowable zones within protected areas not covered by prior vested rights.

The program integrates and unifies all people-oriented forestry activities of the Integrated Social Forestry program, Community Forestry Program, Coastal Environment Program, and Recognition of Ancestral Domains. CBFM's features are as follows:

a. *Security of Tenure.* The CBFM Agreement entitles forest communities to use and develop the forestland and resources for 25 years.
b. *Social Equity.* Social justice is a basic principle underlying CBFMs granting forest communities comprehensive rights to use and develop forest resources.
c. *DENR and LGU Partnership.* DENR and LGUs provide technical assistance to CBFM participants to help them achieve sustainable forest management.
d. *Investment Capital and Market Linkage.* CBFM helps participants access investment capital, identify markets, and build marketing capabilities.

There are at least four types of livelihood run by the CBFM: grain processing, lending, consumer store, and selling/marketing of lumber from planted trees. The CBFM now has its own infrastructure, such as a CPEU multi-purpose building with a kitchen and two bedrooms, a consumer store building, multi-purpose pavement, and a small water impounding system (SWIS).

So far, the CBFM's financial performance has been well thought of, utilizing funds from different sources, and paying on time without fail. This successful financial performance has been achieved by the existing management led by chairman Rizaldo Manaeg, with good policy-making and good attitudes based on persistent social capital. Division of labor is practiced, with day-to-day business activities, for instance, being run and managed by the project manager, salespeople, and other staff.

At subdistrict level, the Tambon Council (TC) is regarded as a CSO and officially recognized by the government with the 1995 enactment of the Tambon Administrative Unit Bill. The Council consists of the *kamnan* (Tambon chief), the village heads, the health officer, one 'knowledgeable' person per village, and the secretary. The social capital of the *tambon* is derived from the local community that manages the forest and formulates rules that are observed by its members. This could be considered as best practice since it is a more effective and efficient way of conservation than if practiced by government officials. Through this practice, neglected responsibilities and illegal concessions can be avoided since the community will benefit from conservation in the long run.

20.5 Conclusions and Recommendations

20.5.1 Conclusion

Local community forest management systems in the Southeast Asian region have been proven to contribute to NRM and CBFM because of their interdependent relationship. Local knowledge can be transformed into social capital through encouragement of sustainable forest development and river watershed management transformed from sustainable development. From generation to generation, social capital that includes local knowledge, myths, traditions, and customs in the regions studied has shown how effective CBFM is in protecting and managing forests in Southeast Asia. Civil society also plays an important role in bridging the interests of multiple stakeholders, enabling the various parties involved to combine traditional and modern approaches and technologies in implementing CBFM.

20.5.2 Recommendations

Based on case studies in Thailand and the Philippines, it is recommended that governments in Southeast Asia take into account the lessons learned over the past decades regarding NRM and CBFM, while considering incorporating some of the principles into their policy and regulatory frameworks. However, many social and cultural mechanisms are more or less the same throughout Southeast Asia, so the transfer of learning is not too difficult. The only question that remains is whether there is the political will from a government that has yet to implement NRM and CBFM principles. Indonesia can learn from the experiences of Thailand and the Philippines in terms of recognizing and appreciating social capital such as local knowledge, myths, and customs that have long existed throughout the archipelago. The fact that Indigenous and ethnic communities in Indonesia have a long tradition of sustainable forest management means that they can be considered as part of the policy infrastructure and CBFM system to stop the rapid rate of deforestation which certainly degrades ecological sustainability.

References

Ayuthaya PNN (1997) Community forestry and watershed networks in Northern Thailand. In: Philip H (ed) Seeing forests for trees, environment, and environmentalism in Thailand. Silkworm Books, Bangkok

Bourdieu P (1986) The forms of capital. In: Richardson J (ed) Handbook of theory and research for the sociology of education. Greenwood Press, Westport CT

Blaikie P, Brookfield H (1987): Land degradation and society. London: Methuen

Bromley DB (1993). Reputation, image and impression management. John Wiley & Sons

Buergin Reiner (2000) Hill Tribes' and Forests: Minority Policies and Resource Conflicts in Thailand. Socio-Economics of Forest Use in The Tropics and Subtropics. Freiburg: Albert Ludwigs Universität.

Carney D Ed (1998) Sustainable Rural Livelihoods: What Contribution Can We Make? DFID, London.

Coleman J (1988) Social capital and the creation of human capital. Am J Sociol 94(supplement):S95–S120

Coleman J (1990) Foundations of social theory. Harvard University Press, Boston, Massachusetts

Dacawi R (1982) The Ifugao way of forest conservation. Philippine Upland World 1(2):14–15

Dasgupta P, Serageldin I, Eds, Social Capital: A Multifaceted Perspective, World Bank, Washington DC, 40–59

Derrida J (1982) Myth and language. University of Chicago Press, Chicago

Deepa, Narayan (1999) Cents and Sociability: Household Income and Social Capital in Rural Tanzania The World Bank https://doi.org/10.1596/1813-9450-1796

Deepa, Narayan Lant, Pritchett (1999) Cents and Sociability: Household Income and Social Capital in Rural Tanzania Economic Development and Cultural Change 47(4) 871–897 https://doi.org/10.1086/452436

Dheeraprasart V (2005) Until no trees remain: illegal logging in the Salween Forest. In: After the Logging Ban. Politics of forest management in Thailand. Foundation for Ecological Recovery, Bangkok

Etzioni A (1995) The spirit of community: rights, responsibilities and the communitarian Agenda. Cambridge University Press, London

Feyerabend G, Tarnowski CB (2005) Participatory democracy in natural resource management: a'Columbus's Egg'? In: Brosius, J.P., Tsing, A.L. and Zerner, C. (eds) Communities and Conservation: Histories and Politics of Community-based Natural Resource Management. AltaMira Press, Walnut Creek, CA, 69–90

Fisher RJ (2000) Decentralization and devolution in forest management: a conceptual overview. In: Enters T, Victor M, Durst P (eds) Decentralization and devolution of forest management in Asia and the Pacific. RECOPTC Report No, Bangkok, p 18

Flora JL (1998) Social capital and communities of place. Rural Sociol 63(4):481–506

Forestry Statistics of Thailand (2007)

Grootaert C (1998) Social capital: the missing link. In: World Bank Social Capital Initiative Working Paper, No 5

Habermas J (1987) Theory of communicative action: critique of functionalist reason, vol II. Polity Press, Oxford

Hirsch P (1998) The Politics of Environment in Southeast Asia: Resources and Resistance. Routledge, London and New York

Ian, Scoones Ian, Scoones Sustainable Livelihoods and Rural Development Methods for Livelihoods Analysis Fernwood Publishing Practical Action Publishing 98–108

Jacobs J (1961) The life and death of great American Cities. Random House, London

Kaiyoorawong S (2005) Conserving Thailand's forest: legal perspectives. In: After the Logging Ban. Politics of Forest Management in Thailand. Bangkok: Foundation for Ecological Recovery

Klock J, Tindungan M (1995) The past and the present: a meeting of forces for a sustainable future. In: Forest, trees and people newsletter. Issue no. 29. FTP Program Network

Laungaramsri P (2001) Redefining nature, Karen ecological knowledge and the challenge to the modern conversation paradigm

Lasco RD (2006) Environmental impact of community-based forest management in the Philippines. Int J Environ Sustain Dev 5(1):46

Milton, K (1996) Environmentalism and Cultural Theory, Exploring the Role of Anthropology in Environmental Discourse. London&New York: Routledge

Narayan D, Pritchett Lant (1996) Cents and Sociability: Household Income and Social Capital in Rural Tanzania. Economic Development and Cultural Change, 47(4) (July 1999), 871–897. The University of Chicago Press

Ngidlo RT (1998) Conserving biodiversity: the case of the Ifugao farming system. PCARRD, Los Baños, Laguna

Nyunt KM (2005) The impact of trade liberalization on the forestry sector in Thailand. PSDR LIPI, Jakarta

Ostrom E (1998) Social capital: a fad or fundamental concept? Center for the study of institutions, population and environmental change, Indiana University, USA Palmer, I

Paterson, Matthews (2000) Understanding Global Environmental Politics: Domination, Accumulation, Resistence. New York: St. Martin's Press

Philip, Hirsch Carol, Warren The Politics of Environment in Southeast Asia Dams resources and the politics of environment in mainland Southeast Asia Routledge 69–84

Prasert (1991) "The impact of tourism to hilltribe society: Case study in Pang Ma Pha District, Mae Hong Son Province". Research report

Prasit (1991) Who cares? Northern Thailand. Thailand: World Agroforestry Centre (ICRAF)

Pretty J (1998) The living land: agriculture, food and community regeneration in Rural Europe. Earthscan Publications, London

Putnam R (1995) Bowling alone: America's declining social capital. J Democr 6(1):65–78

Putnam Robert D, with Leonardi R, Nanetti RY (1993) Making democracy work: civic traditions in modern Italy. Princeton University Press, Princeton

R., Chambers Piers, Blaikie Harold, Brookfield (1988) Land Degradation and Society The Geographical Journal 154(1) 114-https://doi.org/10.2307/633496

Reksohadiprojo (2000) Kajian Identifikasi Potensi dan Permasalahan hutan Kemasyarakatan Di Desa Soakonora Kabupaten Halmahera Barat Propinsi Maluku Utara. Universitas Nuku

Röling NR, Wagemakers MA (eds) (1997) Social learning for sustainable agriculture. Cambridge University Press, Cambridge

Rondolo MT (2001) Fellowship Report. Tropical Forest Update. 11, 4. ITTO, Japan

Rowley G (1999) The Tragedy of the Common Waters: Towards the Deepening Crisis within the Jordan Basin. The Arab World Geographer, 2, 26–40

Sandler T (1997) Global challenges: An approach to environmental, economic and political problems

Scoones I (1998) Sustainable Rural Livelihoods: A Framework for Analysis. IDS Working Paper 72. Institute of Development Studies

Seri P (ed) (1993) Indigenous knowledge and rural development. Amarin Printing Group, Bangkok

Shalardchai R. Pa mai Kap Karn Pattana Chonnabot (2003) Forestry and Rural Development. Bangkok: Public Policy Studies, Social Science Association of Thailand

Tongpan S et al (1990) Deforestation and poverty: can commercial and social forestry break the vicious circle? Paper presented at the 1999 TDR Year End Conference. Bangkok, Thailand

Uphoff N (2000) Understanding social capital: learning from the analysis and experience of participation. In: Paper of Seminar Staff, Mansholt Institute, Wageningen, 13 September

Woolcock M (1998) Social capital and economic development: towards a theoretical synthesis and policy

Yonariza (2009) The effect of the logging Ban in Thailand's National Forest on Timber Tree Domestication. In: Confluences and Challenges in Building the Asian Community in the Early 21st Century. The Work of the 2008/2009 API Fellows. The Nippon Foundation, Tokyo

Yong Ooi Lin C (2003) Flowed over: The Babagon Dam and the Resettlement of the Kadazan Dusuns in Sabah. Centre for Orang Asli Concerns, Subang Jaya

Yos S (2003) Biodiversity: local knowledge and sustainable development. Regional Centre for Social Sciences, Chiang Mai

Yos S (2008) Flexible peasants: reconceptualizing the third world's rural types. Regional centre for social science and sustainable development (RCSD), Faculty of Social Sciences, Chiang Mai University, Chiang Mai

Zaenuddin D (2002) Establishing Golden Paradigm for New Indonesia. Jurnal Masyarakat dan Budaya 4(2)

Zulueta, F.M and Liwag, Dolores B (2001) Social Problems and Issues in the Philippines. Manila: National Book Store

Document

SMI (2022) Sustainable Development: Definition, Concept and Goals, https://ptsmi.co.id/pembelian-berkeuntungan

Bappenas, SDGs Goals, https://sdgs.bappenas.go.id/

Part VI
Afterword

The Sustainable Development Paradigm in Relation to the Socio-Ecological Crisis 21

Henny Warsilah and Agus Heri Purnomo

Abstract

As we have seen in the narratives presented in *Socio-Ecological Crisis of Natural Resources from a Sustainable Development Perspective,* the management of natural resources in Southeast Asia and Indonesia can be categorized into three major problems: (1) economic growth and development continuing to rely on the availability of primary natural resources, such as land.

As we have seen in the narratives presented in *Socio-Ecological Crisis of Natural Resources from a Sustainable Development Perspective,* the management of natural resources in Southeast Asia and Indonesia can be categorized into three major problems: (1) economic growth and development continuing to rely on the availability of primary natural resources, such as land, forests, peat swamps, karst, mining, sea, and energy; (2) industrial growth continuing to take place in urban areas, especially Java Island and Outer Islands of Java, leading to industrial pollution combined with urban factors such as waste and vehicle emissions; and (3) high economic growth resulting in the capitalization of natural resources and triggering conflicts in land and forest use, along with increased access to other natural resources.

The twentieth century marked the emergence of sustainable development as a developmental direction. According to Susan Baker (2006), the term 'sustainable development' entered the public discourse in 1980 when the Union for the Conservation of Nature and Natural Resources presented its World Conservation Strategy (IUCN 1980). This agenda aimed to achieve sustainable development through the conservation of living natural resources. However, the focus primarily addressed ecological sustainability, without explicitly connecting sustainability to broader social and economic concerns. Baker further emphasizes that it was not until 1987, with the publication of the World Commission on Environment and Development (WCED) report *Our Common Future*, that the explicit relationship between the social, economic, and ecological dimensions of development was highlighted (WCED 1996).

Sustainable development, as the seventh point of the UN Millennium Declaration, seeks to ensure environmental sustainability by incorporating the principles of sustainable development into efforts to reduce the loss of environmental resources. This includes reducing the loss of

H. Warsilah (✉) · A. H. Purnomo
Research Centre for Society and Culture - National Research and Innovation Agency, Jakarta, Indonesia
e-mail: hennywarsilah@gmail.com

H. Warsilah and A. H. Purnomo (eds.), *Approaching Natural Resources for Sustainable Development in Indonesia*, Sustainable Development Goals Series, https://doi.org/10.1007/978-981-96-6913-4_21

biodiversity, the proportion of land area covered by forests, the consumption of ozone-depleting substances, and the proportion of bird and fish caught within safe biological limits. Other aspects include the proportion of total aquatic resources used, the proportion of terrestrial and marine protected areas, and the proportion of species threatened with extinction.

Alex Evans (2014) analyzes the modest ambition of the Sustainable Development Goals (SDGs) in an ecological context. These goals range from ending poverty and reducing inequality within and between countries, to fostering better governance and peaceful societies, taking action on climate change, restoring ecosystems, and making significant shifts towards sustainable consumption and production.

The deforestation in Indonesia is part of a socio-ecological crisis triggered by an environmentally unsustainable development orientation, which fundamentally contradicts the objectives of sustainable development. The reality on the ground demonstrates structural impoverishment, displacement from living spaces, and human rights violations against local communities. These communities have not been included in the development process and, instead, have been expelled from the ancestral lands that have sustained their livelihoods.

The Food Estate initiative in Central Kalimantan's Gunung Mas Forest and the MIFEE mega-agro-industry, for example, look inconsistent and diverge from the forest moratorium policy. The revised policy clause aims to encourage community participation in land management through a 12.7-hectare community land program (see mongabay.co.id, 27/5). In another example, the plan to construct a cement factory is the root cause of the ecological injustice experienced by the residents of Rembang, where the establishment of a cement factory could potentially threaten their fundamental right to access clean water sources. The voices of concern and rejection have been vociferously expressed by Rembang residents, particularly women. This underscores the significance of the Rembang case in the socio-ecological crisis of natural resources.

The narratives presented in this volume highlights a considerable gap between environmental preservation and economic development (Martinuzen 1997). These two aspects, economics and environment, often cannot coexist harmoniously; instead, there is a trade-off, necessitating the selection of one aspect as a priority (Clark 2001). The socio-ecological crisis is not merely an 'impact' but a 'condition' inherent in capitalism. This implies that the crisis itself has become an intrinsic part of each cycle of capitalist development (Batubara and Handriana 2021). The term 'socio-ecological crisis' has been extensively developed and utilized by non-governmental organizations to illustrate the interconnectedness of social and ecological problems in Indonesia.

The dilemma of sustainable development lies in simultaneously achieving high levels of economic growth without causing environmental harm. If the leniency in natural resource management prevails, production potential may increase, but the risk of environmental damage will escalate. Conversely, stringent environmental protection regulations could deter investors. This underscores the crucial role of the government in establishing appropriate regulations for natural resource management.

In terms of recommendations, several approaches can be adopted to mitigate environmental damage and conflicts over natural resource management. As exemplified by Stiglitz (2000), the public and private sectors can collaboratively address externalities through legal frameworks. To respond to externalities, the government must assume a pivotal role in regulating and curbing environmental damage. In practice, this involves enforcing regulations such as pollution standards, efficiency benchmarks, and implementing environmental taxes (Clark 2001).

Currently, the government is required to take a more serious approach to addressing the ecological damage caused by natural resource exploitation and eradicating rent-seekers.

The suggested actions presented in this section are as follows:

a) The responsibility of managing natural resources will become more challenging if population growth remains uncontrolled. Therefore, population control should be a top priority to minimize the region's vulnerability to the five types of air-related hazards.
b) The National Policy on Natural Resource Management outlines the strategic direction for natural resource management from 2010 to 2030. Relevant ministries and agencies should follow up on the strategic plan documents within their respective areas of responsibility as part of the National Medium-term Development Plan (RPJM).

Policies concerning natural resource management at the provincial level should be promptly formulated, referring to the National Policy on Natural Resource Management, while considering regional conditions and issues. Similarly, natural resource management policies at the district/city level should be developed based on the provincial-level policies.

Regular monitoring and evaluation of the implementation of natural resource management policies should be conducted annually to assess the achievements and shortcomings of the set goals, identify the root causes of any failures, and simultaneously provide feedback to formulate corrective actions aimed at achieving maximum success.

For Indonesia, lessons can be learned in terms of overcoming the socio-ecological crisis by following the example of other Asian countries such as Japan and Malaysia in three key respects. First, by tightening regulations on sustainable natural resource management cultivation. Second, cultivating natural resources that are the mainstay for improving the country's economy and the needs of the community on a mass scale in accordance with the relevant regulations. Third, providing stable opportunities to harvest natural resources through education for children in schools and families about the terrestrial (forest) and marine environments to encourage community attitudes and behavior towards natural resource conservation, and to avoid the socio-ecological crisis of marine resources.

References

Baker S (2006) Sustainable development. Routledge, New York, p 2006

Batubara and Handriana (2021) Socio-Ecological Crisis and Capitalism. Yogyakarta: WALHI

Clarke G (2001) From ethnocide to ethnodevelopment? Ethnic minorities and indigenous peoples in Southeast Asia. Third World Q 22(3):413–436

Evans A (2014) Sustainable development goals. Eight Ways to Make Reality Match Ambition. https://www.theguardian.com/global-development/2014/dec/19/sustainable-development-goals-eight-ways-reality-match-ambition

IUCN (1980) World Conservation Strategy: Living Resource Conservation for Sustainable Development. IUCN-UNEP-WWF.

IUCN (ed) World Conservation Strategy: Living Resource Conservation for Sustainable Development. https://www.environmentandsociety.org/mml/iucn-ed-world-conservation-strategy-living-resource-cons

Martinussen J (1997) State, society and market: a guide to competing theories of development. Zed Books Ltd., London

Mongabay.co.id, 27/5. Areal Moratorium Bertambah Hampir 1 Juta Hektar. https://www.mongabay.co.id/2015/05/27/areal-moratorium-bertambah-hampir-1-juta-hektar/

Stiglitz JE (2000) The contributions of the economics of information to twentieth-century economics. Q J Econ 115(4):1441–1478

UNESCO (2012) Culture: a driver and an enabler of sustainable development. UNESCO

WCED (1996) Report of the world commission on environment and development: our common future

World Bank (2009) Investing in a more sustainable indonesia: country environmental analysis 2009. World Bank Group, Jakarta

Glossary and Abbreviations

Acacia mangium A species of flowering tree in the pea family, Fabaceae, which is native to Australia, Indonesia, and Papua New Guinea. It is known for its fast growth and ability to thrive on poor soils.

Adaptive governance A governance approach that focuses on the ability to adapt to complex environmental, social, and economic challenges.

AMDAL Analisis Mengenai Dampak Lingkungan (Environmental Impact Analysis)

ANR Assisted National Regeneration

AWC Australian Wildlife Conservancy

BAPPENAS Badan Perencanaan Pembangunan Nasional (The National Development Planning Agency)

BBC British Broadcasting Corporation

BCC Bird Conservation Council

Beekeeping The practice of maintaining bee colonies, typically for honey production

Bi Bismuth

BKM Badan Keswadayaan Masyarakat (Community Self-Help Agency)

BNPB Badan Nasional Penanggulangan Bencana (National Board for Disaster Management)

BPBD Badan Penanggulangan Bencana Daerah (Regional Disaster Management Agency)

BPDLH Badan Pengelola Dana Lingkungan Hidup (Environmental Fund Management Agency)

BPN Badan Pertanahan Nasional (National Agrarian Bureau)

BPS Badan Pusat Statistik (Central Bureau of Statistics)

BRG Badan Restorasi Gambut (The Peat Restoration Agency)

BRGM Badan Restorasi Gambut dan Mangrove (The Peatland and Mangrove Restoration Agency)

BRIN Badan Riset dan Inovasi Nasional (National Research and Innovation Agency)

BSNP Berbak-Sembilang National Park

BT Butyltin

BUMDes Badan Usaha Milik Desa (Village Government-Owned Enterprise)

CA Corresponding Agency

CBFM Community-Based Forest Management

CCBS Climate, Community & Biodiversity Standards

Cd Cadmium

CEPA Communication, Education, and Public Awareness

CG Collaborative Governance

CGRs Collaborative Governance Regimes

CHL Chlordane

CLC Customary Law Community

CM1 Counter Measure 1 (national effort)

CM2 Counter Measure 2 (with international support)

CNN Cable News Network

CO2 Carbon dioxide

H. Warsilah and A. H. Purnomo (eds.), *Approaching Natural Resources for Sustainable Development in Indonesia*, Sustainable Development Goals Series, https://doi.org/10.1007/978-981-96-6913-4

Cr Chromium
Critical Land Land degradation due to previous industrial activities, requiring significant restoration efforts.
Civil Society Masyarakat Sipil
CSOs Civil Society Organizations
Cu Copper
CVM Contingent Valuation Methods
DBH DR Dana Bagi Hasil Dana Reboisasi (Reforestation Profit Sharing Fund)
DDT Dichloro Diphenyl Trichloroethane
DENR The Department of Environment and Natural Resources
DFID Department for International Development
DKI Jakarta Daerah Khusus Ibukota Jakarta (Special Capital Region of Jakarta)
DMO Domestic Market Obligation
DNA Designated National Authority
DPP Dasar Penghitungan Pajak (Basis for Tax Calculation)
DPRD Dewan Perwakilan Rakyat Daerah (Regional Representative Council)
DRR Disaster Risk Reduction
EAAF East Asian–Australasian Flyway
ECAN The Environment Critical Areas Network
EHRA Environmental Health Risk Assessment
ENPV Economic Net Present Value
ESCOR Economic and Social Research Council
ESDM Energi dan Sumber Daya Mineral (Ministry of Energy and Mineral Resources)
ETS Emissions Trading Systems
Extractive Industries Industries, such as mining, that involve extracting natural resources from the earth.
ESDM Energi dan Sumber Daya Mineral (The Ministry of Energy and Mineral Resources (ESDM))
FAO Food and Agriculture Organization
FGD Focus Group Discussion
FOLU Forestry and Other Land Use
Forex Reserve Foreign Exchange Reserves
FPIC Free, Prior, and Informed Concern
FREL Forest Reference Emissions Level
GDP Gross Domestic Product
GHG Green House Gas
GIS Geographic Information System
GMF Global Maritime Fulcrum
Green Revolution Revolusi Hijau
HCH Hexa Chlorocyclo Hexane
HEIA High External Input Agriculture
Hg Mercury
Huma The term for fields in Baduy or Sundanese in general.
IBA Important Bird Area
ICCPR International Covenant on Civil and Political Rights
IDR Indonesian Rupiah
IDS Indonesia Dragonfly Society
IFCG Integrative Framework for Collaborative Governance
IKN Ibu Kota Nusantara (Nusantara Capital City)
ILMI Inspirator Lebah Madu Indonesia (Indonesian Honey Bee Inspirator)
IMF International Monetary Fund
INDEF Institute for Development of Economics and Finance
IPD Indeks Pembangunan Desa (Rural Development Index)
IPRA Indigenous Peoples' Rights Act
IUP Izin Usaha Pertambangan (Mining Business Permit)
IUPK Izin Usaha Pertambangan Khusus (Special Mining Business Permit)
IUUF Illegal Unreported Unregulated Fishing
IWC International Waterbird Census
JATAM Jaringan Advokasi Tambang (Mining Advocacy Network)
JC Joint Capacity
JMPPK Jaringan Masyarakat Peduli Pegunungan Kendeng (The Kendeng Mountains Concerned Community Network)
Kawalu (or ngawalu) The ceremony after the *huma serang* rice harvest.
KBAK Kawasan Bentang Alam Karst (Map of karst landscape area/areas karst natural landscapes)
KepDirjen Keputusan Direktorat Jenderal (Decision of the directorate general)
KepMen Keputusan Menteri (Ministry Decree)
KKI Komunitas Konservasi Indonesia (Indonesian Conservation Community)
KLHS Kajian Lingkungan Hidup Strategis (Strategic environmental assessment)

Komnas HAM Komisi Nasional Hak Asasi Manusia (National Human Rights Commission)
KPH Kesatuan Pengelolaan Hutan (Forests Management Unit)
KRTB Komunitas Raje Teran Belitung (Raje Teran Belitung Community)
LBH Lembaga Bantuan Hukum Jakarta (The Indonesian Legal Aid Institute)
LEISA Low External Input and Sustainable Agriculture
Leuweung garapan Forest areas that are cleared by the community for huma or farming; agroforestry.
Leuweung tutupan Secondary forest areas that can be cultivated into fields
LGU Local Government Unit
LHK Lingkungan Hidup dan Kehutanan (Environment and Forestry)
LIPI Lembaga Ilmu Pengetahuan Indonesia (Indonesian Institute of Sciences)
LNG Liquefied Natural Gas
Local wisdom The norms or behavior of local communities
LPG Liquid Petroleum Gas
LPM Lembaga Pelayanan Masyarakat (Community Service Institutions)
LTS-LCCR Long-term Strategy for Low Carbon and Climate Resilience
Macapat A traditional Javanese song or poetry
Mass tourism A form of tourism that involves tens of thousands of people
MBT Monobutyltin
MHA Masyarakat Hutan Adat (Customary Forest Community)
MIFEE Merauke Integrated Food and Energy Estate
mm Millimeter
MW MegaWatts
Mn Manganese
MNS Malaysian Nature Society
MOMI Minerba One Map Indonesia (Minerals and Coal One Map Indonesia)
MoEF Kementerian Lingkungan Hidup dan Kehutanan (Ministry of Environment and Forestry)
MoI Mean of Implementation
MP3EI Master Plan Percepatan dan Perluasan Pembangunan Ekonomi Indonesia (Master Plan for The Accleration and Expansion of Indonesian Economic Development)
MRV Measurement, Reporting, and Verification
MS CLC Marga Sarampas Customary Law Community
Muyong system A forest conservation strategy
NCICD The National Capital Integrated Coastal Development
NCSC North Central of Selangor Coast
NDC Nationally Determined Eontribution
NGO Non-Government Organization
Ni Nickel
NRI National Research Institute
NRM Natural Resource Management
NTFP Hasil Hutan Bukan Kayu (Non-timber Forest Product)
OCP Organochlorine pesticides
OTM Olah Tanah Minimum (Limited land processing system)
PAMSIMAS Penyediaan Air Minum dan Sanitasi Berbasis Masyarakat (Community-based water supply and sanitation)
Pb Lead
PCB Polychlorinated Biphenyl
PCSD The Palawan Council for Sustainable Development
PDAM Perusahaan Daerah Air Minum (Regional Drinking Water Company)
PE Principled Engagement
Perhutani State-owned enterprises are in the form of public companies
Perpres Peraturan Presiden (President regulatory)
Perum Perusahaan Umum (Public company)
PLG Proyek Lahan Gambut (Peat land development project)
PLN Perusahaan Listrik Negara (The state electricity company)
PLTHM Pembangkit Listrik Tenaga Mikro Hidro (Micro-hydro Power Plant)
PNBP Pendapatan Negara Bukan Pajak (Non-tax State Revenue)
PNNI The Palawan NGO Network Incorporation
PODES Village potential
Ponors Natural openings (holes) where surface water enters into underground passages.

POP Persistent Organic Pollutant
PPID Pejabat Pengelola Informasi dan Dokumentasi (Information and Documentation Management Officer)
PRCF People Resource and Conservation Foundation
PSKL Perhutanan Sosial dan Kemitraan Lingkungan (Social Forestry and Environmental Partnerships)
PT KAI PT Kereta Api Indonesia (PT KAI)
PT Perseroan Terbatas (Limited liability company)
PTUN Pengadilan Tata Usaha Negara (State Administrative Courts)
PVC Plan Vivo Certificate
PLTU Pembangkit Listrik Tenaga Uap (The steam power plant)
PLTN Pembangkit Listrik Tenaga Nuklir (Nuclear Power Plant)
PLTA Pembangkit Listrik Tenaga Air (Hydro Power Plant)
PLTG Pusat Listrik Tenaga Gas (Gas-fired Power Plant)
PLTGU Pusat Listrik Tenaga Gas dan Uap (Gas and steam power Plant
PLTS Pembangkit Listrik Tenaga Surya (Solar Power Plant)
PLTP Pembangkit Listrik Tenaga Panas Bumi (Geothermal Power Plant)
Reclamation The process of restoring land that has been mined to a more natural or economically usable state
REDD + Reduction emissions from deforestation and forest degradation, role of conservation, sustainable management of forest and enhancement of forest carbon sink
Resource curse A paradox where countries with abundant natural resources tend to have less economic growth and poorer governance than countries with fewer natural resources
Restoration The process of assisting the recovery of an ecosystem that has been degraded, damaged, or destroyed
RPJMD Medium Development Plan/RPJMD
RTB Raje Teran Belitung (a community in East Belitung)
RTRW Rencana Tata Ruang Wilayah (Regional Spatial Planning Policy)
Saba (or nyaba) To go, travel, and visit to get to know other cultures.
Sb Antimony
SBY Susilo Bambang Yudhoyono (4th Indonesian President)
SCCM Sustainable Commodities Conservation Mechanism
SDA Sumber Daya Alam (Natural Resources)
SDGs Sustainable Development Goals
SES Social-Ecological Systems
SHEEP Society for Health, Education, Environment, and Peace
Siji kanggo sak lawase In the context of marriage, polygamous marriages are prohibited.
SLA Sustainable Livelihood Approach
SM Shared Motivation
Social Forestry License A legal permit allowing communities to manage and utilize forest areas sustainably.
SOE State-Owned Enterprise
SSP Stumpage Sales Price
SWIS A Small Water Impounding System
Tata Wayah Seasonal rules
TBT Tributyltin
TC Tambon Council
TEV Total Economic Value
Tin Mining The process of extracting tin from the earth, historically significant in the Bangka Belitung region.
TNFP Thai National Forestry Policy
TNI Tentara Nasional Indonesia (The Indonesian National Armed Forces)
TNKS Taman Nasional Kerinci-Seblat (Kerinci Seblat National Park)
Trigona Beekeeping The practice of keeping Trigona bees, a stingless bee species, for honey production.
Trigona Bees A genus of stingless bees commonly found in tropical and subtropical regions.
UKL Upaya Kelola Lingkungan (Environmental Management)
UN United Nations
UNDP United Nations Development Program
UNDRR The United Nations Office for Disaster Risk Reduction

UNFCCC United Nations Framework Convention on Climate Change

UNGPs United Nations Guiding Principles on Business and Human Rights

UNICEF United Nations International Children's Emergency Fund

UPL Upaya Pemantauan Lingkungan (Environmental Monitoring Efforts)

USD United States Dollar

UU Undang-Undang (Indonesian National Law)

V Vanadium

VCS Verified Carbon Standards

VOC Vereenigde Oostindische Compagnie (Dutch East India Company)

WALHI Wahana Lingkungan Hidup Indonesia (The Indonesian Forum for the Environment)

WHO World Health Organization

WTP Willingness to Pay

YLBHI Yayasan Lembaga Bantuan Hukum Indonesia (The Indonesian Legal Aid Foundation)

Zn Zinc

Zoopolis A concept by Donaldson and Kymlicka referring to a model of human–animal coexistence that respects the sovereignty and rights of different animal communities.

WASH Water Sanitation and Hygiene

μm Micrometer

MIX
Papier aus verantwortungsvollen Quellen
Paper from responsible sources
FSC® C105338

If you have any concerns about our products,
you can contact us on
ProductSafety@springernature.com

In case Publisher is established outside the EU,
the EU authorized representative is:
Springer Nature Customer Service Center GmbH
Europaplatz 3, 69115 Heidelberg, Germany

Printed by Libri Plureos GmbH
in Hamburg, Germany